# DEVELOPMENTS
# IN
# APPLIED
# SPECTROSCOPY
## Volume 2

*A Publication of the Society for Applied Spectroscopy, Chicago Section*

# DEVELOPMENTS IN APPLIED SPECTROSCOPY

## Volume 2

*edited by*

## J. R. Ferraro and J. S. Ziomek

Proceedings of the
Thirteenth Annual Symposium on Spectroscopy
Held in Chicago, Illinois
April 30 - May 3, 1962

*Distributed by*

**PLENUM PRESS**
**NEW YORK**
**1963**

ISBN 978-1-4684-8687-2      ISBN 978-1-4684-8685-8 (eBook)
DOI 10.1007/978-1-4684-8685-8

Library of Congress Catalog Card No. 61-17720

# PREFACE

For the last thirteen years, the Chicago Section of the Society for Applied Spectroscopy has sponsored a symposium in the spring of each year. In this span, the symposia have shown a steady increase in attendance, in the number of papers presented, in the number of sessions, and in the number of days the conference lasted. The duration of the most recent symposium was four days, with sessions devoted to molecular spectroscopy, including infrared, Raman, ultraviolet, and visible, and to X-ray, NMR, emission, and flame spectroscopy, respectively, with a special session devoted to gas chromatography because of its growing interest in applied spectroscopic work.

Another feature of this last symposium was the attempt on part of the Symposium Committee to establish and maintain the scientific level at that of applied physics. This should place the present symposium at a level somewhere between that of the Ohio State symposium and that of the Pittsburgh meeting, thus approaching the level of applied chemical physics.

In addition, the symposium was designed to offer to scientists from other disciplines and students an opportunity to attend introductory panels and lectures and at the same time the mature investigators a meeting ground and the chance to keep abreast of the latest developments in spectroscopy. How well these aims have been accomplished is best attested to by the phenomenal growth of the symposium.

In 1961 the first attempt was made to publish the proceedings of these symposia. The result was Volume 1 of "Developments in Applied Spectroscopy." This book is Volume 2 in this series and contains 41 papers in the various areas of spectroscopy discussed at the 1962 symposium. The contributors to this volume are some of the country's leading spectroscopists. The distribution of the topics is based on the organization of the symposium program.

  The symposium and the proceedings would not have been possible except through the tremendous effort of the Symposium Committee. We wish to acknowledge the contributions of Jay A. Sheinkop, William J. Driscoll, Edward A. Piotrowski, John Danaczko, John P. Kapetan, Elma Lanterman, John E. Forrette, Robert J. Manning, Stuart Armstrong, and Miles Schwartz. In particular, we wish to give special thanks to Dr. Lanterman and Mr. Forrette for reviewing several of the manuscripts pertaining to X-ray spectroscopy and gas chromatography that appear in these proceedings.

John R. Ferraro
Argonne National Laboratory
Argonne, Illinois

Joseph S. Ziomek
Martin-Marietta Corp.
Orlando, Florida

# CONTENTS

vii

## X-RAY SPECTROSCOPY

### Absorption

### Emission

## GAS CHROMATOGRAPHY

### Preparative

### General

# INFRARED AND RAMAN SPECTROSCOPY

# WORLD-WIDE COMMUNICATION
# OF SPECTROSCOPIC INFORMATION

## Forrest F. Cleveland

Illinois Institute of Technology
Chicago, Illinois

There are hundreds of thousands of spectroscopists in the world, and they carry out tens of thousands of investigations and obtain hundreds of thousands of dollars worth of results, but the potential value of all this effort is largely lost because of an archaic and ineffective means of communication of these results.

If spectroscopy is to make its maximum contribution, efficient and effective methods must be devised for the rapid diffusion of new information throughout the domain of spectroscopy, and for the retrieval of the older information buried in the literature.

## THE LANGUAGE PROBLEM

The first difficulty encountered is, of course, the language problem. According to a study made by UNESCO, scientific results are published in at least 34 languages, but half of the value is lost because half of the scientists cannot read the articles in the language of publication; for example, 50% of the results are published in English, but 50% of the world's scientists cannot read English.

At first thought, translations might seem to be the answer. But translations are slow, costly, often erroneous, and for the most part unobtainable. In fact, most of the articles are never translated, and it would obviously require enormous effort and expense to translate material from each of the 34 languages into each of the other 33.

What is needed is an interlanguage which would be the common denominator of those half dozen European languages in which the scientific vocabulary largely originated, and which would be readable at sight by any scientist who has a knowledge

2

of one or more of these languages. Since the scientific vocabulary has now spread throughout the world, such an interlanguage would be quite suitable for world-wide communication. Fortunately, an interlanguage of this type already exists in Interlingua, which appeared in 1951 after 27 years of research and development.

## "SPECTROSCOPIA MOLECULAR"

Impressed by the ease of reading Interlingua and convinced of the necessity of doing something to improve the transmission of spectroscopic information beyond the language barriers, I decided to begin the publication in Interlingua of a small periodical, called "Spectroscopia Molecular," in my own field of research. The first issue was mailed, by a strange coincidence, exactly 10 years ago today, on May 1, 1952. The use of Interlingua assured the easiest readability by the greatest number of spectroscopists.

"Spectroscopia Molecular" now goes to 400 spectroscopists, laboratories, and libraries in 31 countries. The periodical has supported itself by subscriptions and advertisements, and now has a reserve sufficient to guarantee publication for a year in advance. If more than this is received, it is used to improve the periodical.

"Spectroscopia Molecular" contains brief reports of new research results, summaries of significant articles, reports of spectroscopic meetings, reviews of new books, a calendar of future spectroscopic events, details of new instruments, news about spectroscopists, and other news and information of interest and value to workers in this field. The emphasis is on new developments and on keeping the spectroscopists up to date.

Experience with this periodical during its first ten years of monthly publication provides convincing proof of the usefulness of Interlingua for world-wide communication.

## THE CENTRAL LIBRARY

The Society for Applied Spectroscopy itself, through the dedicated effort of Theodore H. Zink, Chairman of its International Activities Committee, is using Interlingua in its Central Library for Spectroscopy (212 Chestnut Hill Drive, Ellicott City, Md., U. S. A.). This library, as one of its services, publishes twice a month the periodical, "Titulos Spectroscopic," a complete list of titles of current spectroscopic books or articles. These titles are sent to the library by spectroscopists in all parts of the world. They are then translated into Interlingua by the Interlingua Division of Science Service (80 E. 11th St., New York 3, N. Y., U. S. A.) and are published within one month after their appearance in the original journal. The language of the article is indicated. If the spectroscopist can read the language, a reprint may be obtained from the author, or a photographic copy of the article can be obtained from the library. If not, an Interlingua translation of the abstract or complete article can be obtained. For faster service, the library is collecting reprints of all spectroscopic articles, and expects soon to be able to prepare lists of references on specific topics.

This service enables the spectroscopist to know what articles are being published and to obtain quickly a copy of any article in a language he can read. Keeping up to date is thus no longer such a formidable task. Spectroscopists are indebted to the Society of Applied Spectroscopy for providing this useful service.


## MULTILINGUAL DICTIONARY

Another use of Interlingua is in a "Multilingual Dictionary for Spectroscopy" I am preparing in collaboration with Associate Editors Bonino (Bologna), Dupeyrat (Paris), Matossi (Freiburg), Morcillo (Madrid), Stammreich (São Paulo), and Wolkenstein (Leningrad); Advisory Councilors Herzberg (Ottawa) and Mulliken (Chicago); and Linguistic Editor Gode (New York). In this, a complete list of spectroscopic terms will be given, each one followed by a concise and correct definition in Interlingua. Under this will be given the corresponding terms in English, Spanish, French, German, Italian, Portuguese, and Russian.

The dictionary will be the equivalent of 21 bilingual dictionaries, and its existence will have a tendency to standardize international usage, which in turn will facilitate world-wide communication in spectroscopy. (Dr. F. O. Holmes, of the Rockefeller Institute for Medical Research, is preparing a similar dictionary for virology.)

## ABSTRACTS IN INTERLINGUA

Next to the medical workers, who are using Interlingua for abstracts in 25 of their journals and who have used this inter-language in the programs of a dozen international meetings, the spectroscopists are the ones who have shown the greatest initiative in the improvement of communication in their field. A natural next step would be the use of Interlingua in the programs of international spectroscopic meetings and for abstracts in the spectroscopic journals. This would extend the range and usefulness of the programs and journals.

## CONCLUSION

The spectroscopist can now read about new developments in "Spectroscopia Molecular" and read the titles of all recent articles in "Titulos Spectroscopic." If any article is important for his work, he can quickly get a copy of it, in a language he can read, from the "Central Library for Spectroscopy."

Soon he will be able to find the exact meaning of any un-familiar term, and the corresponding expression in eight languages, by reference to the "Multilingual Dictionary for Spectroscopy."

Soon, too, he will be able to extract the essence of programs and articles in unfamiliar languages by reading the accom-panying Interlingua abstracts.

When these instruments have been developed to perfection, long strides will have been made toward the improvement of world-wide communication in spectroscopy.

# METHODS OF STORING, RETRIEVING, AND CORRELATING INFRARED SPECTRAL DATA

## Freeman H. Dyke, Jr.

Jonker Business Machines, Inc.
Gaithersburg, Maryland

For analytical chemists, retrieval and correlation of data has always been a major problem. Finding a real solution could greatly reduce the amount of valuable time now necessary for qualitative and quantitative analyses, and enable the correlation of spectral data with chemical, physical, and biological properties. In other words, an effective data retrieval and correlation technique would make analytical chemistry an even more valuable research tool.

The purpose of this paper will therefore be to examine more closely both the problem and the currently used or available solutions for the retrieval of infrared spectral data. This examination will consist of:

1. Defining the ideal data processing equipment for meeting the needs of the analytical chemist.
2. Examining the four methods now most widely used.
3. Describing a fifth and entirely new technique known as "Termatrex."
4. Summarizing a trial evaluation of Termatrex equipment for handling infrared spectral data, conducted by the American Society for Testing Materials.
5. Describing the Termatrex index for X-ray diffraction data as authored by the American Society for Testing Materials and produced by Jonker Business Machines.

Examining the parameters of the ideal retrieval system for spectral data, we find it should tell the chemist exactly which compounds are present and furnish a quantitative analysis of an unknown. To date, however, the ideal equipment has not been developed. Nevertheless, we may reasonably expect to find in equipment today at least the following features:

6

1. Speed in output—A search of collections of up to 250,000 compounds should take only a few minutes.

2. Flexibility of vocabulary—The system should be able to absorb an unlimited number of characteristics. This feature would enable the chemist to search by combinations of functional groups, by analytical data of all types, and by chemical, physical, and biological properties.

3. Flexibility in output—The searcher should receive instantaneous feedback from his data retrieval equipment. Only when he does can he employ the detective-like technique of question—answer, question—answer, question—answer so often necessary in a search of spectral data.

4. Immediate availability—The chemist should have direct access to his data retrieval systems, preferably by having the equipment located right in the laboratory.

5. Direct access to information—After completing a search, the chemist needs direct access to the detailed information on those chemical compounds found to meet the requirements of the search questions.

6. Low cost—The equipment should be priced within the means of most analytical laboratories around the world.

The four data retrieval techniques currently in use are the manual index, the edge-notched card, the punched card, and the computer. Let us now examine the advantages and disadvantages of each of these techniques individually.

Probably the oldest of the data processing methods is the manual index. Examples are the "Sadtler Spectra Finder" for infrared data and the "Fink Index" for X-ray diffraction data.

The manual index is generally an inverted system, in which all data on those compounds having a given characteristic are grouped on a page dedicated to that characteristic. The advantages of the manual index can be summed up as follows:

1. At present, this is the least expensive means of organizing and disseminating an index.

2. After completing a search, the chemist has direct access to the name, and as a rule to some summary information on the compounds meeting the requirements of the search questions.

3. The manual index can be kept in the laboratory where it is immediately available for use.

There are, however, some disadvantages of using a manual index for retrieving and correlating data. Among these disadvantages are:

1.  The limitation of having to restrict the search to one or possibly two characteristics at a time. Even when the data are organized in a table format, it is relatively difficult to search for even as few as four characteristics in one operation.
2.  The difficulty of performing a search when there are a large number of items to be scanned and a great many items listed under each characteristic.
3.  The impossibility of performing negative searches, a screening technique recognized as one of the most valuable methods of conducting a search.
4.  The difficulty of keeping the index current. When new compounds are added to the manual index, the entire system must be updated.

The second method currently in use for storing and retrieving data is the edge-notched card. At one time or another, this system has been applied to most forms of spectral data.

In this system, a separate edge-notched card is allotted to each item in the data collection. These cards may vary in size from three inches to two feet on each edge. The characteristics of the item are coded around this edge by notching.

A search of the system is performed by slipping long needles through the area of the card edge representing the characteristics in the search question and shaking the needles. Those cards which fall off the needles are the items meeting the requirements of the search question. The advantages of the edge-notched card system for data retrieval are as follows:

1.  Direct access can be supplied to a microfilm or reduced copy of the spectrum itself as well as to summary information on the compound. This direct access is the major advantage of this method of retrieval.
2.  Updating the system is quite simple because new cards can be disseminated to use whenever new compounds are added to the data collection.
3.  Direct availability to the chemist is possible because the edge-notched card index can be kept in the analytical laboratory.

The edge-notched card system, however, also has several disadvantages. Some disadvantages are:

1. Limited vocabulary—This is the most serious because coding space around the edge of the card is limited. Once this coding space has been used up, there is no way to expand the system to include additional characteristics. It is possible, of course, to use two or more cards per item, but this increases the search time severalfold and forces a collation of the cards.
2. Limited size of the data collection—A search becomes difficult to perform when there are more than 2000 compounds in the system. For example, it requires nearly two hours to search 10,000 cards unless they are prefiled. Even then, it is impossible to prefile the cards for all types of searches.
3. Limited availability—Storage space in the analytical laboratory can become a real problem when the collection contains 10,000 or more edge-notched cards.

The third method of data retrieval previously mentioned is the punched card system, which has been widely used for a number of years. A pioneer in the development of this search technique as applied to spectral data is Les Kuentzel of the Wyandotte Corporation.

In this system, a punched card is devoted to one item and the characteristics of the item are coded by punching into the face of the card. A collection of data is searched by passing the cards through a sorter, which can be a single or multiple column device. Advantages of using the punched card, whether these are IBM or some other make, are as follows:

1. Retrieval is automatic, eliminating all possibility of errors due to clerical malfunction during output.
2. The system is easily kept up to date. As new compounds are added to the collection, new cards can be disseminated to the users.
3. Direct access to the name of the compound is provided at the end of the search.

Disadvantages of using the punched card system for retrieving and correlating spectral data include:

1. Lack of coding space for additional characteristics. This is probably the greatest single drawback to using

this technique. Because of this limitation, many systems using punched cards for retrieval are compromised in their coding. Again, it is possible to use two or more cards per item, but this not only doubles the searching time, but also forces a collation of cards in addition to the sorting operation itself.

2.  Difficulty on output when collections of data contain more than 20,000 compounds.

3.  Lack of a means by which the chemist can browse the system. In many instances, it becomes necessary to start the search all over again if the original search fails to yield the desired answers. It is possible, however, to whittle down the necessity for searching the entire collection two or more times by using the proper search strategy.

4.  Lack of immediate availability of the search equipment in the laboratory. Generally, those chemists using a punched card system must send the card collection to a central data processing installation for the search to be performed. This is necessary because of the relatively high cost of the sorting equipment.

In recent years, several organizations have programed computers to search infrared spectral data. Among these are the U. S. Air Force at Wright Field, Ohio; the Thiokol Corporation in Camden, New Jersey; and the National Analine Division of the Allied Chemical Corporation in Buffalo, New York. The advantages of using the computer for retrieving and correlating spectral data have been found to include:

1.  Relatively high speed on output. On a 7090 computer, the search for an entire system of 50,000 compounds can be completed in 10 minutes, with as many as 50 questions being asked simultaneously. Of course, on a small computer such as a 1401, this same search would take approximately an hour.

2.  The search yields chemical compounds which come closest to answering the search question. On the 7090 program written by Wright Field, for example, the print-out includes addresses of the 40 compounds coming closest to meeting all the requirements of the search. First listed are those items meeting all the requirements; next appear those items meeting all but one requirement; and so on.

3. The coding space is unlimited. The use of tape makes it possible to add as many characteristics to the system as are desired. Of course, the fact that this is a serial search does mean the search time will be doubled if the number of characteristics is doubled.

4. If the computer is given intensity data, this search method can be used for a combination qualitative and quantitative analysis.

There are some disadvantages to using the computer for retrieving and correlating spectral data. These are as follows:

1. The analytical chemist must have access to a rather large computer.

2. When only one question is asked, the cost of the search itself is relatively high. At Wright Field, the figure is $50.00 per search regardless of the number of questions being asked.

3. There is no means of browsing or "talking to" the system.

4. Limited availability of the computer sometimes means it will be possible to get on the computer only once a day or even once every several days to conduct a search.

These four methods, then, are the ones most widely used today to meet the problem of retrieving and correlating spectral data. There is a fifth technique called Termatrex which is gaining widespread attention. The principles on which Termatrex equipment is based are as follows:

1. Inverted grouping. In most systems, the card or section of tape represents the chemical compound, while the characteristics of the item are coded on the card or tape. In the Termatrex system, this is reversed. The card represents a characteristic and the items are coded onto the card.

2. Allotted space. On any card representing any characteristic, a specific location is allotted to each item. Compound 0000 always appears in the lower left corner of the card; compound 9999 always appears in the upper right corner. When a hole is drilled in the space dedicated to an item, it means, "Yes, this item has the characteristic represented by this card." When a hole has not been drilled in this space, it means, "No, this

item does not have or has not been measured for this characteristic."

3. Superimposition. To perform a search, the user selects only those cards representing the characteristics in his search question. These are then superimposed over a light source. Any light-dots appearing through this selection of superimposed cards represent chemical compounds meeting the requirements of the search. These sources of light are identified by reading off the $x$ and $y$ coordinate location on the card. The number thus obtained is the address of the compound.

4. Random card filing. Each Termatrex card has a tab located in 1 of 100 possible positions and in 1 of 10 possible colors. This feature makes it possible to select and refile cards at random.

Although the full-sized Termatrex card has a capacity of only 10,000 items, these cards can be microfilmed. A strip of 25 microframes would then represent a card having a capacity of 250,000 items. In addition, Termatrex drilling equipment has been developed which will automatically transfer data from IBM punched cards to Termatrex cards. Equipment to become available this summer will automatically read out the addresses from either the full-sized or microfilm version of Termatrex cards.

While employed by E. I. du Pont de Nemours and Company, I worked with a number of analytical chemists in investigating the application of Termatrex equipment to the handling of spectral data. Among these chemists were Naomi Schlecter of the Central Research Department, K. Brandt and T. Mackey of the Textile Fibers Department, D. Johnson and A. Martin of the Plastics Department, and G. Patterson of the Film Department.

Termatrex equipment has been found to have the following advantages for retrieving and correlating spectral data:

1. High speed in output. In less than five minutes, an entire collection (up to 250,000 compounds) can be searched for correlations among such concepts as spectral data, functional groups, elements, radicals, and properties.

2. Flexibility of output. The analytical chemist can browse or "talk to" his data collection. This feature allows him to use his knowledge and ability in manipulating a data processing system to identify chemical unknowns.

3. Availability. The Termatrex installation is so compact it can easily be kept in the laboratory, available for instantaneous searches. There are no delays caused by having to schedule the use of a sorter or computers.
4. Unlimited expansion. Compact, low-cost data entry equipment is available to the chemist who wishes to add compounds and/or properties from his own files to the data system.
5. Unlimited vocabulary. An unlimited number of characteristics may be added to the system. Since Termatrex is an inverted system, incorporating additional characteristics does not increase the search time.

There are, of course, a few disadvantages to the Termatrex system. These are:

1. The entire system must be updated when new items are added to the data collection. This is perhaps the most serious disadvantage of this technique.
2. A search yields only the address of the compounds meeting the requirements of the search question. This represents an indirect access to the name of the compound.

Two programs involving the use of Termatrex cards for retrieving and correlating spectral data have been undertaken by the American Society for Testing and Materials (ASTM).

For handling infrared data, a trial program has been initiated, using data on 4000 chemical compounds. Ten trial decks of Termatrex cards were prepared and distributed to ten organizations for evaluation and comparison to the other four data processing techniques. Those performing this evaluation are:

W. T. Cave
Assistant Director
Research Center
MONSANTO CHEMICAL COMPANY
800 N. Lindbergh Blvd.
St. Louis 66, Missouri

J. T. Thompson
INLAND MANUFACTURING DIV.
GENERAL MOTORS CORPORATION
Dayton 1, Ohio

F. J. Ludwig
TRETOLITE COMPANY
369 Marshall Avenue
St. Louis 19, Missouri

E. C. Dunlop
Central Research Dept.
E. I. DU PONT DE NEMOURS &
    COMPANY, INC.
Experimental Station
Wilmington, Delaware

A. W. Pross
Central Research Laboratory
CANADIAN INDUSTRIES, LTD.
Montreal, Canada

W. Fulmor
LEDERLE LABORATORIES
AMERICAN CYANAMID COMPANY
Pearl River, New York

M. V. Otis
Research Laboratory
TENNESSEE EASTMAN
Kingsport, Tennessee

P. Sadtler
SADTLER RESEARCH LABORATORIES
1517 Vine Street
Philadelphia, Pennsylvania

R. E. Seeber
National Analine Division
ALLIED CHEMICAL CORPORATION
P. O. Box 975
Buffalo 5, New York

W. C. Kenyon
Research Division
HERCULES POWDER COMPANY
Wilmington, Delaware

For X-ray diffraction data, ASTM has authorized Jonker Business Machines to prepare 300 copies of a Termatrex index on the 5698 inorganic compounds in the Powder Data File. This index is scheduled to become available for purchase from ASTM by September 1, 1962.

The X-ray diffraction data index will use the full-size Termatrex card having a capacity of 10,000 compounds. If a system is established for infrared spectral data, it is quite probable this will be done on the microfilm version (Minimatrex) of Termatrex equipment, giving the index a capacity of 250,000 compounds.

Personally, I have come to believe that a combination of manual indexes, computers, and Termatrex indexes will completely meet the needs of the analytical chemist searching spectral data. Further, I predict edge-notched and IBM punched cards will be replaced because of the relatively low speed on output and the lack of expandable coding space.

In summary it can be said with certainty that more effective data retrieval techniques should and will enable the analytical chemist to combine his knowledge and skill with his own data processing system in identifying unknown mixtures and compounds. Furthermore, he will be able to use combinations of analytical techniques more effectively and assist research by correlating spectral data with properties of chemical compounds.

# METHODS OF FAR INFRARED SPECTROSCOPY

## T. K. McCubbin, Jr.

Department of Physics
The Pennsylvania State University
University Park, Pennsylvania

## INTRODUCTION

The first part of this paper will be a discussion of some of the physical limitations encountered in far infrared spectroscopy using the echellette grating monochromator with filters to separate orders. An inclusive discussion of the methods of far infrared spectroscopy using gratings and filters will be omitted because these methods are well known and used in hundreds of laboratories. In about the past 5 years such spectroscopic systems have become available from instrument manufacturers. The last part of the paper will describe some less common systems using interferometric equipment. There will be an account of some new detectors which can be applied to far infrared spectroscopy.

## SOURCES FOR FAR INFRARED SPECTROSCOPY

The thermal sources which are used in infrared absorption spectroscopic investigations emit a continuous spectrum which becomes increasingly feeble as wavelength increases. This fact would seem to place a severe limitation on far infrared spectroscopy. Calculation will show, however, that a $\lambda^{-4}$ decrease in radiant energy for a fixed wavelength passband will not limit the energy available at the detector of a spectroscopic system which is operated at a fixed frequency or wavenumber passband.

Because $\nu = 1/\lambda$, where $\nu$ and $\lambda$ are wavenumber and wavelength, respectively, $\Delta\nu = \Delta\lambda/\lambda^2$. This means that wavelength passband of the spectroscopic instrument must be increased in proportion to $\lambda^2$ if the wavenumber passband is to remain constant. Thus, assuming constant dispersion, the slits, both entrance and exit, should have a width proportional to $\lambda^2$. Since

15

the radiant power transmitted by a monochromator with an entrance slit illuminated by a radiation source having a continuous spectrum is proportional to the square of the slit width, the power transmitting ability of a monochromator used to isolate a fixed wavenumber passband would vary in proportion to $\lambda^4$. This $\lambda^4$ dependency of power transmitted on wavelength would exactly cancel out the $\lambda^4$ factor found in the thermal radiation laws.

It should then be concluded that if a source following the $\lambda^{-4}$ law were used and if dispersion with respect to wavelength were constant, the radiant power available in a constant wavenumber passband would be independent of wavelength. Actual sources do not fall off as rapidly as $\lambda^{-4}$ and actual dispersive systems have a dispersion, $d\theta/d\lambda$, which varies as $\lambda^{-1}$, a fact that can be demonstrated by differentiation of the grating equation. Dispersion will increase as a given grating is turned to higher angles, but at some wavelength it will be necessary to change to a lower order or to use a coarser grating. The coarser grating or the lower order will exhibit lower dispersion.

The wavelength dependency of the minimum wavenumber passband as limited by the dispersion and the source output can be calculated from the data given by Strong [1] on the relationships between source brightness, dispersion, and energy limited spectral resolution. The figure shows the results of such calculations based on the assumption that the sources follow Planck's law.

The obtainable spectral resolution actually falls off more slowly with wavelength than is indicated by these curves. Sources such as the Welsbach mantle and the quartz mercury arc which have low emissivities at wavelengths near the blackbody maximum corresponding to their temperature tend to emit an intense continuous far infrared spectrum. Precise quantitative studies of the spectra of far infrared sources have not to my knowledge been made, and such investigations would seem a worthwhile undertaking.

One can conclude from the above discussion of radiant energy sources that the blackbody fall-off does not impose an extremely severe limitation on far infrared spectroscopy. Even though thermal sources emit but a small fraction of their total output at far infrared wavelengths, there is ample energy for the limited range of wavenumbers encompassed by that spectral region.

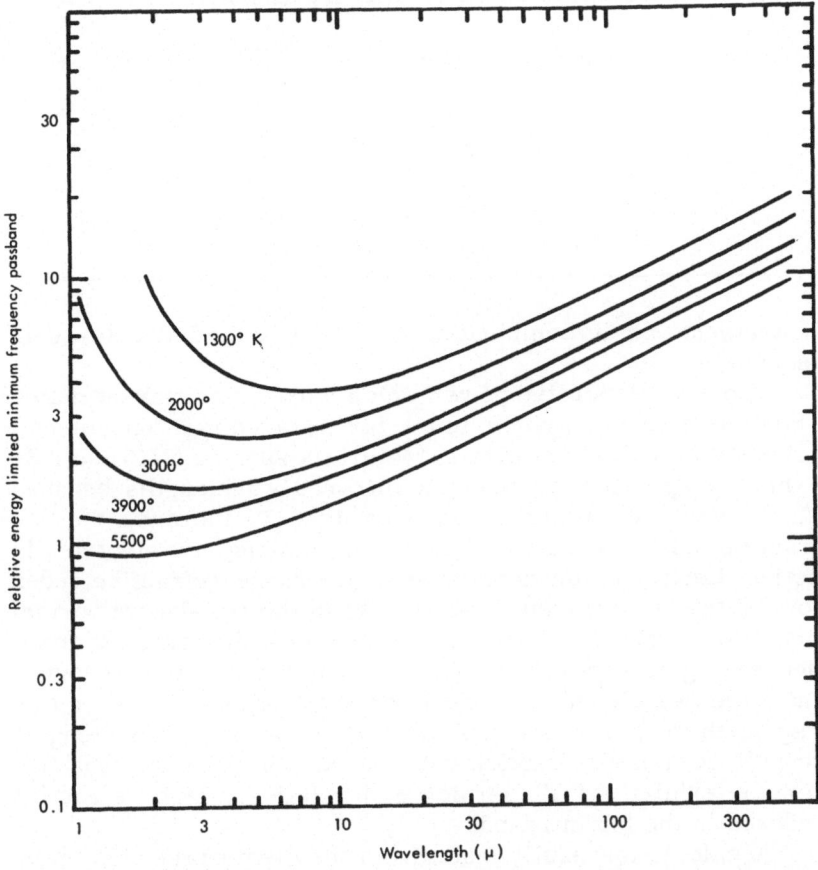

Minimum energy limited spectral resolution using sources which obey Planck's law. Minimum passband can be shown to be proportional to $(\lambda/rT)^{\frac{1}{2}}$, where $\lambda$ is wavelength, $T$ is absolute temperature, and $r = \dfrac{hc}{kT\lambda} \, (e^{-hc/kT\lambda}-1)^{-1}$ the ratio of Planck's law to the Rayleigh-Jeans law.

## ORDER SEPARATION

Spectroscopy at long wavelengths is difficult because the atmosphere is absorbing from roughly 15 to 500 $\mu$ and because prisms or sharp cut-off filter systems for order separation are not available. The atmosphere absorption problem can be solved somewhat unsatisfactorily by flushing the optical path with dry gas. A more satisfactory solution is the use of an evacuated

| Wave-length ($\mu$) | Frequency ($cm^{-1}$) | Order | | | | | |
|---|---|---|---|---|---|---|---|
| | | 1 | 2 | 3 | 4 | 5 | 6 |
| 5 | 2000 | 1 | 3 | 3.2 | 2.7 | 1.5 | 0.88 |
| 10 | 1000 | 1 | 5.2 | 11.3 | 15.2 | 18.1 | 17.0 |
| 30 | 333 | 1 | 7.1 | 21 | 43 | 72 | 108 |
| 100 | 100 | 1 | 7.5 | 24 | 56 | 104 | 172 |
| $\infty$ | 0 | 1 | 8 | 27 | 64 | 125 | 216 |

instrument or a combination of flushing and double-beam technique.

The low dispersive power which makes transparent materials unsuitable as prisms in the far infrared is a fundamental property of optical materials. Materials such as NaCl and KBr exhibit dispersion at medium infrared wavelengths because short-wave electronic resonances tend to increase the refractive index toward the ultraviolet, and longwave electrically active lattice resonances tend to decrease refractive index toward the far infrared. This results in the well-known normal dispersive behavior of optical materials with refractive index decreasing as $\lambda$ increases. Far infrared wavelengths are longer than the wavelengths of the lattice vibrations. Thus, in this region there are no strong absorptions at longer wavelengths to pull down the refractive index. Therefore, transparent materials exhibit low dispersion and relatively high refractive indexes in the far infrared.

A calculation easily made from the elementary diffraction grating equation shows that separation of grating orders requires a resolving power, $R = \lambda/\Delta\lambda$ or $R = \nu/\Delta\nu$, approximately equal to $m + 1$ where $m$ is the order in which the spectrum is to be observed. The relatively low degree of discrimination needed to observe the first- or second-order spectrum can easily be obtained by a combination of scattering filters, transmission filters, reststrahlen, and negative transmission or chopper filters.

It is instructive to compute what degree of attenuation of short-wave radiation will be necessary to eliminate false radiation from the grating spectrum [2].*

On the basis of the Rayleigh-Jeans approximation to the blackbody curve, one can readily show that the energy passing

*Discussion and the table from [2].

a given geometrical exit slit increases as the cube of the order. Factors proportional to the power in the various spectral orders calculated for 2000°K blackbody (Planck equation) at several wavelengths are listed in the table. Thus, for example, when the first-order spectrum at 10 $\mu$ is being observed with a 2000°K source, the band of second-order radiation of wavelength near 5 $\mu$ admitted by the slits contains 5.2 times as much power as does the first-order band. The table can also be used to determine the power that must be expected in the overlapping orders when higher-order spectra are of primary interest. For example, with third-order operation at 10 $\mu$, the fourth-order (7.5 $\mu$) spectrum would be 43/21, or twice as intense; fifth-order radiation at 6 $\mu$ would be 72/21, or 3.4 times stronger than that of the third order; and so on.

## INTERFEROMETRIC FAR INFRARED SPECTROSCOPY

The grating-filter methods of Rubens, Czerny, and Randall are still used for most of the measurements made at far infrared wavelengths. In recent years workers in several laboratories have been applying two-beam interferometric techniques to far infrared spectroscopy. One interferometric method applicable to the far infrared is Fourier transform spectroscopy, often called the Fellgett technique because Dr. P. Fellgett called attention [3] to the fact that this method is better than the grating method of spectroscopy in that, for a given time of observation of a given spectral region, more energy will reach the detector.

In Fourier transform spectroscopy the beam of radiant energy is divided into two beams, a variable path difference is introduced into one beam, and the beams are recombined. A Michelson or other type of interferometer can be used. In fact, Fourier transform spectroscopy in a primitive form enabled Michelson to investigate the fine structure of hydrogen lines in the late 19th century.

Michelson analyzed his visibility (or percent modulation) curves using a hand cranked computer which is presently on exhibit at the Franklin Institute in Philadelphia. Present-day Fourier transform spectroscopists use the fastest digital computers, in at least one case feeding the output from the detector to a digitalizer and then directly to punch cards. Note that whereas Michelson recorded only the visibility of the fringes, present-day workers record the intensity of the entire fringe system, thus escaping the possible ambiguities that were discovered by Michelson and Lord Rayleigh.

A major disadvantage of Fourier transform spectroscopy is the necessity of using a computer. As one worker in this field expresses it, it is as if a photographer had to send his film to the Eastman Kodak Company each time he took a picture. The Fourier transform method is being used in the far infrared in the laboratories of Dr. Strong at Johns Hopkins, Dr. Gebbie at The National Physical Laboratory in England, and Dr. Gush at the University of Toronto. There are probably other workers unknown to the author.

Another mode of interferometric spectroscopy is to drive the interferometer with a periodic "triangular-wave" motion. The intensity of the radiant energy will be a periodic function of time and the electrical output from the detector will contain only the drive frequency and its harmonics. A tuned amplifier is used to select the desired harmonic. The infrared wavelength that gives rise to the electrical signal selected by the amplifier is changed by changing the amplitude of the interferometer motion. A two-beam periodically driven interferometric system using a variable depth laminary grating in the zeroth order has been used extensively both with thermal and microwave sources in the millimeter and submillimeter wavelength region by Dr. L. Genzel [4] at Frankfort, Germany. The periodic two-beam system with part of the signal rejected electrically will of course not have the energy advantage that results when the entire fringe system is recorded.

A two-beam interferometer can be used instead of filters to separate the spectral orders of a diffraction grating. The radiant energy emerging from the exit slit of a grating monochromator has a spectrum consisting of a series of nearly monochromatic lines of frequency $\nu$, $2\nu$, $3\nu$, etc., where $\nu$ is the first-order frequency. If this radiation is sent to a two-beam interferometer with a variable path which is increasing at a constant rate, each order will be modulated at a frequency proportional to its optical frequency. A frequency selective amplifier can select the desired order and reject the others.

A two-beam interferometer consisting of a laminary grating of variable groove depth was constructed for grating order separation by the author and Dr. J. Strong at the Johns Hopkins University [5]. The interferometric modulation technique is also used by Dr. Genzel and by Dr. Bell at The Ohio State University.

The advantages of using diffraction gratings in a high order at incidence angles near 60°, i.e., the use of gratings as echelles, have recently been demonstrated for the near and medium infrared regions [6, 7]. Far infrared filter systems lack the selectivity needed to separate grating orders higher than about the second. It might be possible to gain the advantages of echelle use of the gratings in the far infrared using interference modulation for order selection.

## CRYOGENIC DETECTORS FOR THE FAR INFRARED

Recent work with low temperature infrared detectors may soon lead to vast improvements in the signal-to-noise ratio of far infrared spectra and to increased spectral resolution. Some recently reported cryogenic detectors are the carbon bolometer of Boyle and Rodgers [8], the gallium doped germanium bolometer of Low [9], and the superconducting tin bolometer of Bloor and others [10, 11].

Detectors operating at liquid helium temperatures are being developed which may be orders of magnitude better than room temperature detectors such as the Golay cell. The minimum detectable power of such detectors used with spectrometers will be background limited, i.e., the unavoidable statistical fluctuations in the temperature of an object coupled by thermal radiation to its environment may be their major source of noise. It will be necessary to prevent all thermal radiation except the signal that is intended to be detected from reaching the detector, for the unwanted input would both increase the radiation noise and raise the bolometer's temperature. A cooled longwave pass filter (such as a 1-mm thickness of crystal quartz) can prevent the thermal radiation from the spectrometer exit slit jaws from reaching a far infrared detector. Cryogenic detectors can be more easily used in the far infrared than in the region of medium wavelengths around 10 $\mu$, for in that region a narrow band transmission filter at liquid helium temperature would be necessary to keep the room temperature radiation from reaching the detector. The narrow band filter would have to be changed as the spectrum is scanned.

A further consequence of background limited detection is that interference and multislit modulation systems and Fourier transform systems in which the detector is always irradiated with a large signal lose many of their advantages when they are used with such detectors.

## CONCLUSIONS

Scientific workers who wish to make far infrared spectro-graphic measurements can readily use the grating spectrometer with filters to separate orders, for this method is extensively described in the scientific literature, and critical components or even the entire system can be purchased commercially. In contrast to infrared spectroscopy at shorter wavelengths, only a narrow wavenumber region in the far infrared can be scanned before filters must be changed. Variations in the background level which result with single-beam systems because of atmospheric absorption and filter characteristics make it advisable to develop double-beam or evacuated systems.

Interferometric far infrared techniques are still at the developmental stage.

Cryogenic detectors for the far infrared are being developed which should operate with much better signal-to-noise ratio than can be achieved using room temperature detectors.

## REFERENCES

1. J. Strong, J. Opt. Soc. Am. 39, 320 (1949).
2. R. C. Lord and T. K. McCubbin, J. Opt. Soc. Am. 47, 689 (1957).
3. P. Fellgett, J. Phys. Rad. 19, 187 (1958); Thesis, Cambridge University (1951).
4. L. Genzel, J. Mol. Spectroscopy 4, 241 (1960).
5. J. Strong, Concepts of Classical Optics, W. H. Freeman and Co., San Francisco (1958), p. 431.
6. D. H. Rank et al., J. Opt. Soc. Am. 50, 821 (1960).
7. T. K. McCubbin, R. P. Grosso, and J. D. Mangus, Applied Optics 1, 431 (1962).
8. W. S. Boyle and K. F. Rodgers, J. Opt. Soc. Am. 49, 66 (1959).
9. F. J. Low, J. Opt. Soc. Am. 51, 1300 (1961).
10. D. Bloor et al., Proc. Roy. Soc. A260, 513 (1961).
11. D. H. Martin and D. Bloor, Cryogenics 1, 159 (1961).

# THE ADVANTAGES OF INTERFERENCE SPECTROSCOPY OVER CONVENTIONAL INFRARED SPECTROSCOPY

## Sol M. Norman

Block Associates, Inc.
Cambridge, Massachusetts

The interferometer spectrometer has certain advantages over conventional spectrometers. These include: 1000 times greater sensitivity, a very large entrance hole instead of slits, and a very rapid scan—up to 6 spectra per second.

These advantages permit it to overcome certain current classes of problems in spectroscopy and also permit it to be used for entirely new analytical applications where spectrometers have not been considered feasible before.

Let us consider how the interferometer spectrometer works. First, consider an acoustical analogy, where we want the frequency spectrum of a musical sound. The way we would measure this would be to use a microphone which follows the fluctuations in the sound wave and converts it to a fluctuating electrical signal. This signal is then recorded on tape and fed into an audio frequency wave analyzer. The wave analyzer has a tunable electrical filter which tunes through the frequency spectrum and indicates the amplitude of each frequency present in the source.

The interferometer spectrometer system operates in a similar manner with infrared radiation. In order to have a detector equivalent to the microphone, we would need something fast enough to follow the fluctuations in the light waves. Since infrared detectors are too slow for this, we first process the light through a Michelson interferometer before the detector.

The interferometer used is modified by having one of its mirrors mounted on a transducer. This arrangement is shown in Fig. 1. Light entering the interferometer is divided into two beams by the beamsplitter; the two beams are reflected off the mirrors and recombine. The pattern as seen by the detector is a large circular fringe.

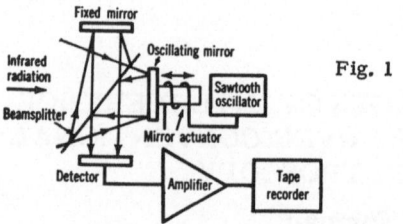

Fig. 1

Let us examine what happens to the brightness of this fringe as the mirror is moved (Fig. 2). Consider that we are looking at a monochromatic infrared source. If we consider our zero position to be where the variable path length is set equal to the fixed path, we have maximum brightness of the fringe. As the mirror is moved, the brightness is reduced, reaching a minimum when the mirror has travelled a quarter of the source wavelength. As the mirror travels further, the brightness increases—going through successive cycles.

If we used another monochromatic source, twice the wavelength of the first, the oscillations in brightness would occur half as frequently.

If we drove the mirror with a sawtooth voltage which produced a linear velocity on the forward motion, the detector would see a monochromatic source as a brightness fluctuation of constant frequency. The exact value of the frequency would depend on the source wavelength and the mirror velocity. For a given mirror velocity, each wavelength in the source would produce a different frequency at the detector. We merely choose the mirror velocity so that the frequencies produced are within the response capabilities of the detector.

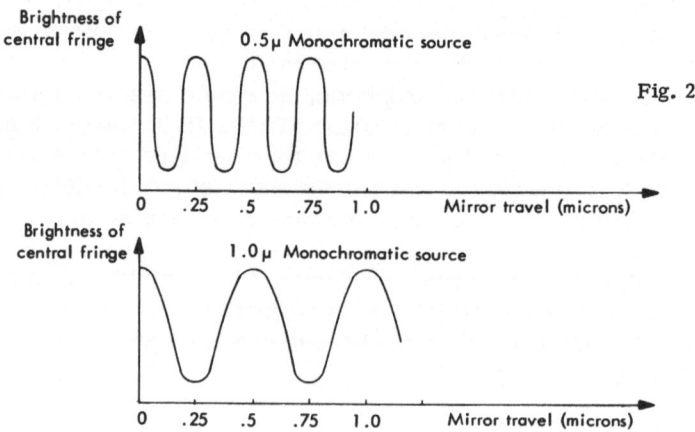

Fig. 2

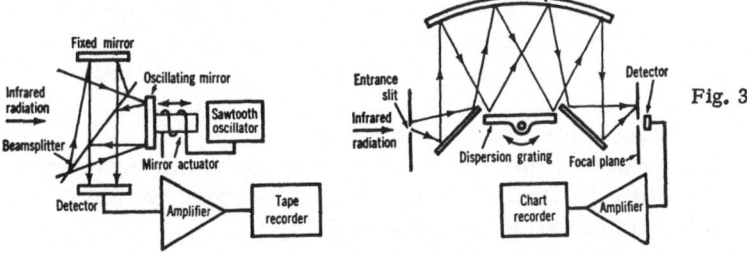

Fig. 3

The detector sees the superposition of all these frequencies—just as the microphone did with the sound frequencies.

Now we can handle our infrared spectral information just as we did the sound spectrum. We can feed it directly to an audio wave analyzer, or we can record it on tape and play it through the wave analyzer at our leisure, or we can handle it other ways if desired.

Let us compare the interferometer spectrometer with a conventional instrument to see how it achieves its gain in sensitivity (Fig. 3). The conventional instrument uses an entrance slit, a dispersive element, and an exit slit. The entrance slit immediately limits the amount of light which can be accepted by the instrument. Also, the light which does go into the instrument is spread out into a spectrum which is scanned by the exit slit. The detector sees only one wavelength at a time.

In the interferometer spectrometer, however, there is a large entrance hole—much larger than the slit in the conventional instrument—so that much more light can be accepted by it, typically 50 to 100 times as much. Also, the light that does enter is used more efficiently. Instead of looking at only one wavelength at a time, the detector sees all wavelengths simultaneously. For the same scan time and same number of resolution elements, say 400, for example, each wavelength in the interferometer is seen by the detector 400 times longer than in the conventional instrument. For infrared detectors, this will improve sensitivity by the square root of 400, or a factor of 20.

The combination of a larger aperture and more efficient use of scan time results in an over-all advantage of approximately 1000 in sensitivity.

The instrument's ruggedness and reliability stem from its simple construction—it has only one moving part suspended on flexing springs. There are no wearing surfaces anywhere in the instrument.

The scan rate is established by the electronic drive circuitry to the transducer; this requires no gear changes or other troublesome operations—just the changing of a selector switch position. Also there is no order sorting problem, since the output frequencies are not ambiguous.

The nature of the instrument output offers many interesting possibilities. Instead of being limited to a piece of chart paper, it can be presented in a number of ways.

We can use a multiple filter wave analyzer that presents the complete spectrum on the face of an oscilloscope at a rate up to 6 times per second. With this we can watch rapid changes taking place in a process.

Or, if we are interested in what is happening at only two or three wavelengths, we can use fixed electrical filters connected in parallel with the output of the instrument and monitor these wavelengths, or ratios between them, continuously on a meter. This is much simpler than running off individual sheets of spectra and then comparing them by hand.

Or, if we desire a chart, we can record several spectra on

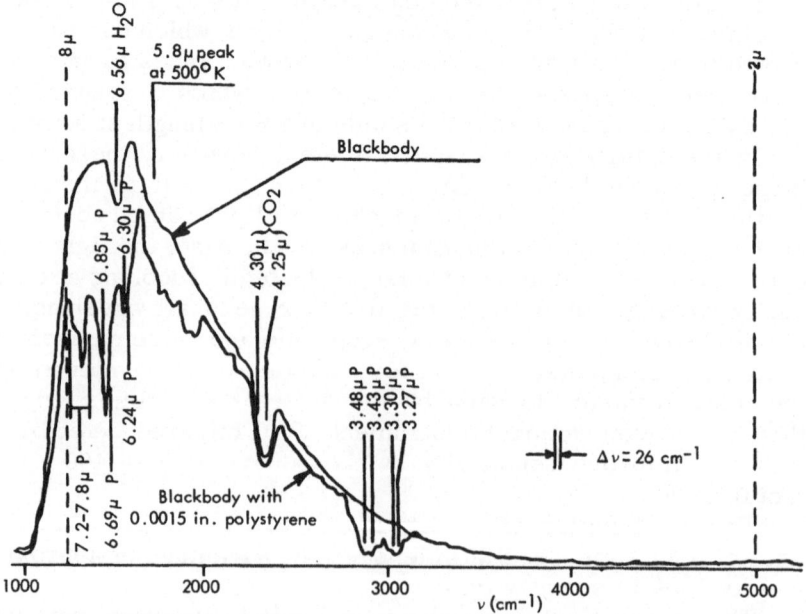

Fig. 4. Spectrum obtained with an I4S interferometer spectrometer under the following conditions: Bolometer detector; 1 cm aperture, $CaF_2$ optics; 4° field of view; 500°K blackbody source; 6-second recording time at 2 scans per second; ~30 cm$^{-1}$ resolution.

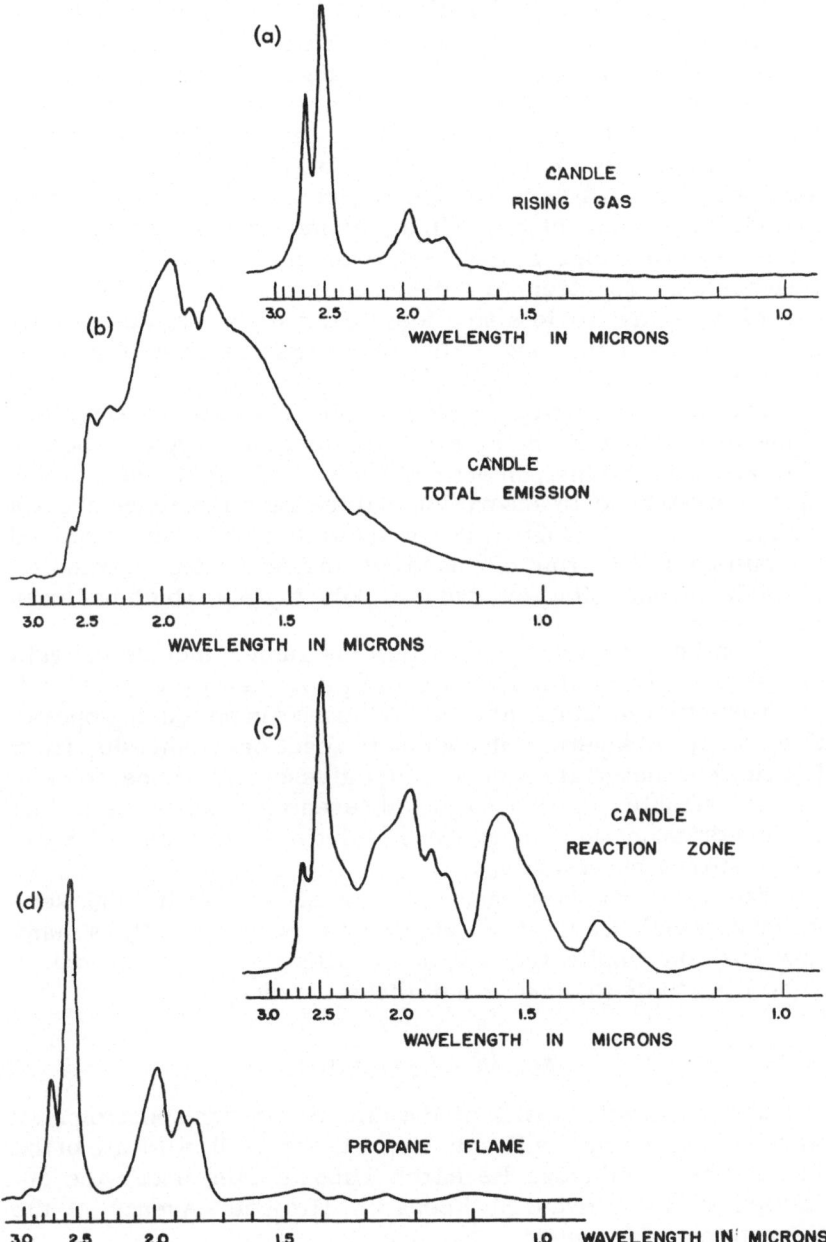

Fig. 5. Candle: (a) Rising gas; (b) total emission; (c) reaction zone. (d) Propane flame.

a tape loop and play it back through a wave analyzer and recorder, which will draw out the spectrum.

Also if one is interested in storing, cataloging, and comparing large numbers of spectra, these can be stored on tape and processed electronically to obtain the desired information.

The first uses for the rapid scan interferometer spectrometer were in military and geophysical applications. These included a variety of classified emission measurement problems and use under severe environmental conditions. An interferometer spectrometer designed for a space vehicle application is the world's smallest infrared spectrometer—only 1 in. diam. × 5 in. long, with an electronics control unit 4× 8 × 1 in.

The interferometer spectrometer has only recently become available for commercial applications. Typical spectra obtained with the instrument are shown in Figs. 4 and 5, which show polystyrene in absorption and the reaction zones about a candle flame. Although it is not shown here, we have obtained the emission spectrum of a hand held in front of the instrument, as well as emission spectrum of polyethylene which had been warmed to about 100°F.

Promising applications for this instrument include analysis of emission from films, control of pilot plant processes such as orientation in films, analysis of furnace gases during operation, study of kinetics of chemical reactions, emission from fluorescent materials such as laser glasses, microspectroscopy, reststrahlen spectroscopy, differential absorbance measurements such as for aqueous solutions, and use with gas chromatography systems.

The interferometer spectrometer, because of its high sensitivity, rapid scan, small size, and ruggedness, offers many new possible applications. It is a challenge to spectroscopists as well as to us to make use of these advantages.

## ACKNOWLEDGMENT

Certain developments of the interferometer spectrometer were sponsored by the Geophysics Research Directorate of the Air Force Cambridge Research Laboratories under the direction of the Advanced Research Projects Agency of the Department of Defense.

# NANOSECOND TIME RESPONSES
# OF DIFFERENT INDIUM ANTIMONIDE
# PHOTODETECTORS

## Heinz Fischer and Harold Rose

Air Force Cambridge Research Laboratories
Hanscom Field
Bedford, Massachusetts

## OBSERVATIONS

The time response of an InSb PEM Photodetector* [1] was recently observed to be approximately 18 nsec [2] by means of a new type nanosecond light source [3]. This response is considerably faster than that claimed by the producer and agrees with predictions derived from steady-state investigations [4]. Spectral measurements with the PEM show a slower decay in the infrared which is possibly the result of a slower IR decay of the detector but may also result from an IR tail of the light pulse, a possibility which has not yet been confirmed (see Fig. 1).

True rise times cannot be measured accurately because the observed rise times are so fast ($\gtrsim 2.5$ nsec) that they are comparable to the rise of the light pulse. On the other hand, observed rise times also depend on the length of the light pulse, as will be explained later.

Studies comparing this PEM with PC cells of fast geometry (linetype) prove the PC to be less sensitive at comparable time response—which can be regulated by doping [4]. The experiments show that the rise time of PC always is slower than that of the PEM even when the decay is faster; this indicates, on the other hand, that the observed rise apparently is not circuit limited. The results are presented in Fig. 2.

## DEDUCTIONS

Figure 3 explains the mechanism of the PEM (Photo-Electro-Magnetic) and PC (Photo-Conductive) detectors. The

*Manufactured by Minneapolis-Honeywell.

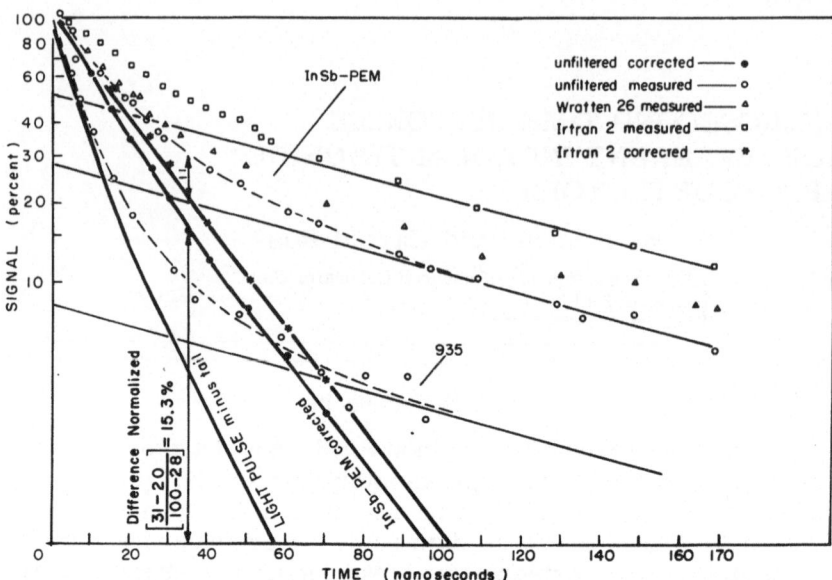

Fig. 1. Decay times of the InSb PEM detector compared with the light pulse as observed with the RCA 935 photodiode.

incident light produces electron—hole pairs which diffuse into the interior. Electrons and holes are bent in opposite directions by an external magnetic field in the PEM detector. Either a net voltage or a current results, depending on connection to an external circuit. In the case of the PC detector the carriers are moved by an external electric field, thus producing a current.

The response of the PC Photodetector is determined by the volume lifetime $r$ exclusively because the PC current presents the sum of all free carriers; it is assumed that the surface recombination velocity is small and that the volume recombination is not due to electron trapping. This is generally accepted for the InSb at room temperatures.

The PEM photocurrent, however, is proportional to the difference of the carrier density at front and backside, and thus depends not only on $r$ but also on the thickness $d$ of the detector—since $d$ affects the carrier difference due to possible reflections from the backside. This effect shows up in amplitude [1], decay, and rise when the thickness comes close to the diffusion length $l = (Dr)^{\frac{1}{2}}$, where $D$ is the ambipolar diffusion constant. As a result the response time of the PEM can be shorter with equal volume lifetime $r$; this goes so far that in one case the decay of the PC was considerably faster (due to

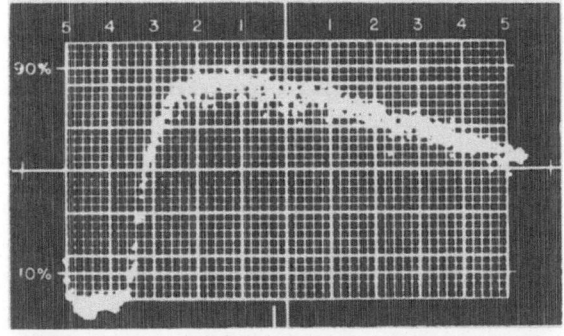

PC#1
Scale = 5 nsec/major div.
10-90% Risetime = 5.3 nsec
Half width = 55 nsec

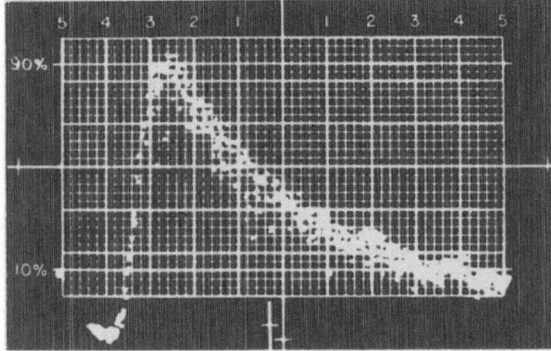

PC#2  (Highly doped)
Scale = 5 nsec/major div.
10-90% Risetime = 3.2 nsec
Half width = 17 nsec

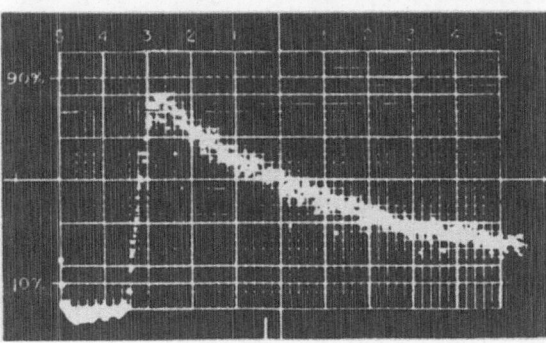

PEM
Scale = 5 nsec/major div.
10-90% Risetime = 2.3 nsec
Half width = 25 nsec

Fig. 2. Time functions of PEM and PC detectors as observed by Lumatron-Sampling-Scope (rise time ~0.3 nsec).

different doping) than that of a PEM and still had a longer rise time (see Fig. 2).

However, observed rise times of the PC and PEM are only partly a detector property since they depend on the duration of the incident light pulse when the light pulse is shorter or comparable to the volume lifetime.

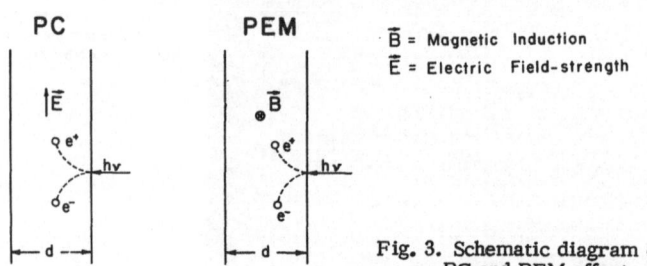

Fig. 3. Schematic diagram of the PC and PEM effect.

A shorter light pulse not saturating the detector results in an apparent faster rise of the photocurrent; however, the amplitude is decreased in comparison to that resulting from a longer light pulse. This decrease of the amplitude is stronger with the PC than with the PEM because the PEM is generally faster and consequently reaches a certain percentage of its saturation level earlier. A detailed theoretical study will be published separately.

## AMPLITUDES

Actual signals as produced by the combination of light source and PEM are given in the table. From these values some judgment of the potential for various spectroscopic applications may be derived; however, detailed knowledge of noise level and its origin will be required for adequate analysis.

The PEM itself ($\sim$20 ohm) is assumed to be "Johnson" noise limited [1]. This may not be of much significance for some kind of "single" pulse nanosecond spectroscopy where more serious noise sources including electric pickups have to be considered. On the other hand, high repetition pulse rates exceeding $10^4$ per sec, which are possibly due to negligible erosion of the electrodes, permit application of sensitive pulse-integrating systems.

Single Pulse Signals as Received through Various Spectral Filters*

| Filter | No filter | Wratten 4 | Wratten 26 | Irtran 2 |
|---|---|---|---|---|
| Range ($\mu$) | — | > 0.46 | > 0.61 | > 1.5 |
| Amplitude (mv) | 750 | 490 | 430 | ≲ 200 |

*Measurements of spectral source emittances are under way.

The light of the SM 4 [3] is focused by means of a Ploetz clamshell surface mirror system [5] of small $f$ number, i.e., large solid angle. The signal from the PEM is fed into a 1-m, 50-ohm cable, terminated at the scope end and connected into a Tektronix 585 which has adequate band width to record true maximum amplitudes; the half-width of this particular light pulse $\sim 7$ nsec.

## SUMMARY

Nanosecond time responses of PEM and PC indium antimonide detectors as determined by means of a new light source promise considerable interest for nanosecond infrared spectroscopy—such as flash photolysis, for example. Light output and cell sensitivity are adequate. Until now, this combination of an ultrafast light source and an infrared detector very nearly as fast has not been available.

## ACKNOWLEDGMENTS

W. B. Rüppel and C. C. Gallagher of AFCRL performed the experiments; C. Hergenrother of Block Associates developed the PC detectors and participated in discussions.

### REFERENCES

1. P. W. Kruse, J. Appl. Phys. 30, 770 (1959).
2. H. Fischer and P. von Thuena, Appl. Optics 1, 373 (1962).
3. H. Fischer, J. Opt. Soc. Am. 51, 543 (1961).
4. R. N. Zitter, A. J. Strauss, and A. E. Attard, Phys. Rev. 115, 266 (1959).
5. G. Ploetz, AFCRC-TN-59-759 (1959).

# A DOUBLE-BEAM AUTOMATIC PRISM-GRATING INFRARED RECORDING SPECTROPHOTOMETER

## Martin H. Gurley III and Edward Merrill

Baird-Atomic, Inc.
Cambridge, Massachusetts

The Baird-Atomic Model NK-3 prism-grating infrared spectrophotometer described in this report provides rapid and accurate quantitative analysis. By adding a grating monochromator to the basic NK-1 spectrophotometer, greater resolution has been provided, while retaining the simplicity and ease of operation inherent in all Baird-Atomic infrared instruments.

On the NK-3, spectra is presented on a strip chart recorder with no breaks or overlap. A pen on the margin marks the frequency intervals and a frequency indicator on the instrument rotating with the cams gives a continuous frequency reading. By employing the variable scan speed of the instrument and speed changes in the recorder, a variety of spectra presentations are available.

The unbroken spectra presentation is achieved by a simple, yet functional indexing mechanism. (Figure 1 illustrates the electrical block diagram.) The region covered is from 5000 $cm^{-1}$ ($2.0\mu$) to 400 $cm^{-1}$ ($25\mu$) with change points at 2500, 1250, and 700 $cm^{-1}$. These portions of the spectra are covered using two gratings, 150 lines/mm and 40 lines/mm, each in two orders.

This unique indexing mechanism (Fig. 2) consists of a clutch and reversible motor combination to rotate the grating, an AC solenoid (index solenoid) which locks the cam follower to the grating shaft, and a latching relay which operates these components. There also are two slotted drums with movable actuators which operate the indexing microswitches. The operation of the indexing mechanism is straightforward. As the scan approaches an index point, two of the index switches in series close. This action operates the latching relay. The latching relay in turn simultaneously operates the index

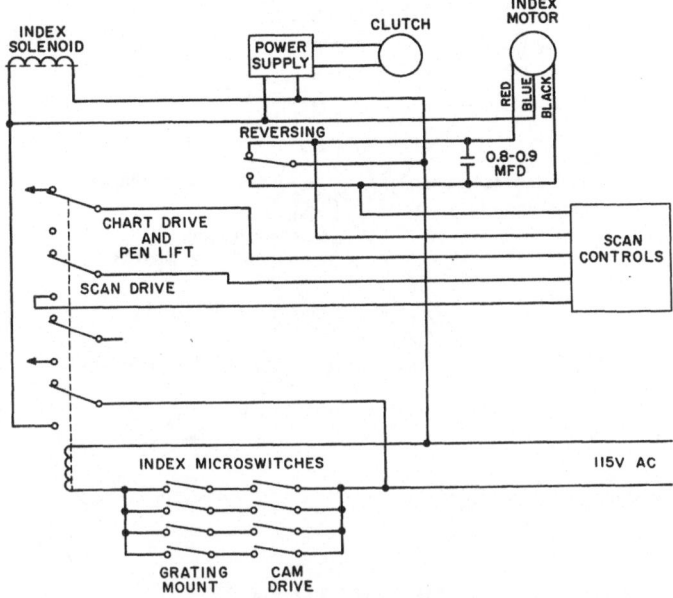

Fig. 1. Indexing mechanism electrical diagram.

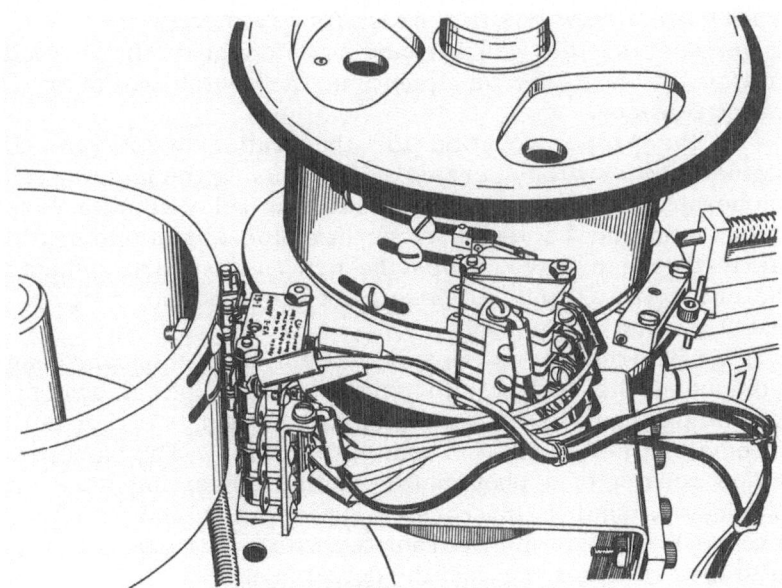

Fig. 2. Indexing mechanism.

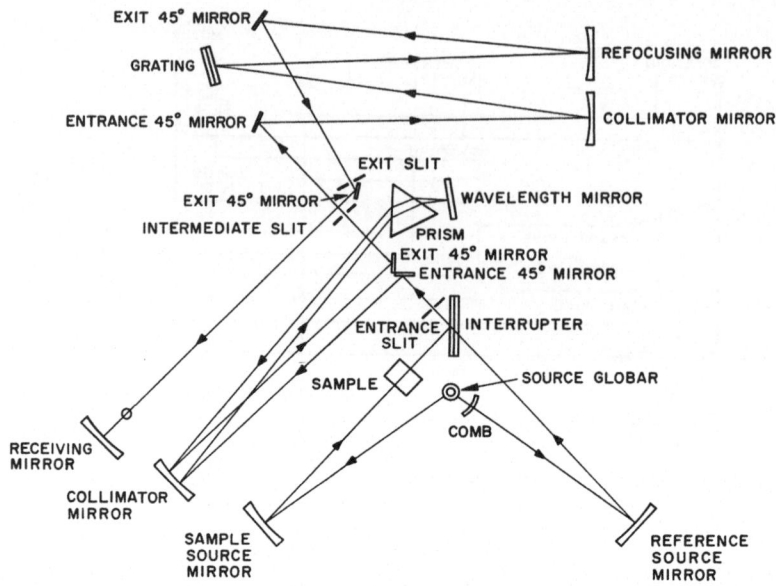

Fig. 3. NK-3 optics.

solenoid, which releases the cam follower from the grating shaft, actuates the motor clutch combination to rotate the grating to a new position or shifts gratings, stops the scan drive, operates the pen lift, and stops the chart drive. As the grating moves to the new position, the switch on the grating mount releases.

At the proper location for the grating to start the next segment, this switch operates again, causing the latching relay to operate. This operation of the latching relay simultaneously releases the index solenoid, stops the motor-clutch combination, starts the scan drive, drops the pen on the chart, and starts the chart drive. This process is repeated at each index point and at the end of the scan.

The instrument is a double-beam dual monochromator type using the optical null principle. The optical system is that of the basic NK-1 with a modified Czerny-Turner monochromator* at the NK-1's exit slit (Fig. 3). The basic NK-1 optics consist of a photometer section containing the source (a globar), source mirrors, sample areas and interrupter (chopper), a prism monochromator with a KBr prism which is used for order sorting, and the detector optics.

*See H. Czerny and A. F. Turner, Z. Physik 61, 792 (1940).

In Fig. 3, the NK-1 exit slit now becomes an intermediate slit in this arrangement. Therefore, the small 45° mirror previously located at the exit slit has been removed and the beam travels to the first 45° mirror (entrance) in the grating monochromator. From there it goes to the 85-cm focal length collimator mirror, then to the grating, and back to the 100-cm focal length refocusing mirror. From this point it goes to the second 45° mirror (exit) and to the new exit slit which is on an extension from the intermediate slit. A 45° mirror behind the exit slit then directs the beam to the receiving mirror and bolometer.

The modified NK-1 serves as photometer and order-sorter. A KBr prism is used as a fore prism with the exit slit of the prism monochromator functioning as an intermediate slit. The prism-grating instrument gives approximately 1 cm$^{-1}$ resolution throughout in a normal scan which takes approximately 20 minutes. The instrument is capable of much greater resolution, 0.25 cm$^{-1}$ at 950 cm$^{-1}$, under more stringent operating conditions. It also is adaptable to other frequency ranges and operating modes, using other prism and grating combinations.

The NK-3 uses two gratings, a 150 line/mm blazed at 6$\mu$ and a 40 line/mm blazed at 22.5 $\mu$. Each grating is used in two orders. By rotating one grating in one direction and the other in the opposite it is possible to "fold" the cam, i.e., it rises as the two orders of the 150 line/mm grating are scanned and falls as the two orders of the 40 line/mm grating are scanned. This tends to reduce the tracking error of the follower and keeps the cam at a more manageable size.

At the present time the NK-3 covers the range from 5000 to 400 cm$^{-1}$, but interchange is possible using different grating-prism combinations. All the accessories available for the NK-1 may also be employed with the NK-3.

# RUBY OPTICAL MASER AS A RAMAN SOURCE [*]

## S. P. S. Porto and D. L. Wood

Bell Telephone Laboratories, Inc.
Murray Hill, New Jersey

A successful Raman source must have high radiance, strict monochromaticity, and a frequency for which the sample is not opaque. The ruby optical maser has from its first demonstration been an obvious possibility for this purpose. This report describes the successful use of this device for exciting Raman spectra.

The ruby optical maser has been described elsewhere [1], and it suffices to describe our particular equipment using the schematic diagram of Fig. 1. The exciting lamp (L) was a G. E. No. FT524 Xenon flash lamp with an American Speedlight Company condenser bank, charging supply, and lamp housing. The ruby rod (R) of 0.05% Cr concentration was obtained from the Linde Company, and the ends were polished flat to about $\frac{1}{5}$ fringe and parallel to about 30 sec of arc. One end of the rod was coated with an opaque silver layer, but the other end had a silver coating of about 25% reflectance. This permitted the total emitted light flux to be as large as possible at the expense of some directionality. A pyrex tube (T) surrounded the ruby rod so that a flow of gas could pass over the rod for cooling as indicated in Fig. 1.

The slightly divergent light beam from the end of the ruby rod was focused with a lens ($L_1$) to a spot several millimeters in diameter on the surface of the pyrex Raman cell. The radiation entering the cell was diffusely reflected many times through the sample by a coating of white $BaSO_4$ deposited on the surface of the cell. The scattered light from the cell was collected by a lens ($L_2$) and brought to a focus on the slit of a high aperture spectrograph built after the design of Bass and Kessler [2]. The maser was operated about three times a minute with a flash energy of about 3000 joules producing flashes having a total duration of about 1 msec. The exposures

*Published in J. Opt. Soc. Am. 52, 251 (1962).

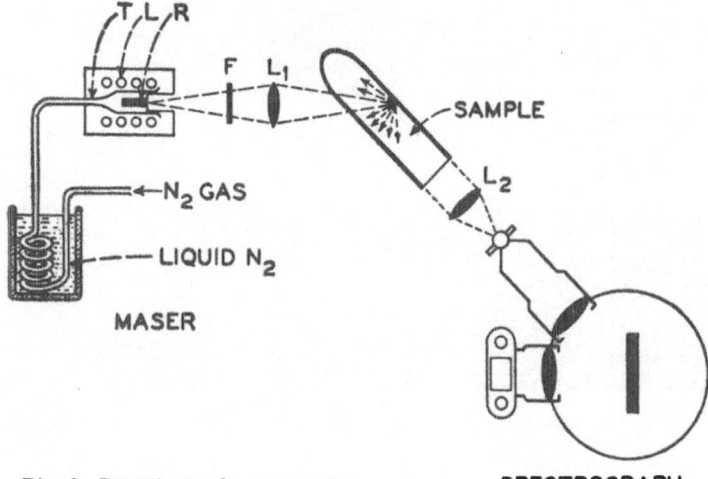

Fig. 1. Experimental arrangement.                    **SPECTROGRAPH**

were recorded on Eastman 35 mm type 1N spectrographic film, and each exposure involved from one to one hundred flashes, depending on the experiment.

The key features of this experimental arrangement which we believe are especially important for the successful observation of Raman lines are: (1) low reflectance on the output end of the ruby rod; (2) the fast spectrograph; (3) the $BaSO_4$ diffuse-reflection coating of the Raman cell.

The spectrograms which were first recorded with this equipment showed many other lines besides the Raman lines. These are shown in the spectrograms of Fig. 2. At the top in the figure (a) is the pure fluorescence spectrum of ruby, showing at E the $R_1$ line at 14,400 used for Raman excitation. There are lattice bands (F) between 14,400 $cm^{-1}$ and 13,670 $cm^{-1}$, and a strong one at 12,930 $cm^{-1}$. These features are present in the spectrum at (b) in Fig. 2 and in the spectrum of the empty cell. In addition to the fluorescence, there are seven lines due to Xe in the exciting lamp of the maser. The line G is due to a grating ghost from the 14,400 $cm^{-1}$ exciting line (E). An unfortunate reflection R is also present in the middle spectrogram of Fig. 2. Since these lines from the empty cell all come from the maser source, we have been able to eliminate them to some extent by using a second-order interference filter having peak transmission at the wavelength of the maser.

A better method for reducing unwanted lines in the spectrum from the ruby maser depends on the fact that the maser light

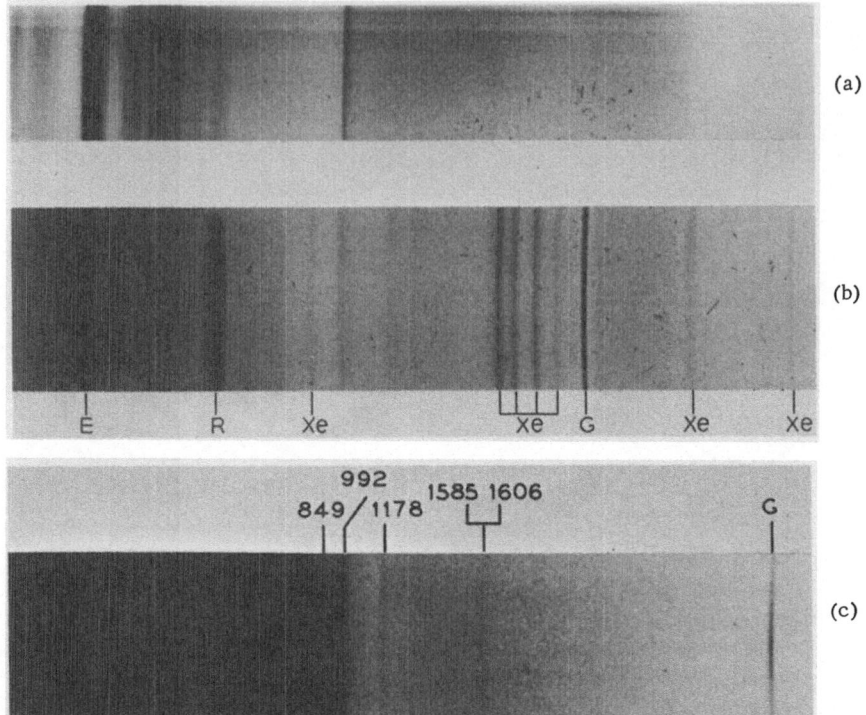

Fig. 2. (a) Fluorescence spectrum of a 0.05% ruby; (b) the Raman picture of $C_6H_6$ obtained by placing the ruby maser close to the Raman cell and with no light filtering; (c) the Raman effect of $C_6H_6$ obtained by an effective filtering of the maser light.

is highly directional, while the fluorescent radiation is not. If the Raman cell is located a considerable distance (in our experiment about 150 cm) from the maser rod, there will be essentially no loss in the maser light, while the fluorescent radiation diverges into a large solid angle and is thereby greatly reduced relative to the other. By this means, we are able to clean up the exciting light without the use of interference filters. Figure 2c shows a Raman spectrogram of benzene recorded this way with 50 flashes of the maser. The observed lines fall at 849, 992, 1178, 1585, and 1606 cm$^{-1}$. The 3050 cm$^{-1}$ line was not observed because of the drastic loss in sensitivity of the 1N film at 8800 A. Also the low lying frequencies were not observed in our experiment because of halation in the film in the vicinity of the exciting line. This can be reduced by filtering the light from the cell through a ruby crystal placed between the Raman cell and the spectro-

graph to reduce the intensity of the exciting line. It is necessary, however, to match the temperature of the filter crystal
to that of the maser rod since there is a relatively large shift
in frequency with temperature [3] of the maser line and the
corresponding ruby absorption line [4]. It is, incidentally, also
necessary to keep the temperature of the maser rod constant
during the many flashes that go to make up a single exposure,
since the Raman lines will also shift with the exciting line. It
is worth noting in this respect that below 77°K the shift of
frequency with temperature is very small and immersion of
the ruby rod in liquid nitrogen would provide better frequency
control than the gas-flow cooling.

Figure 3 shows the Raman effect of both $CCl_4$ and $C_6H_6$
with a Ne reference spectrum. They were taken with the improvements indicated above and by placing an absorber just in
front of the photographic plate so as to absorb the Rayleigh-
scattered maser light.

Even though we have successfully demonstrated that the
ruby maser will work as a Raman source, it is not in its present
form so very much better than other sources of red exciting
lines. We are aware, however, that many improvements may be
made in the experiment, after which this may indeed be a very
important Raman source. The first such improvement would
be to use a flash lamp having higher luminous flux and longer
flash duration for exciting the ruby rod in the maser. There
are many possibilities to be evaluated here. The second improvement might be in sample geometry to produce more
efficient excitation. For example, we have tried making the
second Fabry-Perot plate of the maser cavity highly reflecting
and removing it from the end of the ruby rod to form the rear
surface of the Raman cell. The maser light then traverses the
sample once for each reflection within the mode-selection
cavity. If the reflectance is high and as many as ten reflections
occur, there is a corresponding increase in the probability for
occurrence of the two-photon Raman process. The third improvement is obvious when it becomes feasible to pump the
ruby maser continuously, for then it will be possible to make
much longer exposures than we have used here, but still of the
order of seconds' duration. This will be a very important
Raman source, indeed, and calculation shows that the radiant
flux required for continuous pumping is not by any means
impossibly large and the continuously pumped ruby maser is
surely feasible.

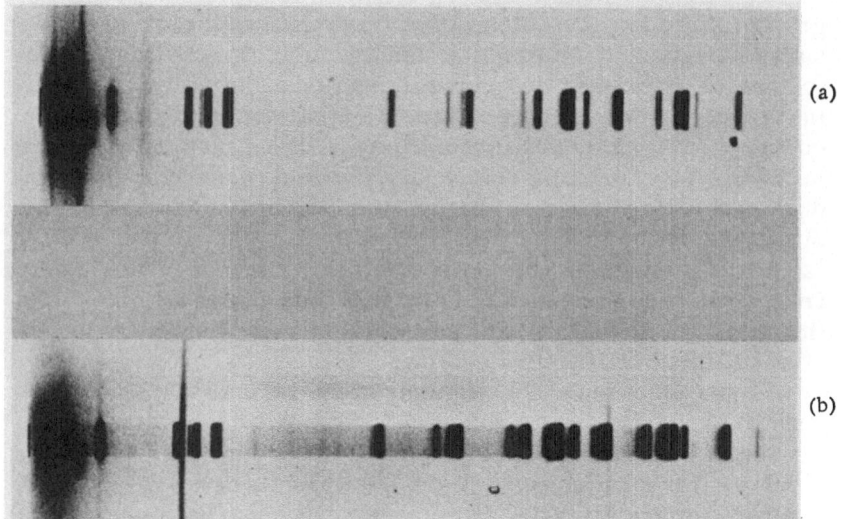

Fig. 3. Raman effect of (a) $CCl_4$; (b) $C_6H_6$ obtained by using the maser at a distance of 1.5 m from the cell and by reabsorbing the maser light in front of the plate.

It hardly needs to be pointed out that this intense, truly monochromatic Raman source at 6940 A in the red is very desirable for studying the vibrations of molecules of absorbing substances, of fluorescing materials, and for low lying frequencies. It is also possible that studies of the coherent Raman effect may also be possible, though this has yet to be demonstrated.

## ACKNOWLEDGMENTS

We are very grateful to W. K. Kaiser for the loan of a power supply; to Miss B. E. Prescott and L. E. Cheesman for general assistance in carrying out the experiments; and to S. Raman, now at Penn State University, for obtaining the pictures in Fig. 3.

### REFERENCES

1. T. H. Maiman, Nature 187, 493 (1960); British Communications and Electronics 7, 674 (1960); R. J. Collins et al., Phys. Rev. Letters 5, 303 (1960).
2. A. M. Bass and K. G. Kessler, J. Opt. Soc. Am. 49, 1223 (1959).
3. J. D. Abella and H. Z. Cummings, J. Appl. Phys. 32, 1177 (1961).
4. O. Deutschbein, Ann. d. Phys. 20, 828 (1934); K. S. Gibson, Phys. Rev. 8, 38 (1916).

# RAMAN AND INFRARED SPECTRA OF THE DICYANAMIDE ION

## Alfred J. Perkins
University of Illinois
Chicago, Illinois

---

The dicyanamide or dicyanimide ion $(N\equiv C-N-C\equiv N)^-$ is isoelectronic with carbon suboxide and therefore might exist in a linear resonance form $(N^-=C=\overset{+}{N}=C=N^-)$ having the same double bond array as the latter. A comparison of the infrared with Raman spectra should show at once whether this ion is linear or bent, since the linear form would have a center of symmetry and there would therefore be no coincidence in absorption between the two types of spectra.

The IR spectrum of this ion has been investigated at the American Cyanamide Corporation laboratories [1], and since this work was completed, Kuhn and Mecke [2] have reported the IR spectra in the NaCl and KBr regions of a number of salts. Since quite low values for bending frequencies for nitrile compounds have been reported [4-6], and the Raman spectra of a solution of the sodium salt showed indications of a broad line at 170 cm$^{-1}$, the infrared measurements were carried out to 100 cm$^{-1}$ on the instrument at MIT built by Lord and McCubbin [7].

## EXPERIMENTAL

Sodium dicyanamide was the principal salt used in this investigation. Two samples were prepared: one from sodium cyanamide and cyanogen bromide according to the method of Madelung and Kern [8] and the other from zinc dicyanamide kindly supplied by the American Cyanamide Corporation. The spectra of the two samples were identical.

Infrared spectra in the CaF$_2$, NaCl, and KBr regions were run on a Baird Associates Model A double-beam instrument using both nujol mulls and KBr pellets. For the spectra at frequencies below 425 cm$^{-1}$, the sodium dicyanamide was dispersed in polyethylene.

43

The Raman spectra were obtained using two instruments: the photographic grating instrument at MIT similar to the one designed by Lord and Miller [9]; and the Cary-81 at Argonne Laboratory, through the courtesy of Dr. J. Ferraro. Polarization measurements were made with each instrument. An extremely weak displacement at 545 cm$^{-1}$ from the exciting line was found on the Cary instrument which did not show up in the photographic work, although long exposures were used. Spectra of the solid sodium and barium salts also were obtained on the Raman instrument at Argonne using the technique of Ferraro et al. [10].

## RESULTS

The results of this experimental work are shown in Table I. The IR data agree very closely with that of Kuhn and Mecke [2]. Note that there are six coincidences between the Raman and infrared, four of the lines being strong and polarized. These coincidences tell us at once that the ion is not linear and leads to a $C_{2v}$ configuration as the only possibility since a $C_{2h}$ structure would have a center of symmetry and thus no coincidences between Raman and IR. For a 5-atom ion of this form, group theory predicts 4 vibrations of species $A_1$, 2 stretching and 2 bending, all active in both IR and Raman, the Raman lines being polarized; one out-of-plane bending vibration of species $A_2$, Raman active only; one out-of-plane bending vibration of species $B_1$, active in both spectra; and 3 vibrations of species $B_2$, 2 asymmetric stretches and one asymmetric bending, all active in both types of spectra.

The assignment of the observed frequencies to the $A_1$ species is apparent on the basis of the Raman polarization data. Also, the $B_2$ stretching frequencies must be at 2174 cm$^{-1}$ and 1340 cm$^{-1}$. This emphasizes the usefulness of Raman spectra in making assignments since Kuhn and Mecke [2], without the benefit of Raman data, assigned the 2232 cm$^{-1}$ frequency to the $B_2$ mode on the basis that the asymmetric vibration is usually of higher frequency than the symmetric one. Here this is not the case. Note also the wide separation of $\nu_2$ and $\nu_8$ and their very strong combination tone.

The assignment of the lines at 517, 526, and 544 cm$^{-1}$ to the other 3 bending vibrations is not straightforward. Had it been possible to get vapor state spectra, band contours would have helped. However, free dicyanamide polymerizes at once to

## TABLE I
### Vibrational Spectrum of $N(CN)_2^-$ Ion

| Raman | | | Infrared | | | |
|---|---|---|---|---|---|---|
| 2 M soln. | Na salt, solid | Ba salt, solid | Na salt | $Pb^{2+}$ salt | $Hg^{2+}$ salt | Assignment |
| | | 160 | | | | |
| 170 pol | 196 | 206 | 196 s | | | $\nu_4(A_1)$ |
| | | | | | 461 | |
| | | | | 496 | 495 | |
| | | | 517 s | 504 | 514 | |
| | | | 526 s | 528 | 527 | |
| 545 | | | 544 m | 545 | 545 | |
| | | | | | 654 | |
| 669 pol | 671 | 670 | 664 s | 687 | 662 | $\nu_3 (A_1)$ |
| | | | | 920 | 916 | |
| 927 pol | 937 | 940 | 931 m | 932 | 938 | $\nu_2 (A_1)$ |
| | | | 1034 vw | | | |
| | | | | | 1308 | |
| | | | 1340 s | 1349 | 1335 | $\nu_8(B_2)$ |
| | | | | | 1398 | |
| | | | 1430 vw | | | |
| | | | | 2120 | | |
| 2175 dep | | 2160 | 2174 vs | 2170 | 2180 | $\nu_7 (B_2)$ |
| 2221 pol | 2225 | 2225 | 2232 vs | 2220 | 2290 | $\nu_1 (A_1)$ |
| | | | 2283 vs | 2280 | | $\nu_2 + \nu_8$ |
| | | | 2740 | | | $\nu_1 + 517$ |
| | | | 3058 | | | |
| | | | 3215 | | | |
| | | | 3546 | | | $\nu_1 + \nu_8$ |

dicyandiamide and at least at present cannot be stabilized. The difficulty is due to the fact that there are too many lines in the infrared spectrum. Some help can be obtained from combination bands. The very weak 1034 cm$^{-1}$ absorption, which is 2 × 517, could be $2\nu_5 (2A_2 = A_1)$. However, the very weak lines at 1430 and 2740 cm$^{-1}$ are apparently combination tones of the 517 frequency with symmetric species. Thus, the 517 cm$^{-1}$ band should not be $\nu_5$ as $A_1 \times A_2 = A_2$ is not supposed to occur. This argument is weakened, however, since $\nu_5$ itself apparently is present because of a decrease in symmetry due to the state of aggregation. On this basis, then, the 517 cm$^{-1}$ line is either $\nu_6 (B_1)$ or $\nu_9 (B_2)$.

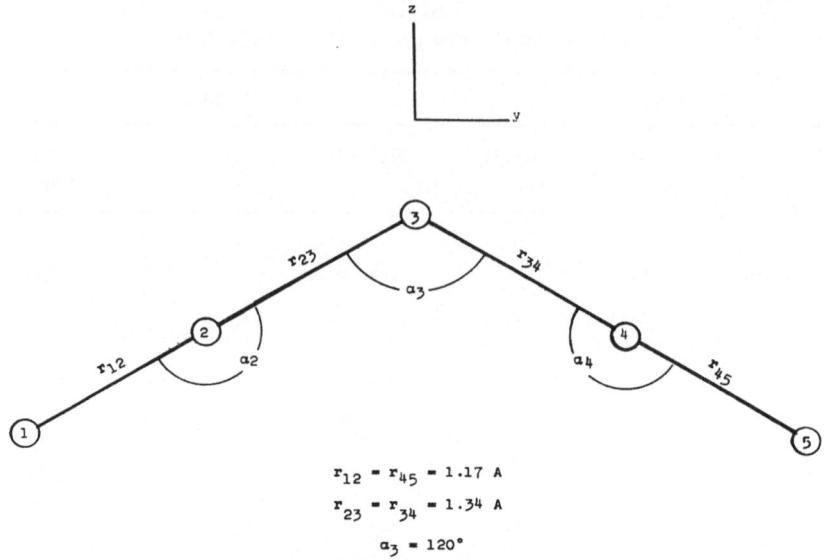

$$r_{12} = r_{45} = 1.17 \text{ A}$$
$$r_{23} = r_{34} = 1.34 \text{ A}$$
$$a_3 = 120°$$

To try to identify these 3 vibrations and obtain some measure of the binding force between the carbon and the central
nitrogen, a normal coordinate analysis was carried out. The
arrangement of atoms used in the calculations is shown in the
figure. The $x$ axis is taken perpendicular to the plane of the
ion. The bond distances were taken from a recent paper by
Miller et al. [11], and are based on a microwave study of
cyanamide and dimethyl cyanamide. They agree well with values
calculated from the bond orders of Davies and Jones [12] for
cyanamide. The angle between the C—N—C bonds is taken as
120°. This assumes $sp^2$ hybridization on the central N atom,
the electrons in the fourth orbital taking part in II bonding resonance or delocalization with the II cloud of the nitrile group.

The symmetry coordinates for this shape of the ion are
shown in Table II. Here the $a'$ indicates out-of-plane bending
about the N—C—N angle. From these coordinates the G matrix
elements were developed. Those for the $A_1$ and $A_2$ species are
shown in Table III, those for the $B_1$ and $B_2$ species in Table IV.

A full quadratic potential function was assumed since the
spacing of the various groups of lines indicated considerable
interaction. The form of the $F$ matrix elements used is shown
in Table V.

In carrying out the numerical computations the nitrile

## TABLE II
### Symmetry Coordinates
( $x$ axis perpendicular to molecular plane)

| $A_1$ species | $B_1$ species |
|---|---|
| $S_1 = \dfrac{1}{\sqrt{2}}(\Delta r_{12} + \Delta r_{45})$ | $S_6 = \dfrac{1}{\sqrt{2}}(\Delta a_2' + \Delta a_4')$ |
| $S_2 = \dfrac{1}{\sqrt{2}}(\Delta r_{23} + \Delta r_{34})$ | $B_2$ species |
| $S_3 = \dfrac{1}{\sqrt{2}}(\Delta a_2 + \Delta a_4)$ | $S_7 = \dfrac{1}{\sqrt{2}}(\Delta r_{12} - \Delta r_{45})$ |
| $S_4 = \Delta a_3$ | $S_8 = \dfrac{1}{\sqrt{2}}(\Delta r_{23} - \Delta r_{34})$ |
| $A_2$ species | $S_9 = \dfrac{1}{\sqrt{2}}(\Delta a_2 - \Delta a_4)$ |
| $S_5 = \dfrac{1}{\sqrt{2}}(\Delta a_2' - \Delta a_4')$ | |

frequencies were factored out by Wilson's method, thus reducing the $FG$ matrix of the $A_1$ species to a $3 \times 3$ and that of the $B_2$ species to a $2 \times 2$. It was further assumed that the force constants for the out-of-plane bending modes were equal to those for the in-plane modes, i.e., that $F_{55} = F_{99}$ and $F_{33} = F_{66}$. When an $F_{55}$ obtained by assigning the 517 cm$^{-1}$ to $\nu_5$ was used to calculate $F_{88}$ and $F_{89}$ using $\nu_9$ as either 526 cm$^{-1}$ or 544 cm$^{-1}$, no real solution to the equations could be obtained. This is partial confirmation of the previous analysis that the 517 cm$^{-1}$ line should be either $\nu_6$ or $\nu_9$. For $\nu_5 = 526$ cm$^{-1}$, a real solution for $F_{88}$ and $F_{89}$ could be obtained only when $\nu_9$ was assigned to the 517 cm$^{-1}$ line. For $\nu_5 = 544$ cm$^{-1}$, real solutions could be obtained for $\nu_9 = 517$ cm$^{-1}$ or 526 cm$^{-1}$. In each of these cases $F_{88}$ came out to be 7.5 ± 0.05 md/A, a value about halfway between those quoted for a C—N single bond and a C—N double bond.

The next step in the calculation was to use these values of $F_{89} = F_{23}$ and the corresponding value of $F_{33} = F_{66}$ to determine the force constants of the $A_1$ species. In this connection, Thompson and Linnett [14] showed that for the type of resonance probably

## TABLE III
### G Matrix Elements

---

$A_1$ species

---

$$G_{11} = \mu_1 + \mu_2$$

$$G_{22} = \mu_2 + 2\mu_3 \cos^2 \frac{\alpha_3}{2}$$

$$G_{33} = \frac{1}{r_{12}^2} \mu_1 + \frac{1}{(r_{12} + r_{23})^2} \mu_2 + \frac{2}{r_{23}^2} \mu_3 \sin^2 \frac{\alpha_3}{2}$$

$$G_{44} = \frac{2}{r_{23}^2} \left( \mu_2 + 2\mu_3 \sin^2 \frac{\alpha_3}{2} \right)$$

$$G_{12} = -\mu_2$$

$$G_{13} = G_{14} = 0$$

$$G_{23} = \frac{\mu_3}{r_{23}} \cos \alpha_3$$

$$G_{24} = -\frac{\sqrt{2}}{r_{23}} \mu_3 \sin \alpha_3$$

$$G_{34} = -\sqrt{2} \left( \frac{\mu_2}{r_{23}(r_{12} + r_{23})} + \frac{2\mu_3}{r_{23}^2} \sin^2 \frac{\alpha_3}{2} \right)$$

---

$A_2$ species

---

$$G_{55} = \frac{\mu_1}{r_{12}^2} + \frac{\mu_2}{(r_{12} + r_{23})^2}$$

---

## TABLE IV
### G Matrix Elements

---

$B_1$ species

---

$$G_{66} = \frac{\mu_1}{r_{12}^2} + \frac{\mu_2}{(r_{12} + r_{23})^2} + \frac{2\mu_3}{r_{23}^2}$$

---

$B_2$ species

---

$$G_{77} = \mu_1 + \mu_2$$

$$G_{88} = \mu_2 + 2\mu_3 \sin^2 \frac{a_3}{2}$$

$$G_{99} = \frac{\mu_1}{r_{12}^2} + \frac{\mu_2}{(r_{12} + r_{23})^2} + \frac{2\mu_3}{r_{23}^2} \cos^2 \frac{a_3}{2}$$

$$G_{78} = -\mu_2$$

$$G_{79} = 0$$

$$G_{89} = -\frac{\mu_3}{r_{23}} \sin a_3$$

---

occurring here, $F_{22}$ should be greater than $F_{88}$, i.e., $f_{r_{23}r_{34}}$ should be positive. Only for the assignment $\nu_9 = 517$ cm$^{-1}$, $\nu_6 = 526$ cm$^{-1}$, and $\nu_5 = 544$ cm$^{-1}$ could reasonable values of both $F_{22}$ and $F_{44}$ be obtained. The other two possible assignments yielded values of below 0.2 for all reasonable $F_{22}$ values.

A set of force constants that will reproduce the frequencies and which seem reasonable are given in Table VI. Since two more force constants than frequencies were used, these numbers are not a unique solution. However, $F_{22}$ cannot be changed much without either making the interaction constants much too large or $F_{44}$ much too small.

## ACKNOWLEDGMENTS

The far infrared spectra and the photographic Raman spectra were obtained at the Spectroscopic Laboratory of the Massachusetts Institute of Technology through the courtesy of

## TABLE V
### $F$ Matrix Elements

| $A_1$ species | $B_1$ species |
|---|---|
| $F_{11} = f_{r_{12}} + f_{r_{12}r_{45}}$ | $F_{66} = r^2(f_{\alpha_2'} + f_{\alpha_2'\alpha_4'})$ |
| $F_{12} = f_{r_{12}r_{23}}$ | **$B_2$ species** |
| $F_{13} = r(f_{r_{12}\alpha_2})$ | $F_{77} = f_{r_{12}} - f_{r_{12}r_{45}}$ |
| $F_{22} = f_{r_{23}} + f_{r_{23}r_{34}}$ | $F_{78} = f_{r_{12}r_{23}}$ |
| $F_{23} = r(f_{r_{23}\alpha_2})$ | $F_{79} = rf_{r_{12}\alpha_2}$ |
| $F_{24} = r\sqrt{2}f_{r_{23}\alpha_3}$ | $F_{88} = f_{r_{23}} - f_{r_{23}r_{34}}$ |
| $F_{33} = r^2(f_{\alpha_2} + f_{\alpha_2\alpha_4})$ | $F_{89} = rf_{r_{23}\alpha_2}$ |
| $F_{34} = r^2\sqrt{2}f_{\alpha_2\alpha_3}$ | $F_{99} = r^2(f_{\alpha_2} - f_{\alpha_2\alpha_4})$ |
| $F_{44} = r^2f_{\alpha_3}$ | |
| **$A_2$ species** | |
| $F_{55} = r^2(f_{\alpha_2'} - f_{\alpha_2'\alpha_4'})$ | |

## TABLE VI
### Force Constants* Based on
### $\nu_5 = 544, \nu_6 = 526,$ and $\nu_9 = 517$

| | |
|---|---|
| $F_{22} = 8.29$ | $F_{55} = 0.614$ |
| $F_{23} = 0.882$ | $F_{66} = 0.505$ |
| $F_{24} = 2.41$ | $F_{88} = 7.62$ |
| $F_{33} = 0.505$ | $F_{89} = 0.882$ |
| $F_{34} = -0.422$ | $F_{99} = 0.614$ |
| $F_{44} = 0.919$ | |

*$F_{22}$ and $F_{88}$ are in md/A; $F_{22}$, $F_{24}$, and $F_{89}$ are in md/rad; $F_{33}$, $F_{34}$, $F_{44}$, $F_{55}$, $F_{66}$ and $F_{99}$ are in md/rad$^2$.

Professor R. C. Lord, Jr., during the author's sabbatical leave. The later Raman spectra of solutions and solids were obtained through the courtesy of Dr. John Ferraro of Argonne National Laboratory, Argonne, Illinois.

## REFERENCES

1. "Sodium Dicyanamide," American Cyanamide Co.
2. M. Kuhn and R. Mecke, Chem. Ber. 94, 3010 (1961).
3. F. Halvorsen and R. J. Francel, J. Chem. Phys. 17, 694 (1949).
4. F. Halvorsen, R. F. Stamm, and J. J. Whalen, J. Chem. Phys. 16, 808 (1948).
5. N. E. Duncan and G. J. Janz, J. Chem. Phys. 23, 434 (1955).
6. F. A. Miller, R. B. Hannan, Jr., and L. R. Cousins, J. Chem. Phys. 23, 2127 (1955).
7. R. C. Lord and T. K. McCubbin, J. Opt. Soc. Amer. 47, 689 (1957).
8. Madelung and Kern, Ann. 427, 1, 26 (1922).
9. Harrison, Lord, and Loofbourow, Practical Spectroscopy, Prentice-Hall, New York (1948).
10. J. R. Ferraro, J. S. Ziomek, and G. Mack, Spec. Chim. Acta 17, 802 (1961).
11. D. J. Miller, G. Topping, and D. R. Lide, Jr., J. Mol. Spect. 8, 153 (1962).
12. M. Davies and W. J. Jones, Trans. Far. Soc. 54, 1454 (1958).
13. Wilson, Decius, and Cross, Molecular Vibrations, McGraw-Hill (1955), p. 74.
14. H. W. Thompson and J. W. Linnett, J. Chem. Soc., p. 1384 (1937).

# THE VIBRATIONAL SPECTRA OF trans-$C_2H_2I_2$ AND trans-$C_2D_2I_2$

## Robert H. Krupp, Edward A. Piotrowski, Forrest F. Cleveland
Spectroscopy Laboratory, Physics Department

## and Sidney I. Miller
Chemistry Department
Illinois Institute of Technology
Chicago, Illinois

## INTRODUCTION

With the availability of the deuterated form of trans-$C_2H_2I_2$, it was decided to examine the vibrational spectrum of trans-$C_2D_2I_2$ and thereby obtain a more complete assignment of the vibrational frequencies of the trans-diiodoethylenes. The only previous spectral work done on the cis- and trans-diiodoethylenes was by Miller et al. [1]. Since their Raman spectrum was obtained with a low resolution spectrograph, the Raman and infrared spectra of the trans-$C_2H_2I_2$ were also reinvestigated.

The rotational isomers cis- and trans-$C_2H_2I_2$ are shown in the figure. Trans-$C_2H_2I_2$ belongs to the point group $C_{2h}$. Six fundamentals are allowed in the Raman, but not in the infrared; of these, five are of the $a_g$ type and one of the $b_g$ type. The remaining six fundamentals, two of the $a_u$ and four of the $b_u$ types, are allowed in the infrared, but not in the Raman.

## EXPERIMENTAL PROCEDURES

The Raman spectra of trans-$C_2H_2I_2$ and trans-$C_2D_2I_2$ were obtained in solutions of $CCl_4$ and $CS_2$, using a two-prism Hilger spectrometer with a dispersion reciprocal of 16 A/mm at 4358 A. Two air-cooled pyrex mercury arcs were used to illuminate the Raman tube. The 4358 A mercury line was isolated by using a solution of rhodamine 5G DN-Extra and paranitrotoluene in ethyl alcohol [2].

Since trans-$C_2H_2I_2$ and trans-$C_2D_2I_2$ quickly decompose when irradiated with light from the mercury arcs, the samples were

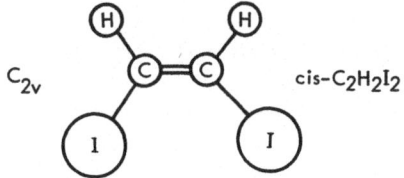

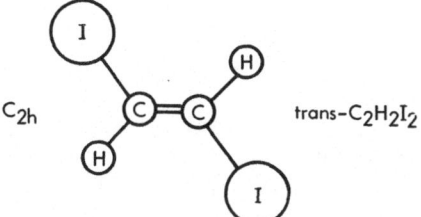

Rotational isomers
cis- and trans-$C_2H_2I_2$.

changed every 30 minutes in the 0.5- to 16-hour Raman exposures. In addition, a few drops of pinene ($C_{10}H_{16}$) were added
to the samples to remove iodine as it was formed. The depolarization ratios were obtained by inserting a split-field
Polaroid between the Raman tube and the slit of the spectrograph, and then measuring the intensities of the two components of the Raman spectrum. A Leeds and Northrup microphotometer was used in intensity determinations.

The infrared spectra of trans-$C_2H_2I_2$ and trans-$C_2D_2I_2$ were
obtained in the range of 400-4000 cm$^{-1}$ with a Perkin-Elmer
Model 21 double-beam infrared spectrophotometer, equipped
with CaF$_2$ and NaCl optics; and a Beckman IR-2 spectrophotometer equipped with KBr optics. As in obtaining the Raman
spectra, solutions of trans-$C_2H_2I_2$ and trans-$C_2D_2I_2$ in CCl$_4$ and
CS$_2$ were used in obtaining the infrared spectra.

## RESULTS

Table I contains the observed Raman and infrared spectral
data of trans-$C_2H_2I_2$. In the first column are the Raman displacements ($\Delta\sigma$) in cm$^{-1}$; in the second column are the semiquantitative relative intensities, in which the strongest line was
given a value of 1000; in the third column are the depolarization
ratios ($\rho$); and in the fourth column are the polarization states
(PS). The fifth and seventh columns contain the infrared wavenumbers in cm$^{-1}$. The intensities listed in columns six and

## TABLE I
### Raman and Infrared Spectral Data and Assignments for trans-$C_2H_2I_2$[*]

| Raman | | | | Infrared | | | | | Tentative theoretical assignment | Type |
|---|---|---|---|---|---|---|---|---|---|---|
| $\Delta\sigma$ | I | $\rho$ | PS | $\sigma$ ($CCl_4$) | $I_e$ | $\sigma$ ($CS_2$) | $I_e$ | $\sigma_c$ | | |
| 154 | 1000 | 0.22 | P | | | | | | $\sigma_5$ | $a_g$ |
| | | | | | | | | ~ 163† | $\sigma_7$ | $a_u$ |
| | | | | | | | | ~ 186† | $\sigma_{12}$ | $b_u$ |
| | | | | 499 | vvw | 499 | vvw | 504 | $\sigma_4 - \sigma_7$ | $A_u$ |
| 555 | 3 | 0.44 | P | | | | | 551 | $\sigma_8 - \sigma_5$ | $B_g$ |
| | | | | 587 | s | 585 | s | | $\sigma_{11}$ | $b_u$ |
| 594 | 11 | 0.79 | | | | | | 593 | $(\sigma_8 - \sigma_5) - \sigma_7$ | $A_g$ |
| 667 | 185 | 0.29 | P | | | | | | $\sigma_4$ | $a_g$ |
| | | | | 694 | vvw | 694 | vvw | 699 | $(\sigma_4 + \sigma_{12}) - \sigma_5$ | $B_u$ |
| 705 | 5 | 0.84 | D | | | | | | $\sigma_8$ | $b_g$ |
| | | | | 740 | vw | 740 | vw | 740 | $\sigma_5 + \sigma_{11}$ | $B_u$ |
| | | | | 784 | w | 783 | w | 779 | $(\sigma_{10} - \sigma_{12}) - \sigma_7$ | $A_u$ |
| | | | | | | 855 | vvw | 853 | $\sigma_4 + \sigma_{12}$ | $B_u$ |
| | | | | 912 | vs | 908 | vs | | $\sigma_6$ | $a_u$ |
| | | | | 1057 | m | 1053 | m | 1064 | $\sigma_5 + \sigma_6$ | $A_u$ |
| | | | | 1129 | vs | 1126 | vs | | $\sigma_{10}$ | $b_u$ |
| 1174 | 5 | 0.74 | P | | | | | 1172 | $2\sigma_{11}$ | $A_g$ |
| | | | | 1229 | m | 1226 | m | 1224 | $2\sigma_8 - \sigma_{12}$ | $B_u$ |
| 1230 | 229 | 0.26 | P | | | | | | $\sigma_3$ | $a_g$ |
| 1329 | 6 | 0.64 | P | | | | | | $2\sigma_4$ | $A_g$ |
| 1542 | 6 | 0.62 | P | | | | | | $\sigma_2$ | $a_g$ |
| | | | | 1605 | s | 1601 | s | 1615 | $\sigma_6 + \sigma_8$ | $B_u$ |
| | | | | 1794 | w | 1791 | w | 1795 | $\sigma_4 + \sigma_{10}$ | $B_u$ |
| | | | | 1815 | w | 1811 | w | 1816 | $\sigma_3 + \sigma_{11}$ | $B_u$ |
| | | | | 2671 | vw | 2668 | vw | 2670 | $\sigma_2 + \sigma_{10}$ | $B_u$ |
| | | | | 3072 | s | 3067 | s | | $\sigma_9$ | $b_u$ |
| 3097 | 6 | 0.53 | P | | | | | | $\sigma_1$ | $a_g$ |
| | | | | 3729 | w | 3726 | w | 3737 | $\sigma_4 + \sigma_9$ | $B_u$ |
| | | | | 3790 | vvw | 3787 | vvw | 3775 | $\sigma_8 + \sigma_9$ | $A_u$ |

[*]$\Delta\sigma$ = Raman displacement in $cm^{-1}$; I = semiquantitative relative intensity; $\rho$ = depolarization ratio; PS = polarization state (P, polarized; D, depolarized); $\sigma$ = wavenumber in $cm^{-1}$; $I_e$ = estimated relative intensity (s = strong, m = medium, v = very); $\sigma_c$ = calculated wavenumber in $cm^{-1}$.
†Deduced from combinations.

eight are estimated. The assignment of the fundamentals was based on the intensity measurements, depolarization ratios, and knowledge of the position of group frequencies.

## TABLE II
### Raman and Infrared Spectral Data and Assignments for trans-$C_2D_2I_2$*

| Raman | | | | Infrared | | | | | Tentative theoretical assignment | Type |
|---|---|---|---|---|---|---|---|---|---|---|
| $\Delta\sigma$ | I | $\rho$ | PS | $\sigma$ (CCl$_4$) | I$_e$ | $\sigma$ (CS$_2$) | I$_e$ | $\sigma_c$ | | |
| 155 | 1000 | 0.24 | P | | | | | | $\sigma_5$ | $a_g$ |
| | | | | | | | | $\sim 158^\dagger$ | $\sigma_7$ | $a_u$ |
| | | | | | | | | $\sim 177^\dagger$ | $\sigma_{12}$ | $b_u$ |
| | | | | | | | | $\begin{cases}391\end{cases}$ | $\sigma_{11}-\sigma_{12}$ | $A_g$ |
| 400 | 6 | | | | | | | $\begin{cases}410\end{cases}$ | $\sigma_{11}-\sigma_7$ | $B_g$ |
| | | | | | | | | $\begin{cases}443\end{cases}$ | $\sigma_4-\sigma_5$ | $A_g$ |
| 455 | 7 | | | | | | | $\begin{cases}445\end{cases}$ | $\sigma_8-\sigma_5$ | $B_g$ |
| | | | | 568 | vs | 568 | vs | | $\sigma_{11}$ | $b_u$ |
| 598 | 112 | 0.36 | P | | | | | | $\sigma_4$ | $a_g$ |
| | | | | | | | | 599‡ | $\sigma_8$ | $b_g$ |
| 633 | 16 | 0.41 | P | | | | | 619 | $\sigma_3-(\sigma_7+\sigma_{12})$ | $B_g$ |
| | | | | 668 | s | 665 | s | | $\sigma_6$ | $a_u$ |
| | | | | 841 | vs | 838 | vs | | $\sigma_{10}$ | $b_u$ |
| 894 | 22 | 0.54 | P | | | | | 899 | $\sigma_2-\sigma_8$ | $B_g$ |
| | | | | 896 | m | 894 | m | 903 | $\sigma_7+\sigma_{11}+\sigma_{12}$ | $A_u$ |
| | | | | 910 | vw | 907 | vw | $\begin{cases}911\\912\end{cases}$ | $\begin{array}{l}\sigma_4+\sigma_5+\sigma_7\\ \sigma_5+\sigma_7+\sigma_8\end{array}$ | $\begin{array}{l}A_u\\B_u\end{array}$ |
| 954 | 325 | 0.20 | P | | | | | | $\sigma_3$ | $a_g$ |
| | | | | 1129 | w | 1126 | w | | $\sigma_3+\sigma_{12}$ | $B_u$ |
| 1184 | 18 | 0.54 | P | | | | | 1196 | $2\sigma_4$ | $A_g$ |
| | | | | 1186 | m | 1183 | m | 1185 | $\sigma_{12}-(\sigma_5+\sigma_7)$ | $A_u$ |
| | | | | 1268 | m | 1265 | m | $\begin{cases}1265\\1266\end{cases}$ | $\begin{array}{l}\sigma_4+\sigma_6\\ \sigma_6+\sigma_8\end{array}$ | $\begin{array}{l}A_u\\B_u\end{array}$ |
| 1498 | 8 | | | | | | | | $\sigma_2$ | $a_g$ |
| | | | | 1523 | m | | | 1522 | $\sigma_3+\sigma_{11}$ | $B_u$ |
| | | | | 2271 | s | 2267 | s | | $\sigma_9$ | $a_u$ |
| 2307 | 8 | 0.41 | P | | | | | | $\sigma_1$ | $a_g$ |
| | | | | 3136 | vvw | 3132 | vvw | 3147 | $\sigma_1+\sigma_{10}$ | $B_u$ |

*Symbols and units as in Table I.
†Deduced from combinations.
‡Calculated from a partial normal coordinate treatment.

The Raman and infrared spectral data of trans-$C_2D_2I_2$ are presented in Table II. The depolarization ratios could not be obtained for the 400, 455, and 1498 cm$^{-1}$ lines, since these were very weak. The 599 cm$^{-1}$ line was not seen directly.

From a partial normal coordinate analysis of the $b_g$ type out-of-plane vibration, in which the $F$-matrix element was transferred directly from trans-$C_2H_2I_2$, the wavenumber 599 cm$^{-1}$ was obtained. Since there is a strong $a_g$ fundamental at 598 cm$^{-1}$, this line at 599 cm$^{-1}$ would not be seen.

Finally, as an aid to the complete assignment of the fundamentals for trans-$C_2H_2I_2$ and trans-$C_2D_2I_2$, the fundamentals of trans-$C_2H_2F_2$, trans-$C_2H_2Cl_2$, and trans-$C_2H_2Br_2$, together with their deuterated analogs, were compared with our values. This information is shown in Table III. In this table the values for trans-$C_2H_2F_2$ and trans-$C_2D_2F_2$ were taken from Craig and Entemann [3]; the values for trans-$C_2H_2Cl_2$ and trans-$C_2D_2Cl_2$ are from Bernstein and Ramsay [4]; and the values for trans-$C_2H_2Br_2$ and trans-$C_2D_2Br_2$ are from Dowling et al. [5].

As may be noted, the fundamentals of the trans-halogen ethylenes show definite progressions, and this information may be used as an aid in assignments. A case in point is that of the 705 cm$^{-1}$ line in trans-$C_2H_2I_2$: there is a more intense line at

## TABLE III
Fundamental Wavenumbers of trans-$C_2X_2Y_2$ (X = H or D; Y = F, Cl, Br, or I)[1]

| Type | $n$ | $C_2H_2F_2$ | $C_2H_2Cl_2$ | $C_2H_2Br_2$ | $C_2H_2I_2$ | $C_2D_2F_2$ | $C_2D_2Cl_2$ | $C_2D_2Br_2$ | $C_2D_2I_2$ |
|---|---|---|---|---|---|---|---|---|---|
| | $\sigma_1$ | | 3073 | 3089 | 3097 | | 2325 | 2323 | 2307 |
| | $\sigma_2$ | | 1578 | 1581 | 1542 | | 1570 | 1531 | 1498 |
| $a_g$ | $\sigma_3$ | | 1274 | 1249 | 1230 | | 992 | 972 | 954 |
| | $\sigma_4$ | | 846 | 746 | 667 | | 765 | 670 | 598 |
| | $\sigma_5$ | | 350 | 216 | 154 | | 346 | 214 | 155 |
| $a_u$ | $\sigma_6$ | 874 | 895 | 898 | 910 | 651 | 658 | 658 | 666 |
| | $\sigma_7$ | 325 | 227[2] | 188[3] | 163[3] | 309 | 221[4] | 183[3] | 158[3] |
| $b_g$ | $\sigma_8$ | 774[5] | 763 | 736 | 705 | 685[5] | 657 | 628 | 599[6] |
| | $\sigma_9$ | 3115 | 3080 | 3085 | 3070 | 2315 | 2285 | 2289 | 2269 |
| $b_u$ | $\sigma_{10}$ | 1274 | 1200 | 1165 | 1128 | 941 | 912 | 858[7] | 840 |
| | $\sigma_{11}$ | 1159 | 817 | 680 | 586 | 1175 | 784 | 648 | 568 |
| | $\sigma_{12}$ | 410[8] | 250[2] | 192[3] | 186[3] | 387[8] | 240[4] | 191[3] | 177[3] |

1 Symbols and units as in Table I.
2 Values from Pitzer and Hollenberg [6].
3 Deduced from combinations.
4 Calculated from Redlich-Teller product rule.
5 Calculated from product rule ratios.
6 Calculated from partial normal coordinate treatment.
   Perturbed by Fermi resonance.
   Calculated from the rotational isomer product rule.

594 cm$^{-1}$ in the Raman spectrum of this molecule that at first was assigned as the $b_g$ fundamental. On using this value as a fundamental and calculating the corresponding line in trans-$C_2D_2I_2$ by a partial normal coordinate treatment, a value of 506 cm$^{-1}$ was determined; this line was not observed. This correlation chart indicated that the 705 cm$^{-1}$ line might be assigned to the $b_g$ fundamental. In a partial normal coordinate analysis using this value, trans-$C_2D_2I_2$ should have a fundamental of the $b_g$ type at 599 cm$^{-1}$; this line would not be seen due to the presence of the strong $a_g$ fundamental at 598 cm$^{-1}$, as previously mentioned.

The infrared bands at 227 and 250 cm$^{-1}$ for trans-$C_2H_2Cl_2$ were observed by Pitzer and Hollenberg [6]. We made use of these values and the Redlich-Teller product rule to obtain 221 and 240 cm$^{-1}$ for the corresponding bands in trans-$C_2D_2Cl_2$. The values for similar fundamentals in the bromine and iodine compounds have been determined by combinations and overtones, both by Dowling and in the present investigation.

In applying the Redlich-Teller product rule to the various types of fundamentals of trans-$C_2H_2I_2$ and trans-$C_2D_2I_2$, a difference of 0.5-1.0% was obtained for the $a_g$ and $b_g$ type vibrations. For the $a_u$ and $b_u$ types, a difference of 0.15-0.8% was obtained.

A normal coordinate analysis is in progress for the eight molecules listed in Table III. This should permit an unambiguous assignment of the fundamental wavenumbers of this group of molecules. A similar analysis of the cis- compounds is also in progress.

## ACKNOWLEDGMENTS

This work was part of a research program which has been aided by grants from the National Science Foundation. The authors are grateful for this assistance.

## REFERENCES

1. S. I. Miller, A. Weber, and F. F. Cleveland, J. Chem. Phys. 23, 44 (1955).
2. J. T. Edsall and E. B. Wilson, Jr., J. Chem. Phys. 6, 124 (1938).
3. N. C. Craig and E. A. Entemann, J. Chem. Phys. 36, 243 (1962).
4. H. J. Bernstein and D. A. Ramsay, J. Chem. Phys. 17, 556 (1949).
5. J. M. Dowling et al., J. Chem. Phys. 26, 233 (1957).
6. K. S. Pitzer and J. L. Hollenberg, J. Am. Chem. Soc. 76, 1493 (1954).

# RAMAN SPECTRAL DATA, ASSIGNMENTS, POTENTIAL ENERGY CONSTANTS, AND CALCULATED THERMODYNAMIC PROPERTIES FOR $CH_2Cl_2$, $CHDCl_2$, AND $CD_2Cl_2$

Brother F. E. Palma, S. M., and K. Sathianandan

Spectroscopy Laboratory, Physics Department
Illinois Institute of Technology
Chicago, Illinois

## INTRODUCTION

Up to the present, no Raman spectral data have been published for the deuterated forms of dichloromethane. Pure samples of $CHDCl_2$ and $CD_2Cl_2$ have recently become available, and the present investigation was undertaken in order to determine the best set of potential energy constants from the vibrational spectra of all three compounds.

## EXPERIMENTAL

The deuterated samples used in this investigation were obtained from Merck, Sharpe & Dohme Company of Montreal, Canada, and had a high degree of isotopic purity.

The Raman spectra were obtained with a Hilger E-612 two-prism spectrograph with glass optics. The reciprocal of dispersion of the spectrograph was 100 $cm^{-1}$/mm at the 4358 A line of mercury which was used to excite the spectra. The spectra were recorded photographically on Eastman 103a-J plates. Exposure times varied from 1 to 16 hours. The spectrum of iron was used as a secondary standard for measuring the wavenumbers, and semiquantitative intensities were obtained by scanning the plates on a Leeds and Northrup microphotometer.

## SYMMETRY PROPERTIES

A substituted methane like $CH_2Cl_2$ can be expected to have a tetrahedral structure. The microwave measurements of

Meyers and Gwinn [1] have shown that the structure is nearly tetrahedral. They measured the H—C—H angle as 112° and the Cl—C—Cl angle as 111°47'.

The CH$_2$Cl$_2$ and CD$_2$Cl$_2$ molecules belong to the point group C$_{2v}$; i.e., they have two mutually perpendicular planes of symmetry and a twofold axis of symmetry. Each compound has nine fundamental vibrations, of which four are type $a_1$, one is type $a_2$, two are type $b_1$, and two are type $b_2$. The spectral lines arising from the $a_1$ vibrations are polarized; the others are depolarized. All the fundamentals are allowed in the Raman, and all but the $a_2$ are allowed in the infrared.

The CHDCl$_2$ molecule has only one plane of symmetry, and hence belongs to the point group C$_s$. It has six $a'$ fundamental vibrations which are polarized and three $a''$ vibrations which are depolarized. All the fundamentals are allowed in both the Raman and the infrared.

## ASSIGNMENT FOR CH$_2$Cl$_2$

The observed Raman displacements, relative intensities, and polarization characteristics of the fundamentals of CH$_2$Cl$_2$ and CD$_2$Cl$_2$ are given in Table I. The assignment of the fundamentals for CH$_2$Cl$_2$ has been discussed at length in the literature by Wagner [2], Corin and Sutherland [3], Herzberg [4], Plyler and Benedict [5], and others. The lower wavenumbers can readily be assigned to the CCl stretching and deformation vibrations. The CH stretching vibrations are at 2986 and 3053 cm$^{-1}$. The 1423 cm$^{-1}$ line is assigned as the CH$_2$ totally symmetric deformation, even though it appears depolarized. The presence of several possible overtone and combination bands of the CCl stretching vibrations is indicated by the peculiar shape of this band in the infrared, and this might explain the apparently high value of the depolarization ratios which have been measured for this line. It is interesting to note that the corresponding line for the CD$_2$ deformation in the completely deuterated compound is strongly polarized, and appears at 1052 cm$^{-1}$, much lower than the CCl stretching overtones.

The torsional vibration of the $a_2$ vibration has been assigned to the 1156 cm$^{-1}$ line. This is forbidden in the infrared, but is observed as a weak band for the liquid. The selection rules are not strictly valid for liquids, however, and this assignment is borne out by the normal coordinate analysis, and by a study of the band contours in the Raman spectrum of the vapor by Welsh et al. [6].

## TABLE I
### Raman Spectral Data for $CH_2Cl_2$ and $CD_2Cl_2$[*]

| Assignment | $CH_2Cl_2$ | | | $CD_2Cl_2$ | | | |
| | $\Delta\sigma_{obs}$ | I | $\rho$ | $\Delta\sigma_{calc}$[†] | $\Delta\sigma_{obs}$ | I | $\rho$ |
|---|---|---|---|---|---|---|---|
| $a_1$ | 285 | 75 | P | | 283 | 494 | P |
| $a_1$ | 703 | 1000 | P | | 677 | 1000 | P |
| $b_2$ | 742 | 10 | D | | 716 | 10 | D |
| $b_1$ | 895 | vw | | 633 | (712)‡ | | |
| $a_2$ | 1156 | 2 | D | 817 | 826 | vw | D |
| $b_2$ | 1265 | vw | | 894 | 953 | vw | |
| $a_1$ | 1423 | 6 | D? | 1006 | 1052 | 6 | P |
| $a_1$ | 2986 | 181 | P | 2111 | 2198 | 199 | P |
| $b_1$ | 3053 | 7 | D | 2158 | 2304 | 4 | D |

*$\Delta\sigma$ is the Raman displacement in cm$^{-1}$, I is the semiquantitative relative intensity, and $\rho$ is the polarization state of the line (P = polarized, D = depolarized).
†Calculated by multiplying the corresponding wavenumber for $CH_2Cl_2$ by $2^{-\frac{1}{2}}$.
‡Calculated from the complete isotopic substitution rule.

## ASSIGNMENT FOR $CD_2Cl_2$

Since the completely deuterated compound has the same symmetry and electronic configuration as $CH_2Cl_2$, one expects the spectra to be similar, except for mass effects due to the deuterium substitution. The $CCl_2$ stretching and deformation wavenumbers are not affected much by this substitution. For the other fundamentals, as a crude approximation one may consider the massive Cl and C atoms to be stationary, while the H or D atoms move in the vibrations. Deuterium substitution should then decrease the $CH_2Cl_2$ wavenumbers by a factor of $2^{-\frac{1}{2}}$ . The results of this very crude calculation are shown in Table I. The assignment for the observed wavenumbers of $CD_2Cl_2$ could then readily be made, except for one $b_1$ fundamental which was not observed because of near coincidence with the strong CCl stretching bands near 700 cm$^{-1}$. An approximate value of 712 cm$^{-1}$ was calculated for this line from the complete isotopic substitution rule [7].

During the course of the present work, Shimanouchi and Suzuki [8], at the University of Tokyo, measured the infrared spectra of these three compounds and published the results of a normal coordinate treatment using a Urey-Bradley potential energy function. They assigned the $CD_2$ deformation vibration to a strong infrared band at 995 cm$^{-1}$. This line, however, can be assigned as a combination band, and in the present work the polarized Raman line at 1052 cm$^{-1}$, which has nearly the same relative intensity as the 1423 cm$^{-1}$ line of $CH_2Cl_2$, has been assigned to the $CD_2$ deformation.

## ASSIGNMENT FOR $CHDCl_2$

The assignments for $CHDCl_2$ were easily made by applying the complete isotopic substitution rules of Brodersen and Langseth [7]. These rules are very useful in a case of this kind where the wavenumbers for each type of vibration of a compound of high symmetry ($CH_2Cl_2$) and those of the completely isotopically substituted compound of the same symmetry ($CD_2Cl_2$) are known, because then all the wavenumbers for each type of vibration for the fundamentals of the partially substituted compound ($CHDCl_2$) can be calculated accurately, solely from the wavenumbers, without using any molecular parameters. It may be useful to discuss here the application of the complete isotopic rule to these molecules as a demonstration of its simplicity, usefulness, and accuracy.

A secular determinantal equation can be written for each fundamental vibration type. For example, if a molecule has three $a''$ fundamentals, the secular equation for the $a''$ vibrations has the form

$$(a'') = \lambda^3 - C_1\lambda^2 + C_2\lambda - C_3 = 0 \qquad (1)$$

Solution of this cubic equation will yield three $\lambda$ values which are related to the wavenumbers by the expression

$$\lambda_i = 4\pi^2 c^2 \sigma_i^2 \qquad (2)$$

where $c$ is the velocity of light in a vacuum and $\sigma_i$ is the $i$th wavenumber of the $a''$ type. The coefficients $C_1$, $C_2$, and $C_3$ are related to the $\lambda$s by the expressions

$$C_1 = \Sigma\, \lambda_i$$
$$C_2 = \Sigma\, \lambda_i\lambda_j \qquad (3)$$
$$C_3 = \Sigma\, \lambda_i\lambda_j\lambda_k$$

If the secular equation is known, it can be solved for the wavenumbers of the $a''$ fundamentals, and, conversely, if the $a''$ fundamentals are known, the secular determinant can be constructed from Eqs. (2) and (3).

The Brodersen-Langseth complete isotopic substitution rules applied to $CHDCl_2$ yield the following expressions:

$$(a'') = \tfrac{1}{2}(a_2)_H(b_2)_D + \tfrac{1}{2}(a_2)_D(b_2)_H = 0 \qquad (4)$$

and

$$(a') = \tfrac{1}{2}(a_1)_H(b_1)_D + \tfrac{1}{2}(a_1)_D(b_1)_H = 0 \qquad (5)$$

Here the parentheses indicate the secular determinants of the various vibration types, and the subscripts $H$ and $D$ refer to $CH_2Cl_2$ and $CD_2Cl_2$, respectively.

For the $a''$ vibrations, the secular determinant is given by a strictly valid rule, which combines the secular determinants for the $a_2$ and $b_2$ fundamentals of $CH_2Cl_2$ and $CD_2Cl_2$. The expanded determinants are polynomials in $\lambda$, of a degree dependent on the number of fundamentals belonging to each type. For example, the expanded determinants for the $a_2$ and $b_2$ vibrations of $CH_2Cl_2$ have the form

$$(a_2)_H = \lambda - C_{1a_2H} \qquad (6)$$

and

$$(b_2)_H = \lambda^2 - C_{1b_2H}\lambda + C_{2b_2H} \qquad (7)$$

The constant coefficients in these expressions are obtained from the wavenumbers by relations similar to Eqs. (3). Expressions similar to Eqs. (6) and (7) are obtained from the wavenumbers for $CD_2Cl_2$. When these relations are substituted into Eq. (4) and the polynomials are expanded and simplified, a cubic equation with the same form as Eq. (1) is obtained. Solutions to this equation are the three $\lambda$s corresponding to the wavenumbers for the three $a''$ fundamentals of $CHDCl_2$.

The approximate rule for the $a'$ vibrations, Eq. (5), gives good results when written in three separate parts for the CH stretching wavenumbers, for the CD stretching wavenumbers, and for the remaining lower wavenumbers. The first two parts simply give the average of the symmetric and non-totally symmetric CH and CD stretching wavenumbers. The last part yields a fourth degree equation which can be solved for the four remaining wavenumbers of the $a'$ type fundamentals.

In the first column of Table II are listed the wavenumbers calculated from the complete isotopic rule. In the remainder of the table are given the observed Raman displacements, relative intensities, and states of polarization. The agreement between the observed and calculated wavenumbers is very good,

## TABLE II
### Raman Spectral Data for $CHDCl_2^*$

| Assignment | $\Delta\sigma_{calc}$† | $\Delta\sigma_{obs}$ | I | $\rho$ |
|:---:|:---:|:---:|:---:|:---:|
| $a'$ | 284 | 283 | 136 | P |
| $a'$ | 681 | 682 | 1000 | P |
| $a''$ | 725 | 725 | 20 | D |
| $a'$ | 773 | 779 | 16 | P |
| $a''$ | 884 | 886 | 1 | D |
| $a''$ | 1220 | 1221 | 1 | D |
| $a'$ | 1277 | 1276 | 1 | P |
| $a'$ | 2252 | 2246 | 59 | P |
| $a'$ | 3038 | 3019 | 66 | P |

*All symbols as in Table I, unless otherwise indicated.
†Calculated from the complete isotopic substitution rule.

even for the $a'$ type vibrations, for which the rules are only approximate. Apparently, in these approximations, the effects of the anharmonicity must largely cancel each other.

## NORMAL COORDINATE TREATMENT

After the preliminary assignment was completed, it was confirmed by a normal coordinate treatment, carried out with a general quadratic valence force potential energy function, according to the Wilson FG matrix method, using wavenumbers corrected for anharmonicity by the method of Dennison [9].

## THERMODYNAMIC PROPERTIES

The heat content, free energy, entropy, and heat capacity for $CH_2Cl_2$, $CD_2Cl_2$, and $CHDCl_2$ were calculated to a rigid rotor, harmonic oscillator, ideal gas approximation, for 12 temperatures from 100° to 1000°K, using the infrared wavenumbers observed for the vapor. The authors hope to publish these thermodynamic properties, and also the potential energy constants obtained in the normal coordinate analysis, in the near future.

## ACKNOWLEDGMENT

This work was part of a research program which has been aided by grants from the National Science Foundation. The authors are grateful for this assistance.

## REFERENCES

1. R. J. Meyers and W. D. Gwinn, J. Chem. Phys. 20, 1420 (1952).
2. J. Wagner, Z. phyzik. Chem. B34, 69 (1939).
3. C. Corin and G. B. B. M. Sutherland, Proc. Roy. Soc. (London) A165, 43 (1938).
4. G. Herzberg, Infrared and Raman Spectra of Polyatomic Molecules, D. Van Nostrand Company, New York (1945), p. 317.
5. E. K. Plyler and W. S. Benedict, J. Research Nat. Bur. Standards 47, 202 (1951).
6. H. L. Welsh et al., Can. J. Phys. 30, 577 (1952).
7. S. Brodersen and A. Langseth, Kgl. Danske Videnskab. Selskab, mat.-fys. Skrifter 1, No. 5 (1958).
8. T. Shimanouchi and I. Suzuki, J. Mol. Spectr. 8, 222 (1962).
9. D. M. Dennison, Revs. Modern Phys. 12, 175 (1940).

# ORTHO-SUBSTITUENT EFFECTS AND THE FUNDAMENTAL NH₂ STRETCHING VIBRATIONS IN ANILINES*

## Peter J. Krueger

Department of Chemistry
University of Alberta
Calgary, Alberta, Canada

## ABSTRACT

The integrated intensities, frequencies, and half-band widths of the fundamental symmetric and asymmetric $NH_2$ stretching vibrations in 32 ortho-substituted anilines, measured in dilute carbon tetrachloride solution, were examined in relation to the corresponding absorption band parameters for 31 meta- and para-substituted anilines, taking into consideration the electronic effects of the substituents. From an almost tetrahedral configuration in $p$-phenylenediamine, the s-character of the nitrogen atom gradually increases as the substituent groups become more electron-withdrawing, with a resulting increase in the HNH angle and the NH force constant. Ortho-substitution in general leads to enhanced HNH angle opening, probably because of intramolecular hydrogen bonding in many cases. The decrease in half-band width for both vibrational modes in ortho-substituted anilines over corresponding values in meta- and para-compounds is ascribed to steric hindrance to solvation of the amino group.

The asymmetric intensities in ortho-substituted anilines are generally increased over corresponding values in meta- and para-compounds, unlike the behavior of the symmetric mode. These results are consistent with a vibrational mechanism taking into account the following factors for each mode: (i) the direction of the transition moment; (ii) the extent of nitrogen lone pair and aromatic $\pi$-electron participation; and (iii) the direct field effect of the ortho-substituent.

*This paper is to be published in expanded form in the Canadian Journal of Chemistry.

## INTRODUCTION

The vibrations localized in characteristic functional groups attached to the aromatic ring have been studied extensively in recent years as a function of electronic changes in the molecule. The frequencies of absorption maxima and the absorption intensities have been correlated with the Hammett $\sigma$ constants [1, 2] of the substituents for the -OH stretching vibration in phenols [3]; the $-C\equiv N$ stretching vibration in benzonitriles [4-6]; the $C=O$ stretching vibration in benzaldehydes, ethyl benzoates, and acetophenones [7, 8]; the -NH stretching vibration in $N$-methylanilines [9]; the two $-NH_2$ stretching vibrations in anilines [5, 9-12]; and the C-O-C stretching vibration in anisoles [13].

For meta- and para-substituted benzene derivatives the frequency/$\sigma$ correlations may be expressed by a Hammett-type equation [14]:

$$\nu = \nu_0 + \rho\sigma \tag{1}$$

where $\nu$ is the observed group frequency in the substituted compound, $\nu_0$ is the group frequency of the "parent" compound in the series, and $\rho$ expresses the sensitivity of the frequency to substituent effects. These plots sometimes exhibit slight curvature as the frequency rises or falls with increasing electrophilic character of the substituents. A similar linear relationship has been sought [15] between $\nu$ and the electrophilic substituent constants $\sigma^+$ [16, 17]. Rao and Venkataraghavan [14] have recently shown that a statistical evaluation of 18 sets of frequencies indicates that both $\sigma$ and $\sigma^+$ correlate the frequencies equally well.

The integrated band intensities ($A$) of vibrations in functional groups in meta- and para-substituted benzenes have been correlated linearly through a logarithmic function so that

$$\log A = \log A_0 + \rho'\sigma \tag{2}$$

or alternatively

$$\log A = \log A_0^+ + \rho^+\sigma^+ \tag{3}$$

Statistical evaluation has indicated that the latter correlation is slightly better [14].

On the basis of a simple molecular orbital model, Brown [18] has proposed that the quantity $A^{\frac{1}{2}}$ should be linearly related to the appropriate substituent constants $\sigma$ or $\sigma^+$. Statistically this correlation appears to be as good as the correlations with $\log A$ [14]. Brown further suggests that the success of these correlations indicates that the changes in electron distribution which occur in the molecule during vibrational distortions closely parallel those which occur in the formation of the transition state during chemical reactions.

In a previous publication, Krueger and Thompson [5] showed that for a wide range of vibrational types the $\log A$ and $\nu$ values also correlated well with the inductive $(\sigma_I)$ and resonance $(\sigma_R)$ parameters deduced by Taft [19a] on the basis that

$$\text{para-}\sigma_{\text{Hammett}} = \sigma_I + \sigma_R \tag{4}$$

as a first approximation. These relationships could be expressed explicitly as

$$\log A = \log A_0 + a\sigma_I + \beta\sigma_R \tag{5}$$

and

$$\nu = \nu_0 = a'\sigma_I + \beta'\sigma_R \tag{6}$$

where the $a$s and $\beta$s could be interpreted as the relative susceptibility of the correlated quantity to the inductive and resonance effects. The significance of this has been discussed.

The Hammett treatment of kinetic data is restricted to meta- and para-substituents to avoid complications due to the effect of neighboring groups, although Farthing and Nam [20] have attempted to extend it to ortho-substituted compounds. Krueger and Thompson [5] previously reported some spectroscopic measurements on ortho-substituted benzonitriles, anilines, phenols, ethyl benzoates, and benzaldehydes, and attempted to relate these values to those for meta- and para-substituted compounds. The substituent constants derived by Taft [19b] for ortho-substituents were used in integrating all the data, since these $\sigma_{\text{ortho}}$ values are on the same scale as the Hammett $\sigma_{\text{meta}}$ and $\sigma_{\text{para}}$ values. The additivity principle proposed by Jaffe [2] was used as a first approximation to get a $\Sigma\sigma$ value for compounds with several substituents. This was reasonably successful for the sterically favorable benzonitrile series, but a number of unexplained anomalies remained for the phenols and anilines.

The data available for the aniline series have now been re-examined, and some conclusions have been drawn concerning the nature of the amino group and its response to ortho-substitution. Further high resolution infrared measurements have also been made on selected model compounds.

## EXPERIMENTAL METHODS

A Perkin-Elmer 12C Single-Beam Spectrometer was employed for the earlier measurements, using a LiF prism to improve the resolution in the 3 $\mu$ region. Standard absorption lines of water vapor and ammonia were used as frequency calibrants. The effective slit width was calculated to be about 8 cm$^{-1}$ in this region. Quartz absorption cells up to 5 cm in length were used to keep the maximum aniline concentrations below 0.01 M in carbon tetrachloride. Measurements were made over a wide concentration range at several path lengths, the band areas determined by numerical integration, and the apparent intensities extrapolated to zero peak absorbance to eliminate slit effects [21]. Since the true shape of liquid phase absorption bands is still in doubt, no "wing corrections" [22] were applied. Integration limits were set at those points where the recorder noise level became comparable with the residual absorbance. Due to slight overlapping of the two bands, a small amount of graphical separation was necessary.

The high resolution measurements were made with a Beckman IR-7 Spectrophotometer with a NaCl foreprism and a 75 lines/mm grating blazed at 12 $\mu$. The instrument was calibrated as previously described [23]; the accuracy of the frequency measurements is believed to be limited only by the widths of the bands themselves. These measurements were made in the fourth grating order, where the slit width was calculated to be about 1 cm$^{-1}$ in this region. A pair of matched 2-cm quartz cells was used. The use of a programed slit, double-beam operation, linear absorbance, and linear frequency (cm$^{-1}$) recording permitted direct evaluation of band areas by numerical integration. Slit width effects were found to be negligible in most cases (the calculated slit width is less than 5% of almost all the band widths measured), and the $\nu$, $A$, and $\Delta\nu_{1/2}^{a}$ values reported are the averages of at least two completely independent measurements, except in a few cases where the small amount of material available precluded this.

Most of the compounds were commercial products, purified by recrystallization, sublimation, or fractional distillation under reduced pressure. A few were synthesized following standard procedures in the literature. "Spectrograde" carbon tetrachloride was used, and the solutions were made up volumetrically. In the high resolution measurements the cell temperature was held at 25°C.

## RESULTS

All the compounds investigated, together with the corresponding substituent constants, are listed in Table I. The detailed experimental results and the calculated parameters derived from them will be published elsewhere. All the results are presented graphically. Points of special interest are numbered in the diagrams, the numbers being those of the correspondingly numbered substituents in Table I.

For the purposes of this discussion the total electronic effect of the aromatic ring substituents on the $NH_2$ group is considered to be proportional to the additive sum of the substituent constants ($\Sigma\sigma$) for meta- and para-substituents, and for ortho-substituents, as originally defined by Hammett [1] and Taft [19b], respectively. Wherever possible the more recent $\sigma$ values from the critical analysis by McDaniel and Brown [24] were used. A few constants were taken from the review by Jaffe [2]. For a number of ortho-substituents, $\sigma$ values were estimated on the basis that $\sigma_{ortho} \approx \sigma_{para}$, since no kinetic data for them were available in the literature. The validity of this approximation is substantiated by the work of Taft and his associates.

Figure 1 shows a plot of $\log_{10} A$ vs. $\Sigma\sigma$ for the asymmetric vibration; Fig. 2 shows the analogous data for the symmetric vibration. Considering that this presentation adjusts all the data for the varying electronic effects involved, the general conclusion can be drawn that for ortho-compounds the symmetric intensities are not significantly different from those for meta- and para-compounds over most of the range. For the asymmetric vibration the intensities for ortho-compounds are about 30 to 40% higher than the corresponding values for meta- and para-compounds. Only where the adjacent substituents are alkyl groups (and $NH_2$) is there no apparent increase in the

# TABLE I
## Substituted Anilines and Substituent Constants

| No.[1] | Substituent(s) | $\Sigma\sigma$ [2] | No.[1] | Substituent(s) | $\Sigma\sigma$ [2] |
|---|---|---|---|---|---|
| 1 | $4\text{-}N(CH_3)_2$ | $-0.83$ | 33 | $4\text{-}I$ | $+0.18$ |
| 3 | $4\text{-}NH_2$ | $-0.66$ | 34 | $2\text{-}Cl$ | $+0.20$[4] |
| 4 | $2\text{-}NH_2$ | $(-0.7)$[3] | 35 | $2\text{-}Br$ | $+0.21$[4] |
| 5 | $2,4,6\text{-}tri\text{-}C(CH_3)_3$ | $(-0.6)$[3] | 36 | $2\text{-}I$ | $+0.21$[4] |
| 6 | $2,4,6\text{-}tri\text{-}CH_3$ | $-0.51$[4] | 37 | $4\text{-}Cl$ | $+0.227$ |
| 7 | $2,4,6\text{-}tri\text{-}CH(CH_3)_2$ | $-0.45$[3] | 38 | $4\text{-}Br$ | $+0.232$ |
| 8 | $2\text{-}OCH_3$ | $-0.39$[4] | 39 | $2\text{-}F$ | $+0.24$[4] |
| 9 | $2\text{-}OCH_2CH_3$ | $-0.35$[4] | 40 | $3\text{-}F$ | $+0.337$ |
| 10 | $2,6\text{-}di\text{-}CH_3$ | $-0.34$[4] | 41 | $3\text{-}Cl$ | $+0.373$ |
| 11 | $2,5\text{-}di\text{-}C(CH_3)_3$ | $(-0.3)$[3] | 42 | $3\text{-}COCH_3$ | $+0.376$ |
| 12 | $2,5\text{-}di\text{-}OCH_3$ | $-0.27$[4] | 43 | $3\text{-}Br$ | $+0.391$ |
| 13 | $4\text{-}OCH_3$ | $-0.268$ | 44 | $2,4\text{-}di\text{-}Cl$ | $+0.43$[4] |
| 14 | $4\text{-}OCH_2CH_3$ | $-0.24$ | 45 | $3\text{-}CF_3$ | $+0.43$ |
| 15 | $3,4\text{-}di\text{-}CH_3$ | $-0.239$ | 46 | $4\text{-}SCN$ | $+0.52$ |
| 16 | $2,5\text{-}di\text{-}OCH_2CH_3$ | $-0.20$[4] | 47 | $2\text{-}CF_3$ | $(+0.54)$[3] |
| 17 | $2\text{-}C(CH_3)_3$ | $(-0.2)$[3] | 48 | $3\text{-}CN$ | $+0.56$ |
| 18 | $4\text{-}C(CH_3)_3$ | $-0.197$ | 49 | $2,4,6\text{-}tri\text{-}Cl$ | $+0.63$[4] |
| 19 | $2\text{-}CH_3$ | $-0.17$[4] | 50 | $2\text{-}CN$ | $+0.64$[5] |
| 20 | $4\text{-}CH_3$ | $-0.170$ | 51 | $4\text{-}N=N\text{-}C_6H_5$ | $+0.640$ |
| 21 | $4\text{-}CH(CH_3)_2$ | $-0.151$ | 52 | $2,4,6\text{-}tri\text{-}Br$ | $+0.65$[4] |
| 22 | $2\text{-}CH(CH_3)_2$ | $(-0.15)$[3] | 53 | $4\text{-}COOCH_2CH_3$ | $+0.678$[6] |
| 23 | $2\text{-}CH_2CH_3$ | $(-0.15)$[3] | 54 | $2\text{-}COOCH_2CH_3$ | $(+0.68)$[3] |
| 24 | $4\text{-}C_6H_5$ | $-0.01$ | 55 | $3\text{-}NO_2$ | $+0.710$ |
| 25 | $2\text{-}C_6H_5$ | $(-0.01)$[3] | 56 | $4\text{-}COCH_3$ | $+0.874$[6] |
| 26 | $H$ | $0$ | 58 | $4\text{-}CN$ | $+1.00$[6] |
| 27 | $4\text{-}CH_2CN$ | $+0.01$ | 59 | $2,6\text{-}di\text{-}CN\text{-}3,5\text{-}di\text{-}CH_3$ | $+1.14$[5] |
| 28 | $3,4\text{-}(CH)_4$ | $+0.042$ | 60 | $2\text{-}NO_2$ | $+1.22$[4] |
| 29 | $2,3\text{-}(CH)_4$ | $(+0.04)$[3] | 61 | $4\text{-}NO_2$ | $+1.27$[6] |
| 30 | $4\text{-}F$ | $+0.062$ | 62 | $2,6\text{-}di\text{-}Cl\text{-}4\text{-}NO_2$ | $+1.67$[4,6] |
| 31 | $3\text{-}OCH_2CH_3$ | $+0.1$ | 63 | $2,6\text{-}di\text{-}NO_2$ | $+2.44$[4] |
| 32 | $3\text{-}OCH_3$ | $+0.115$ | 64 | $2,4\text{-}di\text{-}NO_2$ | $+2.49$[4,6] |
|  |  |  | 65 | $2,4,6\text{-}tri\text{-}NO_2$ [7] | $+3.71$[4,6] |

[1] Compounds 2 [$2\text{-}N(CH_3)_2$] and 57 ($2\text{-}COCH_3$) were omitted from this table to keep the numbering identical with that used in some unpublished studies on the corresponding overtone bands in this laboratory.
[2] Unless otherwise indicated, the values used are from the critical analysis of Hammett substituent constants by McDaniel and Brown [24].
[3] Estimated on the basis that $\sigma_{ortho} \approx \sigma_{para}$.
[4] Involving $\sigma_{ortho}$ value(s) according to Taft [19b].
[5] From a correlation of quadrupole resonance frequencies [25].
[6] Involving special Hammett substituent constants for anilines according to Jaffe [2].
[7] Extremely insoluble in carbon tetrachloride; frequencies measured by Dyall [26] in $C_2Cl_4$ were used.

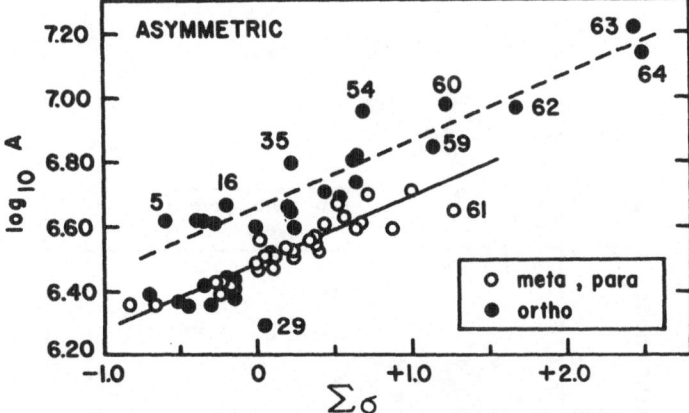

Fig. 1. Correlation of the intensity of the asymmetric $NH_2$ stretching vibration in substituted anilines with the electronic nature of the substituents. The units of $A$ are $cm \cdot mole^{-1}$ ($log_e$).

asymmetric intensities, and a slight decrease in the symmetric intensities.

The frequency/$\Sigma\sigma$ correlations are shown in Fig. 3. Because of the two interacting vibrational modes, and the possibility that the geometry of the amino group may change in the aniline series, these are best discussed in terms of the HNH angle and the NH stretching force constant.

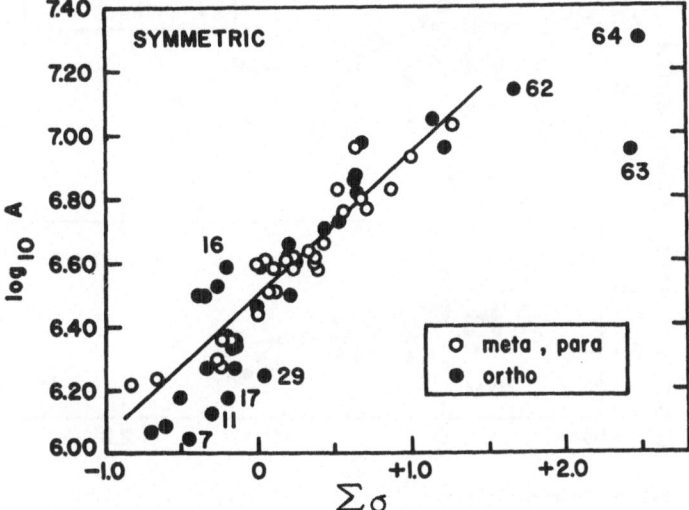

Fig. 2. Correlation of the intensity of the symmetric $NH_2$ stretching vibration in substituted anilines with the electronic nature of the substituents.

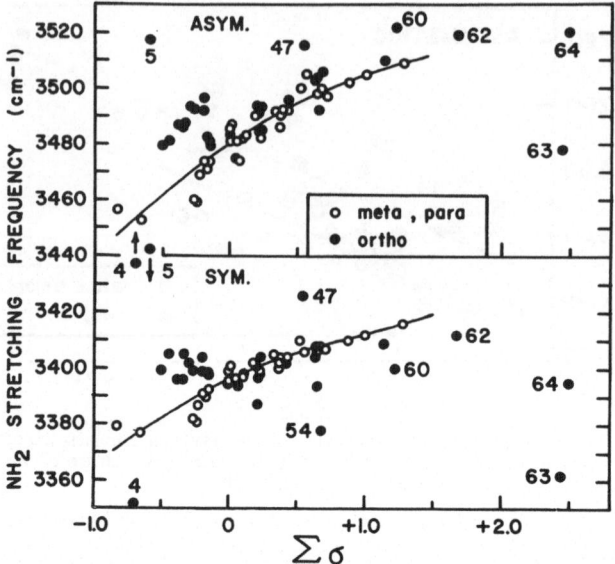

Fig. 3. Dependence of the $NH_2$ stretching frequencies on the electronic nature of the substituents in substituted anilines.

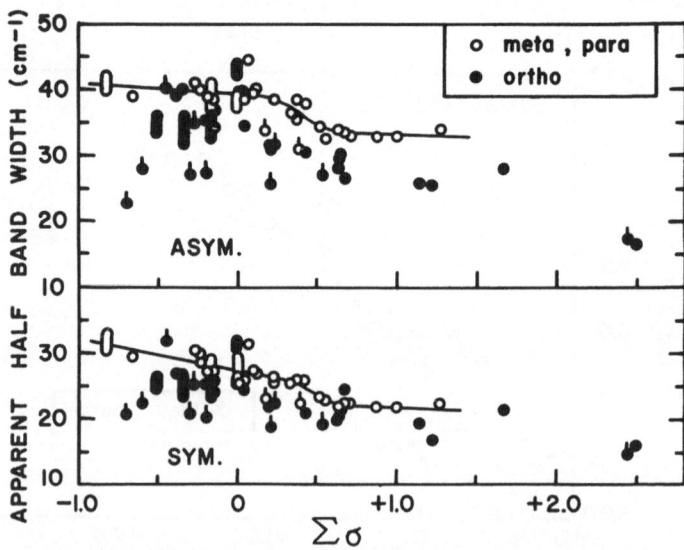

Fig. 4. Relationship of $\Delta \nu_{1/2}^a$ to the electronic nature and position of the substituents in substituted anilines. High resolution measurements indicated by points with vertical tails, and by bottom ends of oval points.

Our attention was first directed to a re-examination of the available data by a comparison of the apparent half-band widths ($\Delta\nu_{1/2}^a$), which are shown in Fig. 4 as a function of $\Sigma\sigma$. In general, both $NH_2$ bands are significantly narrower in ortho-substituted anilines than in meta- and para-substituted anilines. Since most of the band widths shown in Fig. 4 were obtained with an effective slit width of about 8 cm$^{-1}$, they will be somewhat wider than the limiting true widths, but relative values should still be correct. The extent of this distortion is also portrayed in Fig. 4, as indicated by the elongated points corresponding to measurements repeated with the better resolution of a grating spectrophotometer.

## DISCUSSION

An interpretation of these observations is advanced along the lines of a simplified model for the two vibrational modes illustrated schematically in Fig. 5. The essential difference between these vibrational modes was previously pointed out by Orville-Thomas et al. [27].

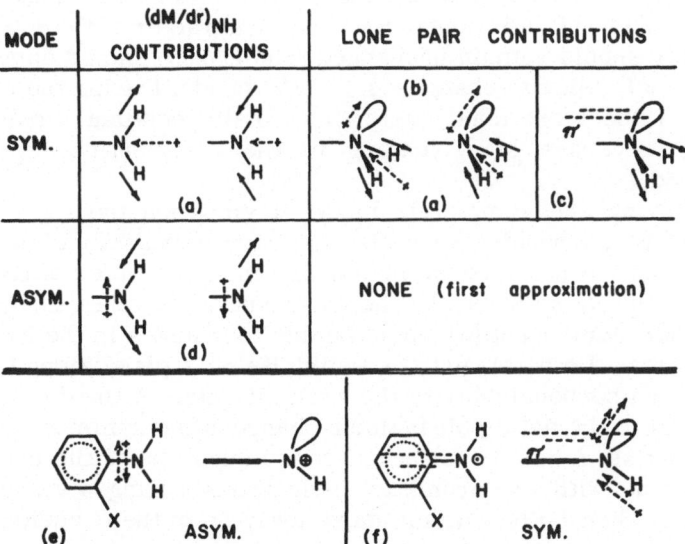

Fig. 5. A simple vibrational mechanism for the $NH_2$ group. (a–d) Contributions to the transition moment. (e, f) Orientation of the components of dipole moment change with respect to the aromatic system and ortho-substituents. Solid arrows refer to atomic motion; dotted arrows to dipole moment.

To a good approximation the only nuclei involved in the $\nu_s$ and $\nu_{as}$ vibrations are the nitrogen and hydrogen atoms. Thus, changes in NH bond lengths during the vibration can be used to define the normal coordinate $Q$. The intensity of a fundamental infrared absorption band is proportional to $(\partial M/\partial Q)^2$, i.e., to the square of the variation of molecular dipole moment $M$ with respect to the normal coordinate during the vibration. The dipole moment of the amino group in its equilibrium configuration can be separated into two NH bond moments, and a third component due to the atomic dipole of the lone pair electrons. Coulson [28] has recently emphasized the importance of lone pair electron contributions to molecular dipole moment changes during vibrations, as well as the unexpected contributions that may be due to hybridization changes as the atoms move.

As the H atoms oscillate in the asymmetric mode the change in dipole moment is perpendicular to the symmetry axis of the $NH_2$ group, and parallel to the plane of the aromatic ring,* as summarized schematically in Fig. 5e. The decrease in s-character of one nitrogen bonding orbital as that NH bond lengthens is exactly compensated by the increase in s-character of the other NH bond as it contracts. The s-character of the lone pair orbital should remain unchanged, and $(\partial M/\partial Q)$ should depend on the small polarity changes in the NH bonds. During the second half of the vibrational cycle this small resultant dipole moment reverses sign, leading to the contribution shown in Fig. 5d.

During the symmetric mode of vibration the s/p ratio of the nitrogen bonding orbitals will vary in phase, and this will give rise to a compensating change in s-character of the lone pair orbital, as shown by the contribution sketched in Fig. 5b. Another component (a) would be due to changes in the NH bond moments. Both (a) and (b) would lie in a plane along the CN bond and perpendicular to the aromatic ring. A third contribution (c) to the net dipole moment change in the symmetric mode will arise from a variation of conjugation of the lone pair electrons with the aromatic $\pi$-electrons during a vibrational cycle. When both H atoms move away from the N nucleus, the

---

*On the basis of the absorption intensity of the 234 m$\mu$ ultraviolet band in substituted anilines, Wepster [29] has concluded that the lone pair electrons of the amino group are oriented for maximum overlap with the aromatic $\pi$-electrons even with bulky tert-butyl groups in the 2,6-positions. Essery and Schofield [30] support this view on the basis of limited infrared measurements on a few ortho-alkyl anilines.

s-character of the lone pair increases and it cannot combine as effectively with the $\pi$-orbitals (these have odd symmetry, whereas an s-orbital has even symmetry with respect to the plane of the ring). When the H atoms approach the N nucleus, the increase in p-character of the lone pair leads to more extensive interaction with the $\pi$-electrons. Thus, the $\pi$-electron component of the permanent dipole of the aromatic part of the molecule contributes to the transition moment of this vibration as it varies in phase with the symmetric vibration, to an extent determined by the hybridization of the N atom; i.e., it leads to a very small contribution when the N atom is effectively $sp^3$, and to a maximum when it is effectively $sp^2$. Figure 5f summarizes the contributions to $(\partial M / \partial Q)_s$.

A comparison of Fig. 5e and f shows that a substituent ortho to the $NH_2$ group might be expected to enhance $A_{as}$ by a direct field effect if it were strongly electrophilic or nucleophilic, since the sign of the dipole that leads to contribution d reverses at every half-cycle, and since the substituent group is almost directly in line with the direction of the transition moment. In the symmetric mode the ortho group is not intimately involved because it is well removed from the transition moment direction, and would likely only influence $A_s$ by the inductive/resonance mechanism operating through the aromatic carbon skeleton and the $\pi$-electrons, respectively. This factor also contributes to $A_{as}$, and is accounted for in the comparison on the $\sigma$ basis. These views are borne out by Figs. 1 and 2. For the $\log_{10} A_{as}/\Sigma\sigma$ correlation, the ortho-compounds in which intensity enhancement is not observed are those where the substituents are $2,4,6$-tri-$CH_3$, $2,4,6$-tri-$CH(CH_3)_2$, $2,6$-di-$CH_3$, $2,5$-di-$C(CH_3)_3$, $2$-$C(CH_3)_3$, $2$-$CH_3$, $2$-$CH(CH_3)_2$, and $2$-$CH_2CH_3$. This is in harmony with the relatively low polarizing ability of alkyl groups. Some of the other deviations from the established patterns remain unexplained, but the effect of ortho-substitution on the mechanical and electrical anharmonicity of these vibrations is being investigated in an attempt to provide some further information on the interaction mechanisms [31].

The HNH bond angle $(\theta)$ and the NH stretching force constant $(k)$ can be calculated from the two absorption frequencies of a primary amine by means of the valency force field equations of Linnett [32]:

$$4\pi^2 \nu_s^2 = k \left[ 1/m_H + (1 + \cos\theta)/m_N \right] \qquad (7)$$

$$4\pi^2 \nu_s^2 = k \left[ 1/m_H + (1 - \cos\theta)/m_N \right] \qquad (8)$$

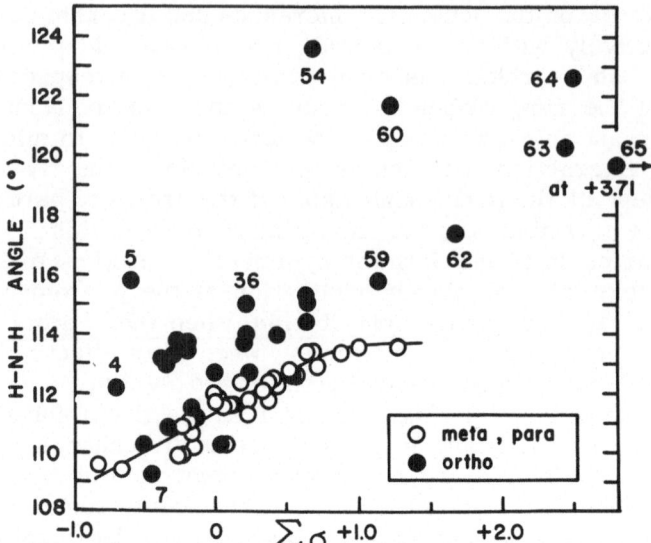

Fig. 6. HNH bond angle in substituted anilines and its dependence
on the electronic nature and position of ring substituents.

where $m_H$ and $m_N$ refer to the masses of the hydrogen and
nitrogen atoms, respectively. This treatment assumes that the
nitrogen and hydrogen atoms move along the NH bond directions,
and that the stretching force constant is much larger than the
deformation and interaction force constants. McKean and
Schatz [33] have shown that this would be a valid assumption
for the ammonia molecule. Mason [34] has successfully applied
these equations to a wide range of primary amines, and has
found $\theta$ to vary from 104° in $LiNH_2$ to about 119° in hetero-
aromatic compounds like 2-aminopyrimidine, 4-aminopyri-
midine, and 3-amino-1,2,4-triazine. Although aniline itself
was included in his study ($\theta = 111.8°$), conclusive evidence for
substituent effects on the hybridization state of the $NH_2$ group
in the aromatic amine series has not been reported.

Figure 6 shows that for meta- and para-substituted anilines
$\theta$ increases linearly with increasing electron-withdrawing
power of the substituents, from a value of 109.6° in $N,N$-di-
methyl-p-phenylenediamine and 109.4° in p-phenylenediamine
(almost exactly $sp^3$) to a limiting value of 113.6° which is
reached in p-aminobenzonitrile and p-nitroaniline. This is in
agreement with the predicted change in lone pair conjugation
with the aromatic $\pi$-electrons. Ortho-substituted anilines show
extensive HNH angle opening, except in those cases where the

substituents involved are one or two methyl, ethyl, or isopropyl groups, or the $CF_3$ group. Undoubtedly there are a number of factors responsible for this increase in $\theta$. For those cases in which the angle is close to or greater than 120°—as in 2,4,6-trinitroaniline (119.7°), 2,6-dinitroaniline (120.3°), 2-nitro-aniline (121.7°), 2,4-dinitroaniline (122.6°), and ethyl ortho-anthranilate (123.6°)—intramolecular hydrogen bonding is suggested, and similar interaction of a weak nature may account for the behavior of ortho-CN, halogen, -OR, and $-NH_2$ groups. Weak bonding of an amino hydrogen atom to the $\pi$-electrons of an adjacent phenyl group can also be envisaged.

The increased HNH angles for 2-nitro-, 2,4-dinitro-, 2,6-dinitroaniline, and 2,4,6-trinitroaniline are interesting in view of some recent controversy in the literature as to whether or not intramolecular hydrogen bonding is present in the first compound. Lutskii and Alexseeva [35], Mortiz [36], and Farmer and Thompson [37] have concluded that there is intramolecular hydrogen bonding in 2-nitroaniline; Dyall [26] claims that solvent effect studies show it to be absent in 2-nitro- and 2,4-dinitroaniline, but present in 2,6-dinitro- and 2,4,6-trinitro-aniline. Since the largest HNH angle found in this work (123.6°) occurs in ethyl ortho-anthranilate, in which hydrogen bonding is strongly favored because of the formation of a six-membered chelate ring, the enhanced angles in the nitroanilines also suggest a similar interaction mechanism.

The compound 1-naphthylamine (No. 29) appears to behave in an anomalous manner in that both $A_s$ and $A_{as}$ are smaller than expected, as is the calculated HNH angle. This suggests that the conjugation of the lone pair electrons with the aromatic $\pi$-electrons is disturbed by some interference due to the peri-H atom. Elliott and Mason [38] have postulated "bending" of the nitrogen bonding orbital due to repulsion, and a slight twisting of the $NH_2$ group to reduce the conjugation. Narrower band widths for this compound support this view due to hindrance to solvation of the $NH_2$ group, as will be described later. 2-Naphthylamine (No. 28) behaves like a normal meta-substituted aniline.

A weakness in the calculation of $\theta$ for ortho-substituted anilines from equations [7] and [8] may arise in the underlying assumption that the two NH bonds are equivalent. However, the fact that anilines which are symmetrically substituted in the 2,6-positions (compounds 5, 49, 52, 59, 62, 63, and 65) and should thus meet this requirement lead to similarly enhanced

$\theta$ values provides some proof for the validity of the method. This variation in HNH angle in the aniline series can be further confirmed by applying the same type of calculation to the limiting cases of a primary alkyl amine $(sp^3)$ and a primary acid amide $(sp^2)$. Using the data of Orville-Thomas et al. [27] for methylamine in $CCl_4$ solution, the calculated HNH angle is 103.8°; and for acetamide in dilute $CHCl_3$ solution, the measurements of Davies [39] lead to 120.3°.

For ortho-halogenated anilines, the HNH angle is found to increase in the order F < Cl < Br < I. Intramolecular hydrogen bonding of the OH group in phenols to ortho-halogen atoms is well known. Baker and Kaeding [40] have shown that the order of increasing hydrogen bond strength in the ortho-phenol series is really I < F < Br < Cl, which can be attributed to both the varying size of the halogens and an "orbital-orbital repulsive interaction" which increases in the order of Cl < Br < I. These authors have also concluded that Badger's rule [41] does not apply in cases where the interacting groups are not free to take up their preferred orientation or interacting distance. In the aniline series the HNH. . . F interaction may be extremely weak because the small size of the fluorine prevents the H atom from getting close to its lone pair orbitals. The increase in size of the halogens appears to more than offset the corresponding decrease in electronegativity.

The concept of intramolecular hydrogen bonding would account for the "normal" HNH angles in anilines with ortho-methyl, -ethyl, and -isopropyl groups. The large angle of 115.8° in 2,4,6-tri-tert-butylaniline could arise as the $NH_2$ group is forced into greater planarity with the aromatic ring by the bulky tert-butyl groups which flank it (see Fig. 10a). This could account qualitatively for the marked increase in $A_{as}$ relative to $A_s$ in terms of NH bond moment contributions, the lone pair conjugation being small because of the electron-donating characteristics of the three tert-butyl groups. To a lesser extent this could also account for $A_{as}$ increases in 2-tert-butyl- and 2,5-di-tert-butylaniline.

The calculated NH stretching force constant increases smoothly with $\Sigma \sigma$ for meta- and para-substituted anilines (Fig. 7). The complex deviation of ortho-substituted compounds is a direct reflection of the dependence of the force constant on the effective charge on the nitrogen atom, as well as on its hybridization state [32, 42].

Once the bond angle $\theta$ is known, the contribution of 2s and

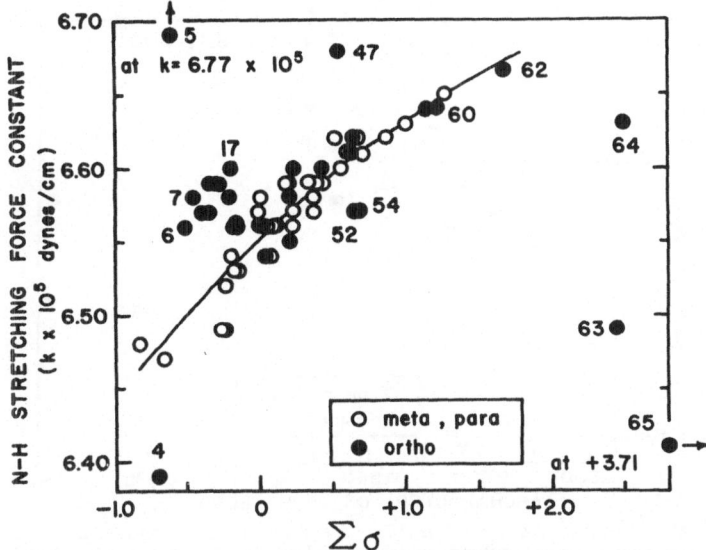

Fig. 7. Dependence of the calculated NH stretching force constant on the electronic nature and position of ring substituents in anilines.

2p orbitals to the hybrid nitrogen bonding orbitals may be calculated [34]. The coefficient $b$ in the hybrid orbital

$$\psi_{\text{hybrid}} = b\psi_{2s} + \sqrt{(1-b^2)} \cdot \psi_{2p} \tag{9}$$

can be used as a measure of s-character, and may be calculated from

$$b^2 = -\cos\theta/(1 - \cos\theta) \tag{10}$$

provided the two NH bonds are equivalent and the hybrid orbitals of N binding the H atoms are orthogonal. For meta- and para-anilines, $k$ is a linear function of s-character (Fig. 8). This is in excellent agreement with Mason's calculations [34] based on Slater orbitals which show that the overlap integral (measuring the strength of the NH bond, as does the force constant) for a hybrid orbital of nitrogen according to equation (9) and a $1s$ orbital of hydrogen increases linearly with $b$. Fig. 8 also shows that changes in the HNH angle alone cannot account for the difference between the NH stretching force constants of ortho-anilines and those for meta- and para-anilines. For a given HNH angle, an increase in the effective charge on the nitrogen atom should decrease its electronegativity and hence lower the NH stretching force constant and

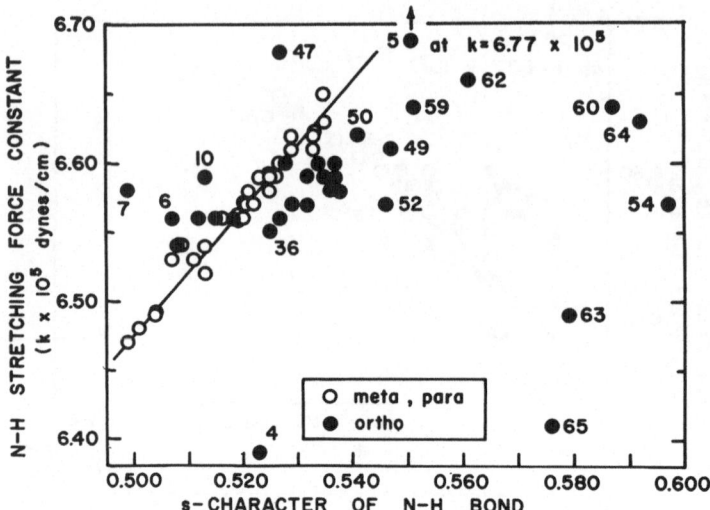

Fig. 8. Relation between the NH stretching force constant and the s-character of the hybrid orbitals bonding these atoms in substituted anilines. A $b$ value of 0.500 corresponds to sp³ hybridization of the N atom.

vice versa. Farmer and Thompson [37] have indicated that hydrogen bond formation with an amino hydrogen atom must induce a negative charge on the nitrogen atom. This would account for those points in Fig. 8 which are below the meta-para line.

The points above this line correspond to ortho-alkyl substituents (and o-CF₃). Since the bulkiest ortho-alkyl groups lead to the greatest increase in the NH force constant, "internal specific solvation" of the amino group is suggested, whereby the effective dielectric environment of the amino group has acquired a larger hydrocarbon nature as CCl₄ molecules are crowded out. This should raise the NH force constant. The NH force constant for aniline in n-hexane does show a slight increase, whereas for 2,6-dimethylaniline in n-hexane it is identical with that calculated from frequencies obtained in CCl₄. Changes of force constant of less than 1% due to this effect would account for almost all the positive deviations observed.

Crystallographic data available are in agreement with the concept of increasing sp² character of the amino nitrogen atom as ring substituents become more electrophilic. The C-N bond lengths [43] in p-hydroxyaniline, p-iodoaniline, and the sym-trinitrobenzene complex of p-nitroaniline are 1.47, 1.43, and 1.37 A, respectively, denoting increasing bond order. All three are classed as planar, with $C_{2v}$ symmetry. The maximum displacement from the mean plane in the $p$-nitroaniline complex is given as 0.06 A. Since the X-ray method is not suitable for the determination of H-atom positions, these results are not at variance with the ideas advanced here. Recently, Ritschl [44] has assigned the C-N stretching frequency in ten anilines, and has shown that it increases linearly with the Hammett $\sigma$ values of the substituents, also implying increasing bond order.

The integrated intensity ($A$) of a fundamental infrared absorption band depends on the dipole moment change with the normal coordinate of the vibration ($\partial M / \partial Q$):

$$A = (N\pi/3c)(\partial M / \partial Q)^2 \tag{11}$$

where $N$ is the Avogadro number and $c$ is the velocity of light. If only the changes of NH bond moments with bond length ($dM/dr$) contributed to the transition moment, and if the contributions from both NH bond moments were equal and additive, then ($dM/dr$) could be calculated in two independent ways from the experimental intensities, using the expressions

$$(\partial M / \partial Q_s) = (2/\mu_s)^{\frac{1}{2}} \cos (\theta/2) \cdot (dM/dr) \tag{12}$$

$$(\partial M / \partial Q_{as}) = (2/\mu_{as})^{\frac{1}{2}} \sin (\theta/2) \cdot (dM/dr) \tag{13}$$

where the $\mu$s are the effective reduced masses governing the amplitudes of vibration. These simplified expressions arise because the resultant NH bond dipole gradients in the symmetric and asymmetric stretching vibrations lie along the internal and external bisectors of the HNH angle, respectively. Alternatively, equations (12) and (13) can be used to estimate $\theta$ from the experimental intensities through

$$(A_s / A_{as}) = \tan^2 (\theta/2) \tag{14}$$

Mason [34] attempted this for $N$-heteroaromatic primary amines and found that these $\theta$ values never exceeded 92°, and were not even in the same relative order as the very reasonable $\theta$ values calculated from the frequencies. Further, Mason found that the two ($dM/dr$) values that could be derived independently from (12) and (13) did not agree, the calculation based on $A_S$ always giving a larger value. The author concluded that because of the possibility of large lone pair contributions to $A_s$, equation (13) may be a better approximation. It would fail insofar as the asymmetric vibration is anharmonic and has a small dipole moment gradient along the internal bisector of the HNH angle

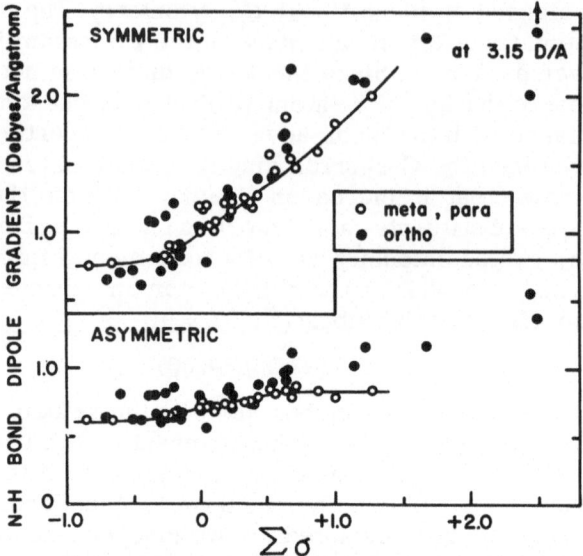

Fig. 9. Substituent dependence of the calculated NH bond moment gradients based on the intensities observed in the two vibrational modes.

and small in-phase pulsations in s–character of the lone pair orbital.

The present measurements and calculations lend strong support to this vibrational mechanism (Fig. 9). For $p$-phenyl-enediamine and $N,N$-dimethyl-$p$-phenylenediamine, $\theta$ values of 109.4° and 109.6°, respectively, indicate virtual $sp^3$ hybridization, since $\theta$ is 106.8° in cyclohexylamine [34] and 106.78° in ammonia [45]. With the lone pair electrons localized on the nitrogen atom and effectively isolated from the $\pi$-electrons, the symmetric intensity is due to variation in the NH bond moments and in the atomic dipole. The NH bond moment component in the asymmetric mode exceeds the corresponding component in the symmetric mode for $\theta > 90°$. Table II gives the relative contributions of these bond moments, assuming a fixed vibrational amplitude and a fixed charge distribution in the bond. The numerical values are normalized to the contributions for $\theta = 90°$, where they are equal for both modes. Since $A_{as} > A_s$ for anilines with large negative substituent constants [9], it appears that the atomic dipole contribution is not large enough to outweigh the deficiency in the NH bond dipole component in the symmetric mode. The calculated values of $(dM/dr)$ for these compounds from equations (12) and (13) are

TABLE II

Relative Contributions of NH Bond Moments to the Resultant Dipole Moment Change in the $NH_2$ Group, for a Fixed Vibrational Amplitude and Fixed Charges on the N and H Atoms

| $\theta$ (°) | Asymmetric Mode (dipole moment change along external bisector of $\angle HNH$)[*] | Symmetric Mode (dipole moment change along internal bisector of $\angle HNH$) |
|---|---|---|
| 90° | 1.00 | 1.00 |
| $(sp^3)$ 109.4° | 1.16 | 0.83 |
| 113.6° | 1.20 | 0.79 |
| $(sp^2)$ 120° | 1.25 | 0.71 |

[*]In the asymmetric mode this dipole moment change reverses sign after every half-cycle of the vibration.

0.8 and 0.6 D/A, respectively. The enhancement of 0.2 D/A in the value based on $A_s$ reflects the lone pair contribution.

As the substituent constants become more positive and the HNH angle begins to approach the trigonal angle of 120°, the lone pair orbital becomes more extensively delocalized over the aromatic ring. Since the lone pair conjugation will vary during a complete vibrational cycle as already described, this should lead to a gradually increasing $\pi$-electron component as $\theta$ increases. This is no doubt a dominant factor in determining the intensities, and explains the pronounced rise in the $(dM/dr)$ values as calculated from $A_s$ by a method which neglects this contribution (Fig. 9). When $\theta$ is large, the $\pi$-electron component (Fig. 5c) more than outweighs the deficiency in the NH bond dipole contribution in the symmetric mode relative to the asymmetric mode (Table II). These results, as summarized in Table III, also explain the crossing of the log $A_s/\sigma$ and log $A_{as}/\sigma$ correlation lines near $\sigma = 0$ (parent compound, aniline* ) previously reported by us [9].

The calculated $(dM/dr)$ values based on $A_{as}$ are probably reliable as far as order of magnitude is concerned. Figure 9 shows that this value rises slightly as $\theta$ increases, which is predicted (Table II). The bond dipole gradients for ortho-substituted anilines are slightly larger than those for the corresponding meta- and para-anilines, as based on both $A_s$ and $A_{as}$. This is a more sensitive plot than Fig. 2, which does not

*Coulson [46] has calculated the $\pi$-electron density on the nitrogen atom in aniline to be 1.91, using an M.O. method. This is consistent with an HNH angle of about 111.7° as found in this work.

## TABLE III

### Resultant Dipole Moment Change in the $NH_2$ Group

| $\theta(°)$ | $b$ | $\sigma$ | Contributions in Vibrational Modes* | | | Relative Intensities | |
|---|---|---|---|---|---|---|---|
| | | | Asymmetric | | Symmetric | Predicted | Observed $A_{as}/A_s$ |
| 109.4 | 0.499 | $-0.83$ ($p$-$NH_2\,C_6\,H_4\,NH_2$) | $d_{min}$ | $>$ | resultant of $a_{max}+b$ | $A_{as} > A_s$ | 1.30 |
| 111.7 | 0.520 | $0$ ($C_6H_5NH_2$) | $d_{min}+\delta$ | $\approx$ | resultant of $(a_{max}-\delta)+c$ | $A_{as} \approx A_s$ | 1.06 |
| 113.6 | 0.535 | $+1.27$ ($p$-$NO_2\,C_6\,H_4\,NH_2$) | $\underbrace{d_{min}+\delta+\delta'}_{d_{max}}$ | $<$ | $\underbrace{\text{resultant of } (a_{max}-\delta-\delta')+c_{max}}_{a_{min}}$ | $A_{as} < A_s$ | 0.43 |

*Refer to Fig. 5 for the identity of the components.

take HNH angle changes into account. Slight enhancement of $(dM/dr)$ on ortho-substitution would be expected to show up in calculations based on both vibrational modes if intramolecular hydrogen bonding were involved. For 2,6-dinitroaniline, $(dM/dr)$ based on the symmetric mode is unusually low (also see point 63 in Fig. 2), whereas the largest value is obtained from calculations based on the asymmetric mode. This must arise from abnormal charge fluctuations in the doubly-chelated amino group depending on the relative movement of the H atoms with respect to each other and to oxygen atoms.

Further evidence for this vibrational interpretation is provided by the log $A_s$ and log $A_{as}$ correlations with $\sigma_I$ and $\sigma_R$ according to equation (5). For the asymmetric vibration, $(\alpha/\beta)$ was found to be 1.7, whereas for the symmetric vibration this ratio was 1.1, indicating that $A_s$ is much more sensitive to the resonance effect than $A_{as}$.

In liquids and solutions collision broadening is considered to be the dominant factor contributing to vibrational band width, since Ramsay [22] has concluded that natural line widths would be $\sim 10^{-6}$ cm$^{-1}$, and Doppler broadening would lead to widths of only $\sim 10^{-3}$ cm$^{-1}$. The decrease in $\Delta\nu_{1/2}$ as the substituents become more electron-withdrawing can now be attributed generally to a rise in force constant, which makes the solvent perturbation of energy levels less significant. However, such factors as a certain amount of torsional libration of the -NH$_2$ group relative to the aromatic ring, without significant loss of $\pi$-electron overlap, cannot be excluded.* Such torsional motion would no doubt lead to band widening, and would decrease with increasing C-N bond order. Califano and Moccia [48] have assigned some of the additional vibrations of the -NH$_2$ group in a few substituted anilines, and Stewart [49] reports the -NH$_2$ torsional oscillation (twisting of the C-N bond) in the far infrared at $\sim 290$ cm$^{-1}$. Further studies on substituent effects on these vibrational modes are now in progress.

In ortho-substituted anilines the half-band widths are significantly narrower than for the corresponding meta- and para-compounds (Fig. 4). This narrowing is ascribed to a decrease in the number of effective collisions of solvent molecules with the amino group due to the shielding nature of the

---

*Jones et al. [47] have suggested that in ortho-substituted acetophenones the acetyl group may actually be in a state of minimum energy when it is displaced slightly from the plane of the ring.

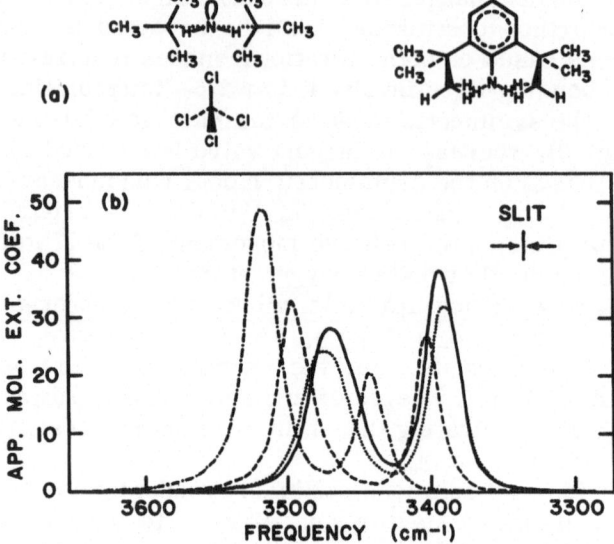

Fig. 10. (a) Steric hindrance to solvation in 2,6-di-*tert*-butyl-aniline. (b) Typical high resolution amine absorption spectra (dilute CCl$_4$ solution).

———————  aniline

·················  4-*tert*-butylaniline

—·——·—  2,4,6-tri-*tert*-butylaniline

— — — —  2-*tert*-butylaniline

adjacent substituents. As expected, 2,4,6-tri-*tert*-butyl, 2,5-di-*tert*-butyl, and 2-*tert*-butylaniline (points 5,11, and 17) show very extensive narrowing. Scale molecular models show that the NH$_2$ group must be vibrating in what is effectively a methyl cage in 2,4,6-tri-*tert*-butylaniline (shown schematically in Fig. 10a). Free rotation of the *tert*-butyl groups is not possible. Wepster and co-workers [50] report that the base strength of 2,6-di-*tert*-butylaniline is lower by $10^3$ than would be expected, and Bartlett et al. [51] found the $pK_A$ of 2,4,6-tri-*tert*-butyl-aniline to be < 2 (in comparison with $pK_A$ values of 4.14, 4.00, and 3.39 for aniline, *p-tert*-butylaniline, and *o-tert*-butylaniline, respectively). Both anomalies were ascribed to steric hindrance to solvation. Very narrow bands are also observed in *o*-phenyl-enediamine.

The narrow band widths for ortho-substituted compounds are in agreement with the concept of intramolecular hydrogen bonding, since the interacting distance involved and the rotational configuration of the amino group is extremely restricted, especially as the lone pair conjugation with the aromatic $\pi$-electrons increases. A further loss of degrees of freedom of the system occurs on hydrogen bond formation [40].

The high resolution measurements indicate that almost all the $NH_2$ bands have some asymmetry on the low frequency side. The significance of this is not yet known. Cabana and Sandorfy [52] have pointed out a similar effect in the $C\equiv N$ stretching vibration in benzonitriles, and have suggested an intramolecular Stark effect where highly polarizable groups are involved.

## ACKNOWLEDGMENTS

Professors B. M. Wepster (Technische Hogeschool, Delft) and L. R. C. Barclay (Mt. Allison University, Sackville, New Brunswick) generously supplied some samples of ortho-substituted amines. The technical assistance of Mrs. G. D. Brown with the high resolution measurements is gratefully acknowledged.

## REFERENCES

1. L. P. Hammett, Physical Organic Chemistry, McGraw-Hill, New York (1940), Chapt. 7.
2. H. H. Jaffe, Chem. Revs. 53, 191 (1953).
3. P. J. Stone and H. W. Thompson, Spectrochim. Acta 10, 17 (1957).
4. H. W. Thompson and G. Steel, Trans. Faraday Soc. 52, 1451 (1956).
5. P. J. Krueger and H. W. Thompson, Proc. Roy. Soc. (London) A250, 22 (1959).
6. M. F. A. E. Sayed, J. Inorg. and Nuclear Chem. 10, 168 (1959).
7. H. W. Thompson, R. W. Needham, and D. Jameson, Spectrochim. Acta 9, 208 (1957).
8. R. N. Jones, W. F. Forbes, and W. A. Mueller, Can. J. Chem. 35, 504 (1957).
9. P. J. Krueger and H. W. Thompson, Proc. Roy. Soc. (London) A243, 143 (1957).
10. M. S. C. Flett, Trans. Faraday Soc. 44, 767 (1948).
11. S. Califano and R. Moccia, Gazz. chim. ital. 86, 1014 (1956).
12. S. Califano and R. Moccia, Gazz. chim. ital. 87, 58 (1957).
13. G. K. Goldman, H. Lehman, and C. N. R. Rao, Can. J. Chem. 38, 171 (1960).
14. C. N. R. Rao and R. Venkataraghavan, Can. J. Chem. 39, 1757 (1961).
15. C. N. R. Rao and G. B. Silverman, Current Sci. (India) 26, 375 (1957).
16. Y. Okamoto and H. C. Brown, J. Org. Chem. 22, 485 (1957).
17. H. C. Brown and Y. Okamoto, J. Am. Chem. Soc. 80, 4979 (1958).
18. T. L. Brown, J. Phys. Chem. 64, 1798 (1960).
19. R. W. Taft, Jr., in M. S. Newman (ed.): Steric Effects in Organic Chemistry, John Wiley & Sons, New York (1956): (a) pp. 594 ff; (b) p. 618.
20. A. C. Farthing and B. Nam in G. W. Bray (ed.): Steric Effects in Conjugated Systems, Butterworth & Co., London (1958), p. 131.
21. R. A. Russell and H. W. Thompson, Spectrochim. Acta 9, 133 (1957).
22. D. A. Ramsay, J. Am. Chem. Soc. 74, 72 (1952).
23. P. J. Krueger, Applied Optics 1, 443 (1962).
24. D. H. McDaniel and H. C. Brown, J. Org. Chem. 23, 420 (1958).
25. P. J. Bray and R. G. Barnes, J. Chem. Physics 27, 551 (1957).

26. L. K. Dyall, Spectrochim. Acta 17, 291 (1961).
27. W. J. Orville-Thomas, A. E. Parsons, and C. P. Ogden, J. Chem. Soc. 1047 (1958).
28. C. A. Coulson, in E. Thornton and H. W. Thompson (eds.): Proceedings of the Conference on Molecular Spectroscopy, Pergamon Press, New York (1959), p. 183.
29. B. M. Wepster, in G. W. Bray (ed.): Steric Effects in Conjugated Systems, Butterworth & Co., London (1958), p. 86.
    B. M. Wepster, in W. Klyne and P. B. D. de la Mare (eds.): Progress in Stereochemistry, Butterworth & Co., London (1958), p. 99.
30. J. M. Essery and K. Schofield, J. Chem. Soc. 3939 (1961).
31. P. J. Krueger, unpublished results.
32. J. W. Linnett, Trans. Faraday Soc. 41, 223 (1945).
33. D. C. McKean and P. N. Schatz, J. Chem. Phys. 24, 316 (1956).
34. S. F. Mason, J. Chem. Soc. 3619 (1958).
35. A. E. Lutskii and V. A. Alexseeva, Zh. obsch. khimii 29, 2992 (1959).
36. A. G. Mortiz, Spectrochim. Acta 15, 242 (1959).
37. V. C. Farmer and R. H. Thompson, Spectrochim. Acta 16, 559 (1960).
38. J. J. Elliott and S. F. Mason, J. Chem. Soc. 1275 (1959).
39. M. Davies, Discussions Faraday Soc. 9, 325 (1950).
40. A. W. Baker and W. W. Kaeding, J. Am. Chem. Soc. 81, 5904 (1959).
41. R. M. Badger and S. H. Bauer, J. Chem. Phys. 5, 839 (1937).
42. R. E. Richards, Trans. Faraday Soc. 44, 40 (1948).
43. L. E. Sutton (ed.): Tables of Interatomic Distances and Configuration in Molecules and Ions, The Chemical Society, London (1958).
44. R. Ritschl, Z. Chemie 1, 285 (1961).
45. G. Herzberg, Infrared and Raman Spectra of Polyatomic Molecules, Van Nostrand, New York (1945), p. 439.
46. C. A. Coulson, Valence, Oxford University Press (1952), p. 246.
47. R. N. Jones, W. F. Forbes, and W. A. Mueller, Can. J. Chem. 35, 504 (1957).
48. S. Califano and R. Moccia, Gazz. chim. ital. 87, 805 (1957).
49. J. E. Stewart, J. Chem. Phys. 30, 1259 (1959).
50. J. Burgers et al., Rec. trav. chim. 77, 491 (1958).
51. P. D. Bartlett, M. Roha, and R. M. Stiles, J. Am. Chem. Soc. 76, 2349 (1954).
52. A. Cabana and C. Sandorfy, Spectrochim. Acta 16, 335 (1960).

# SOLVENT EFFECTS ON THE INFRARED SPECTRA OF ORGANOPHOSPHORUS COMPOUNDS*

## John R. Ferraro

Argonne National Laboratory
Argonne, Illinois

## ABSTRACT

Solvent effects on the $P \to O$ and $[P-O]G$ absorptions in organophosphorus compounds of the type $(GO)_3P \to O$, $(GO)_2G'P \to O$, $[(GO)G'PO(OH)]_2$, $[(GO)_2PO(OH)]_2$, and $G_3P \to O$ are described, in which G is an alkyl or aryl group. The unassociated phosphoryl dipoles are affected by solvents in a manner similar to the carbonyl dipole. The consequences of the solvent effects on the physical properties of these compounds are discussed.

## INTRODUCTION

Interaction of the $P \to O$ dipole in organophosphorus compounds with alcohols and phenols using infrared methods of investigation have been reported [1-7]. Bellamy [8-11] has observed the solvent effects on the infrared spectra of the compounds $POCl_3$ and $(CH_3)_2HP \to O$. The importance of organophosphorus compounds as metallic extractants, where they are used with an organic diluent, motivated further studies as to the effect of these diluents. In the use of compounds of the type $(GO)_3P \to O$, $G_3P \to O$, $(GO)_2G'P \to O$, $[(GO)G'PO(OH)]_2$, and $[(GO)_2PO(OH)]_2$ as metallic extractants, it was apparent that the diluent or solvent used had a depressing effect on the distribution coefficients of the metal extracted [12, 13]. The solvent effects on the $P \to O$ bond in these compounds therefore appeared to be appreciable. Studies of the effects of the various diluents on the phosphoryl stretching vibration of the compounds tri-$n$-octyl phosphine oxide, tri-$n$-butyl phosphate,

Based on work performed under the auspices of the US Atomic Energy Commission.
*Published in the Journal of Applied Spectroscopy.
†G is an alkyl or aryl group.

di(2-ethylhexyl)phosphoric acid, 2-ethylhexyl hydrogen 2-ethylhexylphosphonate, and dibutyl butyl phosphonate would be an approach to determining the magnitude of these effects. This paper reports on such an infrared investigation.

## EXPERIMENTAL

The infrared studies were made with a Beckman Model IR-4 Infrared Spectrophotometer and a Perkin-Elmer Model 221 Infrared Spectrophotometer in matched cells ranging from 0.0125-mm thickness for di-(2-ethylhexyl)phosphoric acid and 0.025-mm for the other compounds. The solution spectra were solvent compensated, and concentrations of 0.05 M were used. The use of higher concentrations was precluded because of possible solute interactions, and the use of greater cell thickness presented the problem of increased losses in energy because of solvent absorption. The solvents used were of the following purity: $n$-hexane, b. p. 68-69°C, Matheson, Coleman & Bell Co.; cyclohexane, spectroscopic grade, Matheson, Coleman & Bell Co.; benzene, spectroscopic grade, Eastman Organic Chemicals; carbon tetrachloride, spectroscopic grade, Eastman Organic Chemicals; chloroform,* reagent grade, Merck & Co.; acetone, reagent grade, Fisher Scientific Co. (dried over anhydrous $CaCl_2$ before use); and absolute methanol, analytical reagent grade, Mallinckrodt Chemical Works. The tri-$n$-octyl phosphine oxide (I) was obtained from Eastman Organic Chemicals and was used without further purification. The tri-$n$-butyl phosphate (II), also from Eastman Organic Chemicals, was purified as previously reported [14]. The purification of di(2-ethylhexyl)phosphoric acid (III) and 2-ethylhexyl hydrogen 2-ethylhexylphosphonate (IV) was also previously reported [15,16]. Dibutyl butyl phosphonate (V) was from Virginia-Carolina Chemical Corp. and was purified in a manner similar to tri-$n$-butyl phosphate [14]. (The above organophosphorus compounds are referred to in the following discussion by their corresponding Roman numerals.)

## RESULTS

The effects on the phosphoryl stretching frequency in these compounds in such solvents as $n$-hexane, cyclohexane, benzene, carbon tetrachloride, chloroform, acetone, and methanol are

*The chloroform was freed from its stabilizer using the method of Halpern [5].

TABLE I
Phosphoryl Absorption of Several Organophosphorus Compounds
in Various Solvents ($cm^{-1}$)

| | Tri-n-octyl phosphine oxide | Tri-n-butyl phosphate | 2-Ethylhexyl hydrogen 2-ethylhexyl-phosphonate | Di(2-ethyl-hexyl)phosphoric acid | Di-n-butyl n-butyl phosphonate |
|---|---|---|---|---|---|
| Liquid | 1170 M | 1280 M<br>1260 M | 1210 S<br>Broad | 1233 S | 1255 M |
| Solid | 1150 M | | | | |
| n-Hexane | 1200 W | 1292 M<br>1272 M | 1221 M | 1233 S | 1255 M |
| Cyclo-<br>hexane | 1200 W | 1290 M<br>1275 M | 1217 M | 1233 S | 1256 M |
| $CCl_4$ | 1170 W | 1278 M<br>1270 M | 1216 M | 1233 S | 1250 M |
| $C_6H_6$ | 1184 W<br>1171 W | 1280 M<br>1260 M | 1209 M | 1230 S | 1252 M |
| $CHCl_3$ | 1144 M | 1260 M<br>Broad | 1209 M | 1233 S<br>1210 M | 1248 M |
| Acetone | | 1280 M<br>1235 M<br>1218 M | | 1240 S<br>1220 M | |
| $CH_3OH$ | 1133 M<br>1110 W | 1255 M<br>Broad | 1200 M | 1230 M<br>Broad | 1220 M<br>1236 M |

tabulated in Table I. The solvent effects increase as the solvent becomes more polar. The ratio of absorbance (P–O)G/absorbance (P → O), using the peak intensities, remains rather constant in less polar solvents. As the solvent becomes more polar, the ratio in (II) and (IV) decreases, while for (V) it remains rather constant in these solvents. For (III) the ratio shows an increase. Although the ratios in peak intensities used in this comparison are at best only qualitative, they do indicate that changes may also be occurring at the (P–O)G bond. In addition there is a tendency for the (P–O)G absorption to broaden and to shift toward higher frequencies in the more polar solvents. This shift is in the same direction as was found for the hydrogen bonding of triethyl phosphate [5].

TABLE II

Values of $\Delta\nu/\nu'$ for Various Organophosphorus Compounds and
Acetophenone with Various Solvents *

| Solvent | Tri-n-butyl phosphate[†] | Di-n-butyl n-butyl phosphonate | Di(2-ethylhexyl)-phosphoric acid | 2-Ethylhexyl hydrogen 2-ethylhexyl-phosphonate | Tri-n-octyl phosphine oxide | Aceto-phenone [8-9] |
|---|---|---|---|---|---|---|
| n-Hexane | — | — | — | — | — | — |
| Cyclo-hexane | 1.5 | 0 | 0 | 3.3 | 0 | 0.6 |
| Benzene | 9.3 | 2.4 | 0 | 9.9 | 13.3 | 4.1 |
| CCl₄ | 10.8 | 4.0 | 0 | 4.1 | 25 | 3.0 |
| CHCl₃ | 24.8 | 5.6 | 2.4 | 9.9 | 46.7 | 8.3 |
| CH₃OH | 28.6 | 15.1 | 2.4 | 20.6 | 55.8 | 4.1 |
|  |  |  |  |  |  | 10.0 |

\* $\nu'$ is the P→O frequency in the n-hexane solution and $\Delta\nu$ is the difference between this frequency in the n-hexane solution and in the other solvents.
†This column shows only the results for the P→O absorption at the higher frequency.

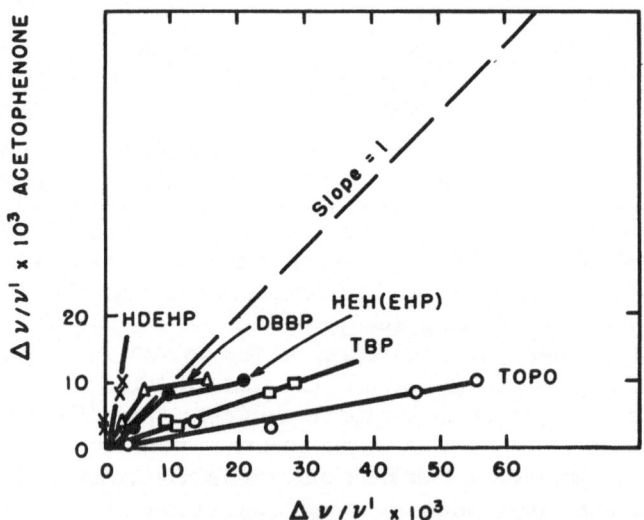

Plot of $\Delta\nu/\nu'$ of acetophenone vs. $\Delta\nu/\nu'$ for the organophosphorus compounds di(2-ethylhexyl)phosphoric acid (HDEHP) di-n-butyl n-butyl phosphonate (DBBP); 2-ethylhexyl hydrogen 2-ethylhexylphosphonate [HEH(EHP)]; tri-n-butyl phosphate (TBP); tri-n-octyl phosphine oxide (TOPO).

Bellamy [8,9] has plotted $\Delta\nu/\nu$ for $X=0$ dipoles vs. $\Delta\nu/\nu$ of acetophenone, obtaining straight lines. The frequency $\nu$ of the gas spectrum was taken as the starting point. For the compounds studied in this work it is extremely difficult to obtain the frequency of the gas spectra, since the liquids (II), (III), (IV), and (V) are very high boiling, and (I) is a solid. Therefore, the frequency in n-hexane ($\nu'$) was used as the starting point. Table II shows the tabulated $\Delta\nu/\nu'$ results, and the figure shows a plot of $\Delta\nu/\nu'$ for the organophosphorus compounds vs. $\Delta\nu/\nu'$ for acetophenone. Straight lines are obtained and slopes of 0.3 and 0.2 are found for (II) and (I), respectively, while a large slope is obtained for (III). For (IV) and (V) the points are better fitted with 2 straight lines, the slope of the initial line being steeper than that of the second. It can be seen from the figure that the P→ O bond in (I) is of greater polarity than (II), which is greater than (IV) and (V). The compound (III), which is a dimer in dilute solutions of n-hexane, cyclohexane, benzene, and carbon tetrachloride [17], is extremely resistant to solvent effects, until one goes to high polar solvents such as acetone, chloroform, or methanol, which will break down the dimer.

## DISCUSSION

The experimental results indicate a difference in the polarity of the phosphoryl dipole depending on the nature of the molecule or the environment around the dipole. The neutral organophosphorus molecules show a much more polar phosphoryl bond than the di(2-ethylhexyl)phosphoric acid type. As expected, (I) shows the most polar phosphoryl bond of the five types of compounds studied. This is in agreement with the infrared results showing a phosphoryl absorption in (I) at a lower frequency than the other compounds. The P → O dipole in (II) appears to behave like (I). Thus, the neutral organophosphorus compounds containing exposed P → O dipoles are more easily affected by solvents. In the plots shown in the figure straight lines are obtained. The results compare favorably with those of Bellamy [9] for similar compounds, such as $(CH_3)_2$ HP → O and $POCl_3$. Compounds (IV) and (V) appear to show different behavior, and have a break in the slope line. Initially, with low polar solvents, the effects are small. However, with a more polar solvent such as alcohol, there is a break and the new line approaches the slope of compounds (I) and (II). This might be caused by a difference in behavior for these solutes in the various solvents from that observed for acetophenone. However, in compound (IV) it is also possible that initially in the low polar solvents the predominant aggregation is dimeric, and the associated $P \rightarrow O$ dipole is thus less available for interaction with the solvent than is the unassociated $P \rightarrow O$ dipole in (I) and (II). Generally, these four classes of compounds appear to have a $P \rightarrow O$ dipole of a polarity somewhere between that of the CO dipole in acetophenone and the SO dipole in dimethyl sulfoxide [9].

The $P \rightarrow O$ stretching frequency in di(2-ethylhexyl)phosphoric acid is independent of the nature of the solvents. This is probably due to the strong dimers formed by this class of acids. Until the ring can be ruptured the phosphoryl bond resists any solvent effects. Polar solvents like chloroform, acetone, or methanol are capable of breaking the ring and of interacting to form hetero intermolecular compounds, causing definite effects on the phosphoryl bond. This behavior is similar to that found for pyrrole [18] and other dimers [19], and is an indication that the $P \rightarrow O$ bond in (III) is only very slightly polar as long as it is part of a strong dimeric ring [15]. Isopiestic molecular weights [17] obtained in various solvents indicated that (III) was predominantly dimeric in dilute solutions of $n$-hexane, cyclohexane, benzene, and carbon tetrachloride, began monomerizing only in chloroform and acetone, and was monomeric in methanol. The molecular weight results parallel the results obtained in infrared. The difference in solvent behavior between (III) and (IV) is probably due to the difference in the stability of the dimers. The acid (IV) dimer, which in very dilute solutions in these solvents appears to break down [17], shows less resistance to solvent effects than (III).

The use of organophosphorus compounds for the solvent extraction of metal cations is well known. Most of the extractions are carried out in an organic phase using a diluent or solvent such as those used in this investigation. It is interesting to note the consequences of these solvent effects on the metal extraction. In the neutral organophosphorus compounds, the more polar phosphoryl bond will interact with the solvent, thus removing the cation extraction sites, which should cause a lower distribution coefficient. An investigation in our laboratory [12, 13] involving organophosphorus compounds of the type described in this paper substantiates this. For example, extraction of promethium into an organic phase of 1,1,1,-trifluoro-3-2'-thenoyl acetone and TBP is highest for cyclohexane and decreases in the order $n$-hexane > carbon tetrachloride > benzene > methyl isobutyl ketone > chloroform. Similar results have been obtained with the acidic type of compounds and the more polar the diluent the lower the distribution coefficients [20, 21].

It is very important then that the distribution coefficients of metal extractions be compared only when the so-called inert diluent or solvent used is the same. To compare distribution coefficients of metallic extractions into organophosphorus compounds when the diluents vary from $n$-hexane to methanol is meaningless.

## ACKNOWLEDGMENTS

The author thanks the following Argonne employees: Mr. George W. Mason, for the purification of di(2-ethylhexyl)phosphoric acid and 2-ethylhexyl hydrogen 2-ethylhexylphosphonate; and Dr. E. Phillip Horwitz, for the purification of di-$n$-butyl $n$-butyl phosphonate.

### REFERENCES

1. W. Gordy and S. C. Stanford, J. Chem. Phys. 8, 170 (1940); 9, 204 (1941).
2. C. S. Marvel, M. J. Copley, and E. J. Ginsburg, J. Am. Chem. Soc. 62, 3109 (1940).
3. L. F. Audrieth and R. J. Steinman, Ibid. 63, 2115 (1941).
4. G. M. Kosolapoff and J. F. McCullough, Ibid. 73, 5392 (1951).
5. E. Halpern et al., Ibid. 77, 4472 (1955).
6. G. Aksnes and T. Gramstad, Acta Chem. Scand. 14, 1485 (1960).
7. T. Gramstad, Ibid. 15, 1337 (1961).
8. L. J. Bellamy and R. L. Williams, Trans. Faraday Soc. 55, 14 (1959).
9. L. J. Bellamy et al., Ibid. 55, 1677 (1959).
10. L. J. Bellamy and P. E. Rogasch, Spectrochim. Acta 16, 30 (1960).
11. L. J. Bellamy and R. L. Williams, Proc. Roy. Soc. (London) A & B 255, 22 (1960).
12. T. V. Healy, J. Inorg. and Nuclear Chem. 19, 314 (1961).
13. T. V. Healy, Ibid. 19, 328 (1961).
14. D. F. Peppard, G. W. Mason, and J. L. Maier, Ibid. 3, 215 (1956).
15. D. F. Peppard, J. R. Ferraro, and G. W. Mason, Ibid. 7, 231 (1958).
16. D. F. Peppard, J. R. Ferraro, and G. W. Mason, Ibid. 12, 60 (1959).
17. J. R. Ferraro, G. W. Mason, and D. F. Peppard, Ibid. 22, 285 (1961).
18. M. L. Josien and N. Fuson, J. Chem. Phys. 22, 1169 (1954).
19. C. G. Cannon, Mikrochim. Acta 2, 555 (1955).
20. C. A. Blake, Jr. et al., Peaceful Uses of Atomic Energy, Proc. United Nations Conf. 28, 289 (1958).
21. D. Dyrrsen and L. D. Hay, J. Inorg. and Nuclear Chem. 14, 1091 (1960).

# NUCLEAR MAGNETIC RESONANCE

# ANOMALOUS CHEMICAL SHIFTS IN THE PROTON MAGNETIC RESONANCE SPECTRA OF THE DIMETHYLCYCLOHEXANES AND RELATED HYDROCARBONS*

## Norbert Muller and William C. Tosch

Department of Chemistry
Purdue University
Lafayette, Indiana

## ABSTRACT

High-resolution proton magnetic resonance spectra were determined for the dimethylcyclohexanes and several related hydrocarbons between −130°C and 130°C. All of the compounds which should undergo rapid ring inversion at room temperature produce spectra which change on cooling because of "freezing out" of this motion. The assumption that appearance of the ring-hydrogen resonances as a relatively narrow band is invariably a symptom of rapid ring inversion is shown to be unfounded. Several of the ring spectra differ drastically from what is predicted using a bond-anisotropy model. A previously unrecognized effect, then, must make a significant contribution to the observed chemical shifts.

## INTRODUCTION

In view of the increasingly widespread use of nuclear magnetic resonance (NMR) spectroscopy as a tool for conformational analysis [1-7], it is important to try to arrive at a detailed understanding of the hydrogen NMR spectra of compounds related to cyclohexane. The spectra of the seven dimethylcyclohexanes which are of interest in this connection have been obtained [3,4] at room temperature and at 40 Mcps. Following the discovery made by Jensen and co-workers [5], and independently in this laboratory, that the axial and equatorial protons of cyclohexane produce separate NMR signals at −95°C,

*This paper is to be published in The Journal of Chemical Physics. This work was supported by a grant from the National Science Foundation and taken from the Ph.D. Thesis of William C. Tosch.

it was decided to reinvestigate the spectra of these compounds at high resolution over a wide temperature range. The purpose of this paper is to present a number of quite unexpected results obtained in the course of this work. Several possible explanations of the observed anomalies were considered and found inadequate, especially when the data were supplemented by including other alkyl-substituted cyclohexanes.

## EXPERIMENTAL

Methylcyclohexane, the dimethylcyclohexanes, and $t$-butylcyclohexane were provided by the National Bureau of Standards. Purified cis -1,1,3,5-tetramethylcyclohexane was kindly made available by Professor N. L. Allinger. The cis - and trans-decalin were prepared from the commercially available isomer mixture by fractional distillation. Chromatographic analysis showed that the isomeric purity of each sample was better than 90%.

The 1,4-methyl-$t$-butylcyclohexanes were prepared by hydrogenation [8] of para-$t$-butyltoluene using W-2 Raney nickel [9] catalyst. The isomers were cleanly separated chromatographically using a 20-ft column of 25% Ucon Polar on 42/60 mesh firebrick.

The other compounds studied and the solvents were commercial materials purified by simple distillation on the vacuum line. The NMR spectra showed no significant traces of hydrogen-containing impurities. Unless otherwise stated, samples were solutions of about 30 mole % of hydrocarbon in carbon disulfide to which a trace of tetramethylsilane had been added to produce the reference signal.

The spectrometer system consisted of a V 4311 high-resolution NMR spectrometer with VK 3506 field stabilizer, a V 4365 field-homogeneity control unit, and a VK 3507 slow-sweep unit.* A V 4340 variable-temperature probe assembly was used as required. Operation of this equipment below the minimum temperature of −60°C specified by the maker constitutes a calculated risk. Most of the data presented here were obtained before thermal stresses resulted in breaking of the Dewar insert. Temperatures were measured with a copper-constantan thermocouple junction built into the Dewar insert.

Peak positions were determined by graphical interpolation between sidebands of the reference signal, produced with a 200J audio oscillator used in conjunction with a 512C electronic

*Manufactured by Varian Associates, Palo Alto, California.

counter.* Most of the measurements were made with an oscillator frequency of 60 Mcps, but several of the spectra were
obtained at 56.4 Mcps, as indicated in the text and figure
captions.

## RESULTS AND DISCUSSION

### Compounds without Axial Substituents

Cyclohexane, bicyclohexyl, and *trans*-decalin. Cyclohexane at room
temperature and 56.4 Mcps produces a sharp line at $-80$ cps
$(-1.42$ ppm). At $-90°C$ the spectrum appears as shown in
Fig. 1a with maxima at $-97$, $-90$, $-69$, and $-63$ cps. It is
probable [5] that the center of the downfield pair of peaks at
$-1.65$ ppm approximately represents the "normal" chemical
shift of equatorial protons and that the normal shift for axial
protons is about $-1.17$ ppm.

Bicyclohexyl consists of two cyclohexyl rings joined so
that each has the other as a rather bulky substituent which
strongly prefers an equatorial position. Then the molecule is
essentially fixed in one conformation even at room temperature.
The spectrum (Fig. 1b) is unaffected by temperature changes
and so similar to the low-temperature spectrum of cyclohexane
as to suggest that the axial-equatorial shift is the only major
type of magnetic nonequivalence for the various ring protons.
In view of the magnitude and variety of the expected indirect
spin-spin couplings, it is not surprising that no distinct spectral
feature attributable to the methine hydrogens can be identified.
This, together with the fact that the two major regions of
resonance partially overlap, makes it impossible to use the
peak areas to test the assumption that the equatorial protons
(outnumbered by axial ones in the ratio of 12:10) indeed produce
the signals at lower fields.

The compound *trans* -decalin is also fixed in a single conformation [1, 6]. Its spectrum at 56.4 Mcps is very similar to
those just described [10].

Methylcyclohexane, *trans*-1,2-dimethylcyclohexane, and *trans*-1,4-di-
methylcyclohexane. The spectrum of methylcyclohexane obtained
in this study agrees precisely with that recently reported by
Anet [11]. The molecules are exchanged rapidly between two
conformations, one with an axial and the other with an equatorial methyl group. More than 95% of the molecules are of the
latter type at any instant, and the time-averaged spectrum is

*Manufactured by Hewlett-Packard Company, Palo Alto, California.

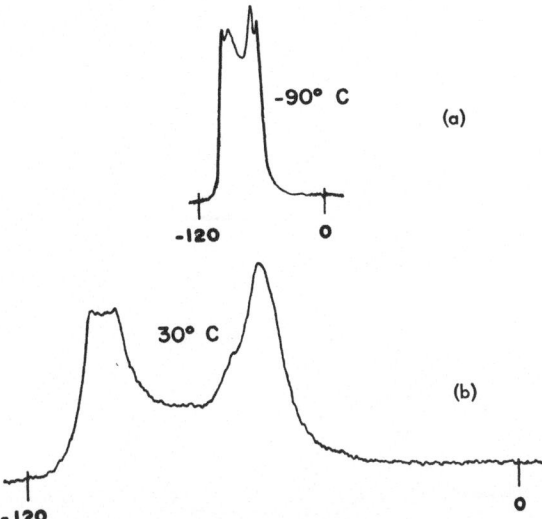

Fig. 1. NMR spectrum of cyclohexane at 56.4 Mcps
and −90°C (a) and bicyclohexyl at 60.0 Mcps (b). In
each spectrum the positions of the tetramethylsilane
resonance and an audio-sideband at −120 cps are
indicated.

essentially that of this conformer. Accordingly the signals due
to the ring hydrogens look much like those of bicyclohexyl or
"frozen" cyclohexane, and there is no appreciable change in
the spectrum with temperature. The methyl doublets of methylcyclohexane and a tetradeuterated species were carefully investigated by Anet [11], who concluded from the intensity ratios
that· the methine proton resonance must lie at −82 cps, although this cannot be verified by direct observation.

The compounds *trans*-1,2- and *trans*-1,4-dimethylcyclohexane
produce rather similar, temperature-independent spectra
(Fig. 2a, b). Essentially all the molecules adopt the conformation having both methyl groups equatorial. The designation *ee*
will be used for such a conformation, as opposed to *ae* when
there is one axial and one equatorial substituent. The ring-
hydrogen spectra are unusual only in that the axial signals are
shifted upfield so that they overlap partially with the methyl
resonances. A striking peculiarity, first noted by Musher [4] is
that the methyl signals are singlets rather than doublets,
although there may be some vestige of a doublet splitting for
the 1,4-compound.

Musher concluded that the methyl-methine spin-spin coupling constant must be nearly zero. To test this, it was decided

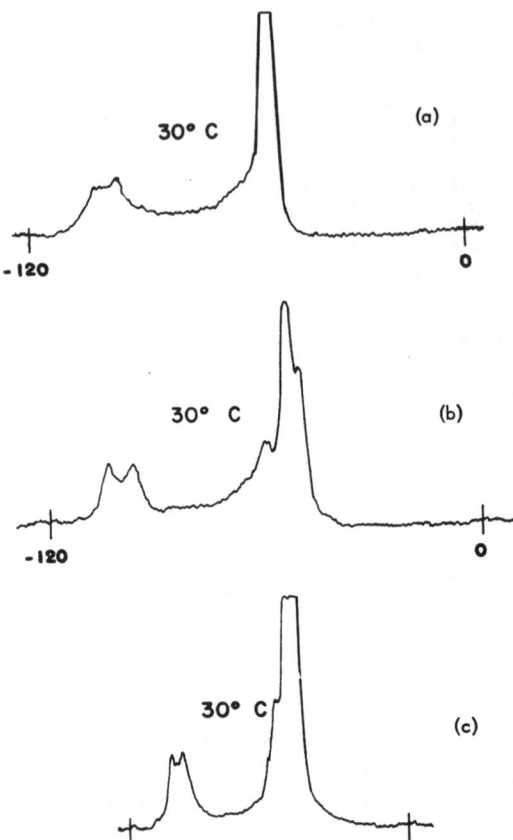

Fig. 2. NMR spectrum of trans-1,2-dimethylcyclohexane at 60.0 Mcps (a); trans-1,4-dimethylcyclohexane at 60.0 Mcps (b); and trans-1,4-methyl-t-butylcyclohexane at 56.4 Mcps (c).

to observe the signals due to naturally present $C^{13}$-methyl groups, occurring about 62 cps upfield from the ordinary methyl signals. If the coupling constant were nearly zero, these $C^{13}$ sidebands should also be singlets. If instead the methyl resonance is unsplit because it and the axial methine proton signal fall at the same field, the sidebands should show a normal fine structure [12].

The $C^{13}$ sideband for the 1,2-compound is about 6 cps wide with a flat-topped contour suggestive of a very poorly resolved doublet. For the 1,4-compound a well-resolved doublet is found with a separation of 5.5 cps. It may be concluded that the methyl-methine coupling constant has a normal value. The distorted shapes of the $C^{13}$ sidebands result from the coupling between the methine hydrogens and other ring hydrogens [11].

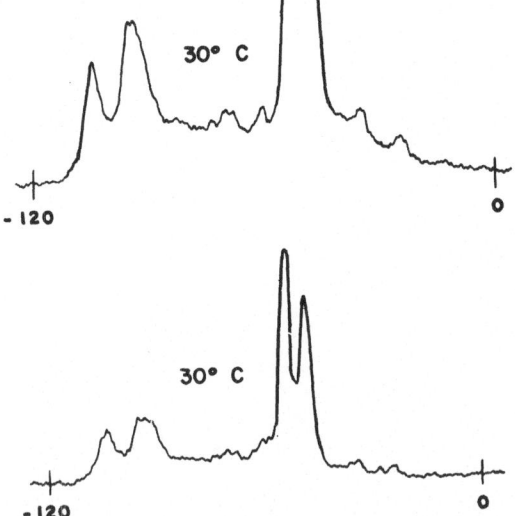

Fig. 3. NMR spectrum of *cis*-1,3-dimethylcyclohexane at 60.0 Mcps and two different levels of amplification.

The resonance of the axial methine proton in each of these compounds must nearly coincide with the methyl resonance. However, a chemical shift near −0.9 ppm for these protons is very surprising.

*trans*-1,4-Methyl-*t*-butylcyclohexane. The spectrum of the lower-boiling isomer of 1,4-methyl-*t*-butylcyclohexane is shown in Fig. 2c. The similarity of this spectrum and that of *trans*-1,4-dimethylcyclohexane helps to establish that the more volatile material is the *trans* isomer, as expected from Allinger's conformational rule [13]. As will be pointed out later, the higher-boiling 1,4-methyl-*t*-butylcyclohexane produces a spectrum quite similar to that of *cis*-1,4-dimethylcyclohexane, further supporting this identification of the isomers.

*cis*-1,3-Dimethylcyclohexane. Figure 3 shows the spectrum of the remaining dimethylcyclohexane with an *ee* conformation. Two features of the spectrum attract attention. First, the methyl resonance is a nicely resolved doublet, showing that once more the axial methine resonance lies downfield, probably beyond −85 cps. Second, the axial ring-proton resonances have components spread over a range extending to about $\frac{1}{2}$ ppm a b o v e the methyl region. These peak positions are undoubtedly affected by spin-spin couplings, but unusually high chemical shift values for one or more methylene protons must also be involved.

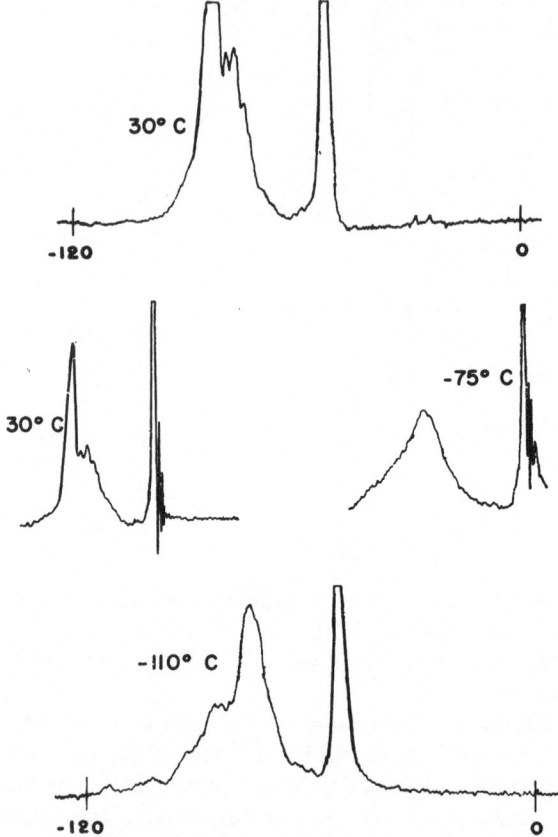

Fig. 4. NMR spectrum of 1,1-dimethylcyclohexane at 60.0 Mcps and three different temperatures. The pair of curves in the center was recorded with faster sweep and reduced amplification.

## Compounds with One Axial Substituent

1,1-Dimethylcyclohexane, cis-1,2-dimethylcyclohexane, and cis-decalin. A molecule with two similar substituents which cannot be simultaneously equatorial may exist in two conformations of equal energy which may be labelled *ae* and *ea*. At room temperature these are rapidly interconverted. Accordingly, 1,1-dimethylcyclohexane (Fig. 4), cis-1,2-dimethylcyclohexane (Fig. 5), and cis-decalin (Fig. 6) each produce a rather narrow ring spectrum at room temperature. Figure 4 also shows that the distinction which exists, in principle, between the hydrogens on carbons 2, 3, and 4 produces no significant difference in chemical shifts.

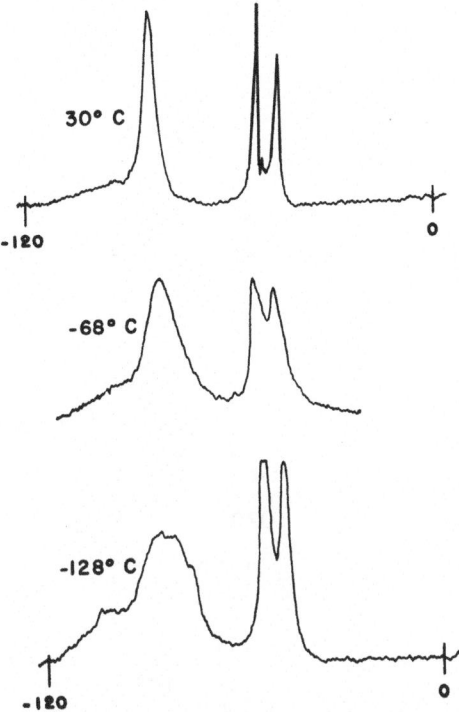

Fig. 5. NRM spectrum of *cis*-1,2-dimethylcyclohex-ane at 60.0 Mcps and three different temperatures.

It was expected that when the temperature was sufficiently reduced, spectra due to the conformationally fixed species would be observed. Indeed, Figs. 4-6 show that the spectra change on cooling, and it is believed that these changes result from conformational "freezing." Contradictory findings have been reported by others [6, 7]. However, Jensen and his co-workers [14] also found temperature variations in the spectra of the *ae* dimethylcyclohexanes and agree with the conclusions reached here: that in each case the barrier for interconversion is of the same order of magnitude as that found for cyclohexane [5], and that below −100°C ring inversion is effectively frozen out.

It follows that for *ae* dialkylcyclohexanes with fixed conformation the ring resonances remain confined within a relatively narrow region around −80 cps, and the "normal" equatorial-proton band near −100 cps and axial band near −65 cps do not occur.

*cis*-1,4- and *trans*-1,3-Dimethylcyclohexane. Figures 7 and 8 show

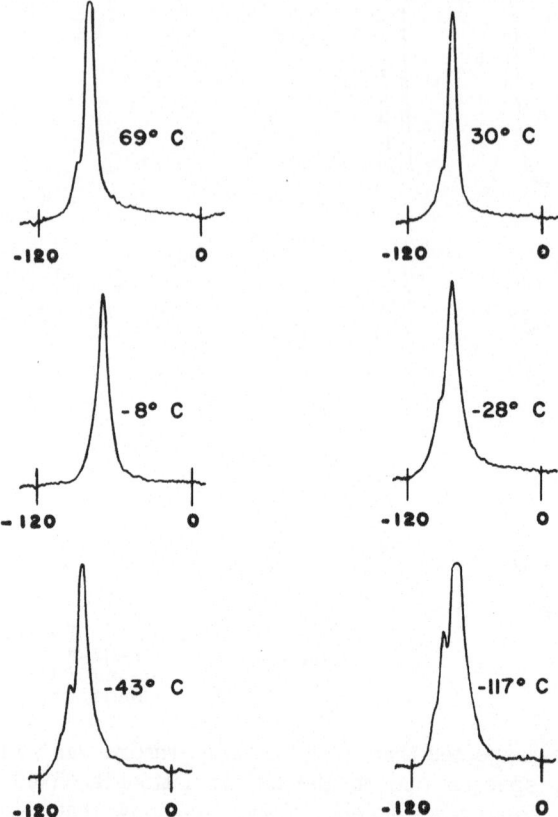

Fig. 6. NMR spectrum of *cis*-decalin at 56.4 Mcps as a
function of temperature.

the spectra of the two remaining *ae* dimethylcyclohexanes at
several temperatures. Ring inversion apparently freezes out
as the temperature is reduced, but again the ring resonances
form a rather narrow band at the lowest temperatures. Some
relatively minor, new features are also apparent.

The ring-hydrogen spectrum of *trans*-1,3–dimethylcyclohexane at room temperature displays a remarkable amount of
fine structure. This structure has disappeared at low temperatures, but then an unusual group of signals is found in the
methyl region. Apparently the methyl groups are magnetically
nonequivalent when the conformation is fixed. This is expected
for all the *ae* isomers but was not found for 1,1- and *cis*-1,2-di-
methylcyclohexane. Interpretation of the signals in the methyl
region is made difficult by the possibility that some of the in-

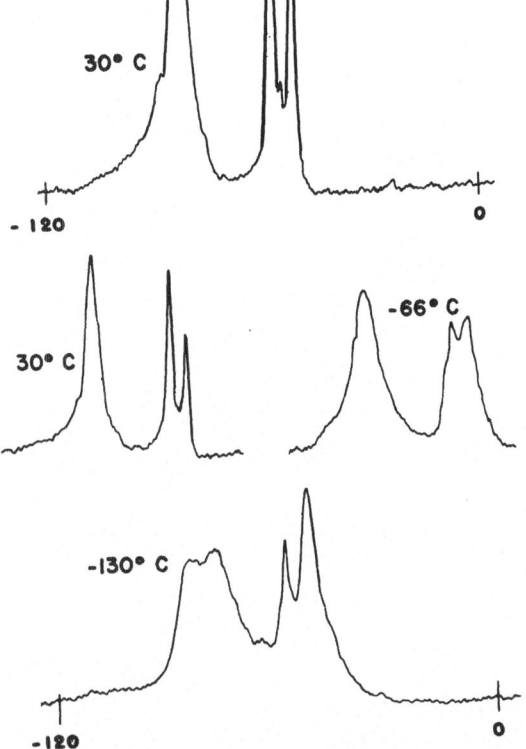

30° C

-120                                    0

30° C

-66° C

-130° C

Fig. 7. NMR spectrum of
cis -1,4-dimethylcyclohex-
ane at 60.0 Mcps and three
different temperatures.

-120                                    0

tensity there may be due to one or two of the ring protons.
This is suggested by the appearance of the spectrum at −78°C,
where the "ring" and "methyl" peaks seem nearly equal in area
although the theoretical intensity ratio is 10:6.

The compound cis -1,4-dimethylcyclohexane also shows an
interesting temperature-dependence for the methyl doublet.
At 25°C the low-field component is the more intense, as expected if the averaged chemical shift of the methine protons
is about −80 cps. At low temperatures, the doublet components
are somewhat broadened, and the order of intensities is
reversed. Apparently the two methyl groups again become
nonequivalent, and one of them (more likely the axial) produces
a doublet while the other produces a singlet like that found in
the trans -1,2- or trans -1,4- isomer. The observed spectrum
suggests that the singlet approximately coincides with the high-
field component of the doublet. It remains unexplained that no
similar effect is found for the cis -1,2-isomer.

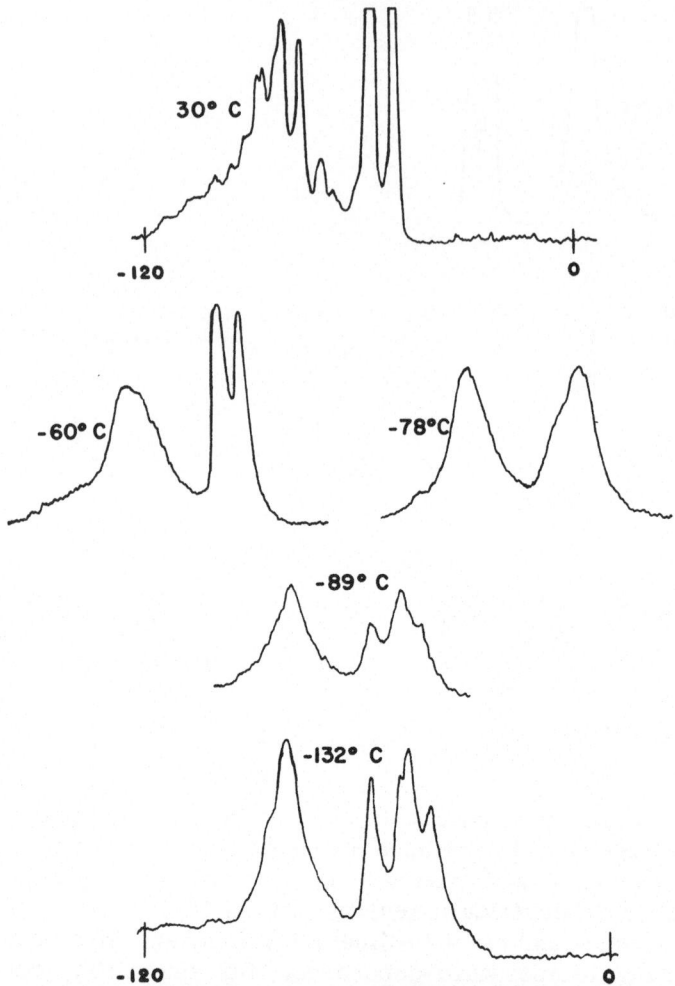

Fig. 8. NMR spectrum of *trans*-1,3-dimethylcyclohexane at 60.0
Mcps as a function of temperature.

*cis*-1,4-Methyl-*t*-butylcyclohexane. The preceding discussion tends
to the conclusion that an unusually narrow band of ring-hydrogen resonances will be encountered whenever a dialkyl
cyclohexane is fixed in an *ae* conformation. The compound
*cis*-1,4-methyl-*t*-butylcyclohexane was prepared in order to
test this generalization.

It has been convincingly argued [15] that a *t*-butyl substituent
forces a cyclohexane ring to adopt the conformation which
permits this bulky group to occupy an equatorial site. There is

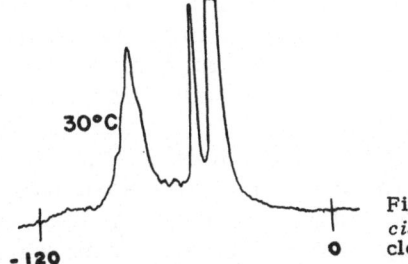

Fig. 9. NMR spectrum of
*cis*-1,4-methyl-*t*-butylcy-
clohexane at 56.4 Mcps.

then no question of ring inversion for either isomer of 1,4-
methyl-*t*-butylcyclohexane. As noted previously, a "normal"
ring spectrum with distinctly separated equatorial and axial
peaks was found for the *ee trans* isomer. The temperature-independent spectrum of the *cis* isomer is shown in Fig. 9. It
closely resembles the anomalous spectrum found for *cis*-1,4-di-
methylcyclohexane at −130°C, except for the obvious difference
in the methyl signals. This result strongly suggests that the
anomalous low-temperature spectra of the *ae* dimethylcyclohexanes and *cis*-decalin should not be ascribed [6, 7] to per-
sistence of rapid ring inversion below −100°C.

    *cis*-1,1,3,5-Tetramethylcyclohexane . This compound was also
studied because it is conformationally fixed at room temperature with one axial methyl group. The spectrum (Fig. 10) shows

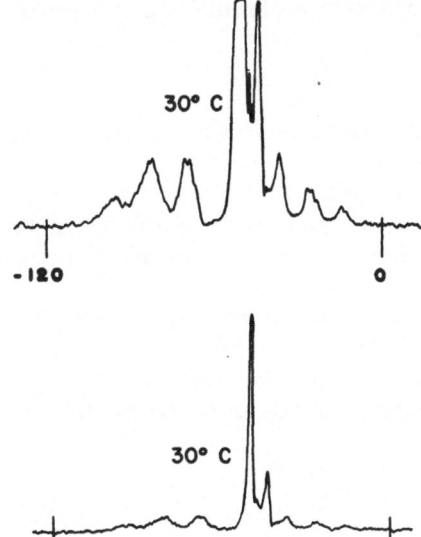

Fig. 10. NMR spectrum of
*cis*-1,1,3,5-tetramethylcy-
clohexane at 56.4 Mcps and
two different levels of amplification.

several unusual features. The two strong, broad peaks centered near −88 and −74 cps do not correspond to the "normal" axial or equatorial hydrogen chemical shifts. Some of the ring-hydrogen signals lie upfield from the methyl resonances, the highest at about −18 cps. Analogous signals were found (see preceding discussion) in the spectrum of *cis*-1,3-dimethyl-cyclohexane. Apart from the unusual effects associated with the presence of an axial methyl group, anomalous upfield shifts for some of the ring protons seem to occur whenever two equatorial methyl substituents are bound to ring carbons separated by one methylene group. Preliminary observations [16] of the spectra of the two hexahydromesitylenes appear to confirm this.

### Intramolecular Origin of the Anomalous Shifts

Early in this study it seemed possible that the observed anomalies might be due at least in part to intermolecular interactions. The spectra of several of the compounds were therefore observed not only in carbon disulfide solutions but also as pure liquids and as dilute solutions in benzene, tetra-methylsilane, and methyl iodide. Although in some cases the solvent change produced an over-all shift of the spectrum by 1 or 2 cps, there was never any detectable change in the relative positions or intensities of the peaks. It was concluded that the environment of the hydrocarbon molecule has no significant effect on the spectrum.

### Theoretical Considerations

The major unexpected results in need of rationalization may be briefly summarized as follows:

1. The "normal" chemical shift difference between axial and equatorial ring protons is drastically reduced in those dimethylcyclohexanes and related hydrocarbons which have one axial alkyl substituent.

2. Very different chemical shifts are found for the axial methine protons in pairs of compounds—such as methylcyclo-hexane and *trans*-1,4-dimethylcyclohexane—for which it would seem intuitively that the environments of these protons must be closely similar.

3. Portions of the ring spectrum for *cis*-1,3-disubstituted compounds are found at fields up to 0.6 ppm above the methyl resonances.

It remains to be investigated whether these phenomena

perhaps follow in a subtle way when existing methods are used to predict the theoretical chemical shifts.

An appealingly simple "bond-anisotropy" model for calculating proton chemical shifts in hydrocarbons was proposed by Bothner-By and Naar-Colin [17] and used by Jackman [2] to rationalize the difference in shielding between axial and equatorial protons in cyclohexane derivatives. This model was used to calculate all ring-proton shifts in methylcyclohexane and the seven dimethylcyclohexanes, assuming for each molecule a single, rigid conformation and including [2] only effects of nearby bonds. The calculations were then repeated including all carbon-carbon bonds in the molecule. Since then, Musher [18] has proposed that this is a better procedure. It was found [10] that the two methods gave only slightly different results, as might be expected because of the occurrence of the factor $1/r^3$ in equation (5) of reference [17].

At best the computed chemical shifts could not accurately reproduce the observed spectra unless allowances were made for the many unequal spin-spin interactions. However, a general rough agreement between the predicted and observed spectra may be expected, and such agreement is found for cyclohexane [2] and methylcyclohexane [10]. For the other compounds there were major discrepancies [10] which may be summarized as follows:

1. No tendency is predicted for the ring-proton resonances of $ae$ dimethylcyclohexanes to be grouped together any more closely than those of the $ee$ isomers.

2. The predicted position of the signal for the axial methine proton is always well downfield from the methyl resonance.

3. None of the ring signals in $cis$-1,3-dimethylcyclohexane is predicted to occur at an exceptionally high field. The same was found for $cis$-1,1,3,5-tetramethylcyclohexane when the calculations were extended to this compound. The observed anomalies thus remain unexplained.

In more recent work on the calculation of anisotropy effects [18, 19], it was suggested that allowance be made for small differences in the inductive effects of a hydrogen and a methyl substituent, and for shielding contributions arising in carbon-hydrogen bonds. The greater flexibility of such a model makes it more difficult to test rigorously, but the data presented here suggest that even with these modifications anisotropy calculations will fail to reproduce the observed shifts. This conclusion may be supported by several qualitative arguments, two of which follow:

1. The spectra of the 1,4-methyl-$t$-butylcyclohexanes (Figs. 2c and 9) show that moving the methyl group on carbon atom 1 ($C_1$) from an equatorial to an axial site results in an upfield shift for the equatorial ring hydrogens which is about the same for those attached to $C_2$ and $C_6$ as for those attached to $C_3$ and $C_5$. The four axial methylene proton signals are shifted downfield, again all about equally. The anisotropy effect of a substituent depends on the inverse cube of the distance between the substituent and the atom affected and on the angle between two appropriately chosen vectors. This makes it very unlikely that a change at $C_1$ will produce equal effects on protons bound to $C_2$ and protons bound to $C_3$. The inductive contribution to these shifts is surely negligible.

2. The compound trans -1,4-dimethylcyclohexane is derived from methylcyclohexane by substituting a methyl group for the equatorial hydrogen originally bound to $C_4$. It was shown previously that this substitution shifts the resonance of the axial methine proton attached to $C_1$ by not less than 20 cps. The newly introduced substituent produces this large shift across a distance of approximately 3.5 A. This could be ascribed to an anisotropy effect only if an unreasonably large value were assigned to the anisotropy in the magnetic susceptibility of the methyl group. Even if this were done, the signals from the ring methylene protons should then be drastically different for the two compounds, whereas in fact they are rather similar.

It thus appears that the bond-anisotropy approach does not make it possible to rationalize all of the data for these cyclic hydrocarbons, and that one must look for some additional factor. Two possibilities suggest themselves: first, for the $ae$ isomers the ring might adopt a boat or skew conformation; second, although remaining essentially in the chair form, the ring may be subject to angle-deformations resulting from steric strain which should be allowed for in the calculations.

A rather considerable body of evidence [20, 21] indicates that the boat or skew conformations are too unstable relative to the chair to be seriously considered. The NMR evidence itself tends to the same conclusion.

The occurrence of deformed chair structures has been the topic of several profitable discussions between Professor E. L. Eliel and the authors. The chief difficulty in using this idea to explain the anomalous spectra is that several of the compounds involved have *ee* conformations and should be virtually strain-free. Even for the *ae* isomers, the following reasoning shows that the chemical shifts are not drastically affected by any ring deformation that may be present:

The repulsion between the axial methyl substituent and axial hydrogens on the same idea of the ring may lead to appreciable deformation in *cis*-1,4-dimethylcyclohexane. In the *cis*-1,2 isomer, if one attempts in a similar way to bend the axial methyl group away from the axial direction, the result is greatly increased interference between the two methyl groups. Thus, the 1,2 isomer should have nearly normal bond angles, while the 1,4 isomer should show the effect of deformation. Since neither compound produces a "normal" spectrum at low temperatures, it seems unlikely that the anomalies arise from ring deformations.

Finally, it may be asked whether postulating ring-current effects [1] would help to resolve the observed anomalies. By modifying the flow of ring-current, a substituent could produce effects on other protons which should be independent of their distance from the point of substitution. However, the attractions of this approach soon fade if one considers, for example, the two 1,4-methyl-*t*-butylcyclohexanes. It would be necessary to suppose that for the *trans* isomer the ring current is nearly the same as that in cyclohexane, but that for the *cis* isomer it is essentially zero. No justification has been found for supposing this. Moreover, even if it is assumed *ad hoc* that an axial methyl group quenches the ring current flow, only one of the three major observed anomalies could be so "explained."

## CONCLUSIONS

1. Of the compounds studied, only 1,1-, *cis*-1,2-, *trans*-1,3-, and *cis*-1,4-dimethylcyclohexane and *cis*-decalin produce spectra which vary with temperature changes. For each of these, the spectra show that the molecules undergo rapid ring inversion at room temperature, but that below −100°C the spectra are those of the conformationally fixed species.

2. The appearance of a relatively narrow region of ring-proton resonances does not, as has been widely supposed, necessarily imply that there is rapid ring inversion.

3. There is no reason to postulate anomalous values for any of the spin-spin coupling parameters in this series of compounds.

4. A number of proton chemical shifts are found which cannot be adequately rationalized with a model based on the recognition only of inductive or bond-anisotropy effects.

5. The observed spectra were essentially unaffected by changes of solvent, and hence the anomalies do not result from intermolecular interactions.

6. The difficulties of interpreting the spectra are not lessened by attempting to allow for hypothetical effects of ring deformations or ring-currents.

7. It follows that a hitherto unrecognized effect contributes significantly to the observed proton chemical shifts in these hydrocarbons. The nature of this phenomenon is not yet understood.

## REFERENCES

1. J. A. Pople, W. G. Schneider, and H. J. Bernstein, High-Resolution Nuclear Magnetic Resonance, McGraw-Hill Book Company, Inc., New York (1959), Chapt. 14.
2. L. M. Jackman, Applications of Nuclear Magnetic Resonance in Organic Chemistry, Pergamon Press, New York (1959), Sect. 7.2.
3. S. Brownstein and R. Miller, J. Org. Chem. 24, 1886 (1959).
4. J. I. Musher, Spectrochim. Acta 16, 835 (1960).
5. F. R. Jensen et al., J. Am. Chem. Soc. 82, 1256 (1960); 84, 386 (1962).
6. W. B. Moniz and J. A. Dixon, J. Am. Chem. Soc. 83, 1671 (1961).
7. R. K. Harris and N. Sheppard, Proc. Chem. Soc. 418 (1961).
8. K. T. Serijan, P. H. Wise, and L. C. Gibbons, J. Am. Chem. Soc. 71, 2265 (1949).
9. R. Mozingo, Org. Syntheses 3, 181 (1955).
10. W. C. Tosch, Ph. D. Thesis, Purdue University (1962).
11. F. A. L. Anet, Can. J. Chem. 39, 2262 (1961).
12. A. D. Cohen, N. Sheppard, and J. J. Turner, Proc. Chem. Soc. 118 (1958).
13. N. L. Allinger, J. Am. Chem. Soc. 79, 3443 (1957).
14. F. R. Jensen, personal communication.
15. S. Winstein and N. J. Holness, J. Am. Chem. Soc. 77, 5562 (1955).
16. N. Muller and O. R. Hughes, unpublished results.
17. A. A. Bothner-By and C. Naar-Colin, Ann. New York Acad. Sci. 70, 833 (1958).
18. J. I. Musher, J. Chem. Phys. 35, 1159 (1961).
19. P. T. Narasimhan and M. T. Rogers, J. Chem. Phys. 31, 1302 (1959).
20. N. L. Allinger and L. A. Freiberg, J. Am. Chem. Soc. 82, 2393 (1960).
21. W. S. Johnson et al., J. Am. Chem. Soc. 83, 606 (1961).

# DETERMINATION OF FAT IN CORN AND CORN GERM BY WIDE-LINE NUCLEAR MAGNETIC RESONANCE TECHNIQUES*

## T. F. Conway and R. J. Smith

George M. Moffett Research Laboratories
Corn Products Company, Argo, Illinois

## INTRODUCTION

In the last ten years, high-resolution nuclear magnetic resonance (NMR) spectroscopy has become an accepted technique for elucidating molecular structure and motion. Yet despite its obvious potential for quantitative analysis, development has been slow. Because more stringent requirements are placed on the instrumentation for quantitative studies, most applications of high-resolution NMR spectroscopy concern structure-type problems. In contrast, commercial wide-line NMR has been used almost exclusively for quantitative analysis [1, 2], because instrumentation and techniques are simpler and more readily adapted to routine measurements. Corn Products Company has used wide-line spectroscopy in analytical research and quality control for about five years; approximately 300,000 samples have been analyzed in the plant control laboratory. The NMR technique offers several advantages: it is fast, nondestructive, precise, and accurate.

This report describes the application of NMR procedures to the determination of fat in corn and corn germ process samples.

Quantitative applications of nuclear magnetic resonance rely on the principle that the absorption signal intensity is proportional to the number of magnetic nuclei. For example, determination of fat in carbon tetrachloride is based on a measurement proportional to the total number of hydrogen nuclei associated with the fat molecules. Referring to Fig. 1, a theoretical NMR absorption spectrum, the area under the absorption mode is proportional to the number of nuclei. This can

*To be published in more detail in The Journal of the American Oil Chemists' Society.

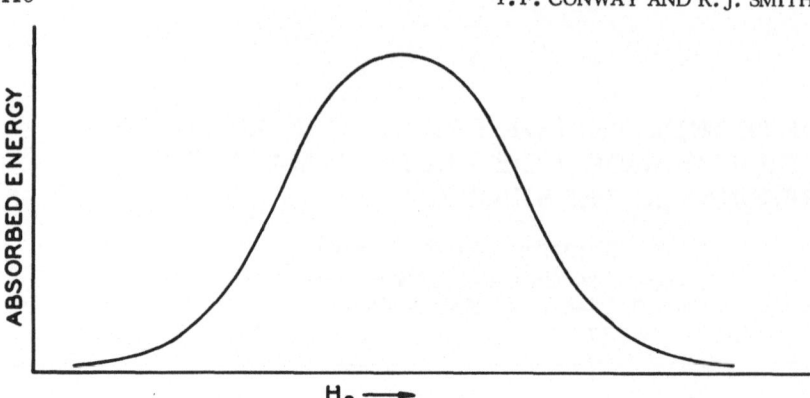

Fig. 1. Theoretical nuclear magnetic resonance absorption spectrum.

be seen from the following equation:

$$\text{Area} = \frac{C}{T} \times \frac{H_1}{r} \times \frac{N}{(1 + \gamma^2 H_1^2 T_1 T_2)^{1/2}} \tag{1}$$

The term $C$ includes the various nuclear and instrumental constants, $T$ is the absolute temperature, $H_1$ is the radio-frequency (r-f) driving field, $r$ is the magnetic field sweep rate, and $N$ is the number of nuclei. The bracketed expression in the denominator is called the saturation factor and shows the dependence on nuclear relaxation times.

Quantitative determination of hydrogen, the principle involved in measurement of fat, has been reported for low- and high-molecular weight materials in solution [3, 4]. The present study extends this type of measurement to corn germ milled in carbon tetrachloride and to corn and corn germ containing less than 5% moisture. Spent corn germ flake samples containing up to 11% moisture were also examined. All experiments were carried out over a sample temperature range of 75° to 80° F.

NMR spectra characteristics depend on the molecular motion within the sample (Fig. 2). For example, fat in corn or corn germ is relatively mobile and behaves like a liquid in that it exhibits a narrow, intense signal. In the nonfat portion (matrix) of the sample, hydrogen nuclei associated with "bound" water, carbohydrates, or proteins are rather rigidly fixed with respect to their neighbors, and they are capable of only restricted movement. Therefore, any given nucleus may be in an effective magnetic environment substantially lower or higher

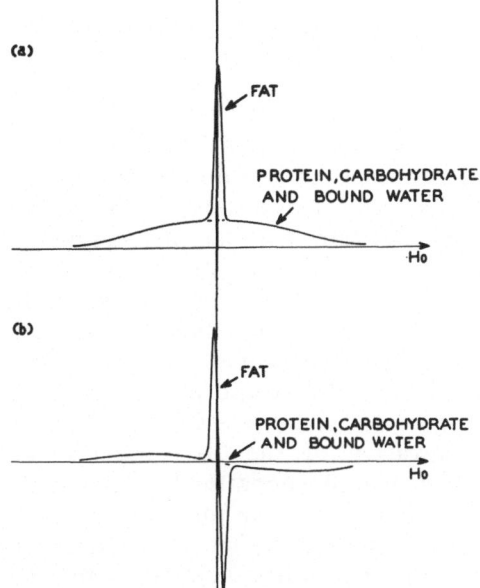

Fig. 2. Proton reso-
nance signals from
fat in dry corn and
corn germ: (a) ab-
sorption; (b) deriva-
tive.

than the applied field. Consequently, signals from the matrix
hydrogen nuclei are weak and very broad, because resonance
is limited to only a few nuclei at a given instant. Proton resonance of such systems gives a greatly broadened signal, and a
narrow line arising from the fat is superimposed thereon. By
measuring the derivative of the absorption signal, the mobile
hydrogen signal (hydrogen in fat) can be analytically separated
from the broadened signal (hydrogen in nonfat matrix), since
the latter is virtually constant over a narrow range of the
applied magnetic field.

## INSTRUMENTATION

The Schlumberger Model 104 NMR Analyzer and Model
104-3 Integrator [5] were employed. These consist of three
major units: the magnet unit, housing the permanent magnet
and associated coils; the console, housing a portion of the
operational control panels, recorder, and electronic circuitry;
and the integrator, which computes the integral signals. Instrument specifications can be described briefly as follows.
The permanent magnet has a field strength of 1717 gauss and a
homogeneity of better than 10 ppm over the sample volume.

Radio-frequency energy is supplied by a crystal-controlled r-f oscillator at 6.880 megacycles per second, with provision to vary the r-f field over a 60-decibel range. Peak-to-peak modulation amplitude can be selected at specific intervals between 5 milligauss and 5 gauss. Sweep time can be varied from 30 seconds to 4 minutes. The electronic integrator is essentially a conventional programmer and operational amplifier. A design feature of the integrator provides a "weight setting" control whereby the sample weight can be dialed into the integrator, and the integral output signal will read out to a common-weight basis. Other design features incorporated in the Schlumberger instrumentation to facilitate use of the integral of the derivative signal for quantitative measurements have been discussed elsewhere [2], along with criteria for selection of instrument parameters.

Operation of the instrument is comparatively simple. A weighed sample contained in a suitable cell is placed in the sample cell cavity in the magnet unit. Instrument parameters are selected, and the "Start" switch is energized. As the instrument automatically "sweeps" through resonance, an NMR derivative signal is recorded. Concurrently, the area under this curve is measured and recorded on the strip-chart recorder at the end of the sweep. The output of the integrator is proportional to the fat content and weight of the sample. Experience has shown that skilled operators are not required for control applications.

## RESULTS AND DISCUSSION

A typical calibration curve for corn oil in carbon tetrachloride is shown in Fig. 3, along with the appropriate instrument parameters. These samples were prepared by dissolving known weights of corn oil in carbon tetrachloride and diluting to solution volumes of 35 ml. Some idea of the precision and accuracy of the method is indicated when application is restricted to solutions of oils having similar hydrogen contents.

We have used several techniques for determining fat in plant process samples. One method applied successfully to corn germ containing less than 5% moisture involves weighing a 3-gram sample into a NMR sample cell and adding 10 ml of carbon tetrachloride. The sample is milled at 45,000 rpm for 3 minutes with a modified Virtis Disintegrator (Fig. 4), and after milling the shaft and blade are rinsed with solvent. After

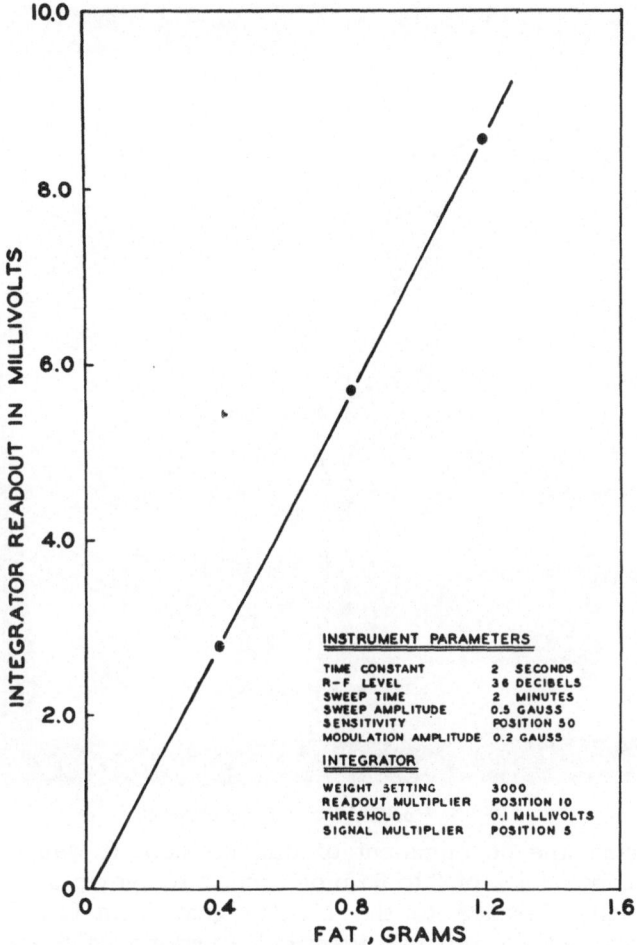

Fig. 3. NMR calibration curve for corn oil dissolved in carbon tetrachloride.

dilution to 35-ml volume with carbon tetrachloride, the sample dispersion is transferred to the instrument. The NMR integral signal is recorded and the fat content is obtained in about 2 minutes. The accuracy of the method is better than 1.0%, relative (95% confidence limits). The relation between NMR signal amplitudes and fat contents for 32 process germ samples is shown in Fig. 5. The specificity of the method is demonstrated by the fact that standard corn oil samples in carbon tetrachloride, examined under identical instrumental conditions, gave precisely the same relationship.

Fig. 4. Modified Virtis Disintegrator.

During the development of this method, it was observed that the presence of 2 to 5% moisture in the germ did not contribute interference to the NMR signal from fat. This was demonstrated by milling identical portions of the 32 germ samples in carbon tetrachloride. Samples in one series were subsequently extracted for 16 hours with the same solvent in a Soxhlet extractor. Milled sample dispersions were compared instrumentally with the extracts; differences between integral signals from equivalent samples in each series were less than 1% relative. This observation is explained by other NMR studies which indicate that the moisture is "bound" tightly to the germ matrix and will not respond under these instrument parameters.

The NMR technique has also been useful in validating extraction procedures for the determination of crude fat. Initial correlations involving NMR signal amplitudes and fat contents

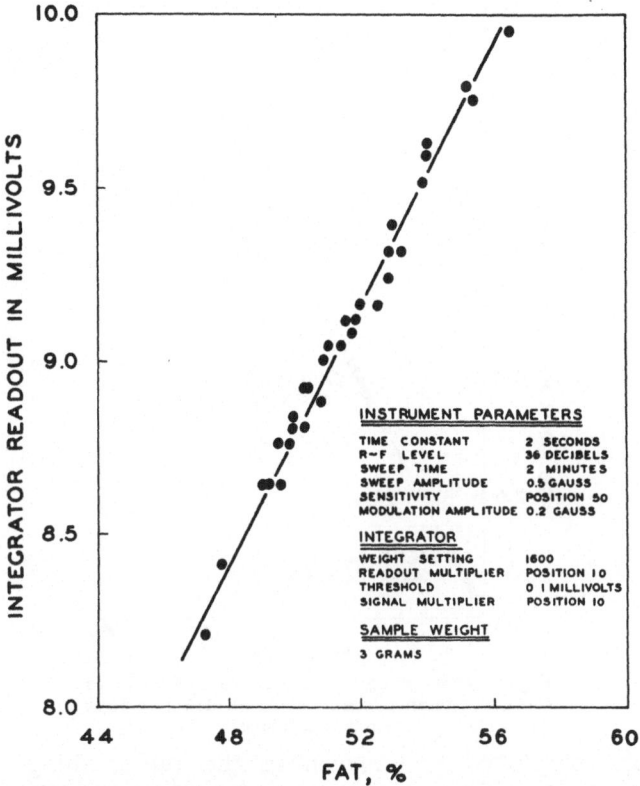

Fig. 5. NMR integral signals vs. fat contents for corn germ
dispersed by high-speed milling in carbon tetrachloride.

were poorer than expected, based on observations with corn
oil–carbon tetrachloride solutions. This anomaly remained
unresolved until NMR examination of the extraction residues
(material remaining after analytical extraction by the standard
procedure) indicated incomplete and inconsistent removal of
fat. A referee extraction method was developed which incorporated wet–milling in carbon tetrachloride followed by
16-hour extraction with the same solvent in a Soxhlet extractor.

Another approach to fat determination involves measurement of NMR signals from predried samples. The relationship
between signal amplitude and fat content for 112 random process samples (germ, expelled germ, and spent germ flake),
dried to less than 3% moisture, is shown in Fig. 6. For comparison, these data were calculated to a common-weight basis
for equivalent instrument parameters. Actual instrument pa-

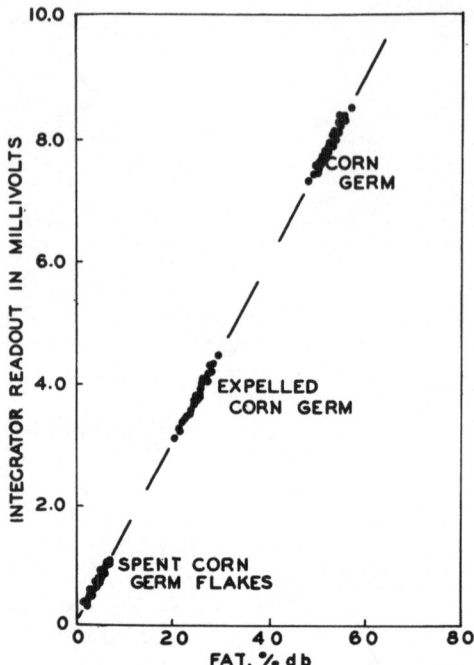

Fig. 6. NMR integral signals vs. fat contents
for corn germ process samples dried to less
than 3% moisture.

rameters employed are shown in the table, along with the
sample weights. These data indicate that the method is accurate
to 1.0, 0.5, and 0.3% fat, absolute (95% confidence limits), for
corn germ, expelled corn germ, and spent corn germ flakes,
respectively. The time required for NMR examination is 2
minutes; the time required for sample drying is 10 minutes at
100°C.

A comparison of the "direct milling" and "predrying"
sample preparation techniques associated with the NMR method
for determination of fat in corn germ indicates the former to
be superior. We attribute this observation to the fact that
milling in carbon tetrachloride places the hydrogen nuclei of
fat molecules in a magnetic environment wherein $T_1$ and $T_2$
(nuclear spin relaxation times) are more nearly alike from
sample to sample. The effect of variations of $T_1$ and $T_2$ on the
quantitative measurement can be seen in Eq. (1). Application of
spin-echo NMR techniques to unmilled corn germ samples has
demonstrated that the spin constants of "mobile" hydrogen in
process samples often vary as a function of sample history.

## Instrument Parameters and Sample Weights

| Instrument parameters | Sample type | | |
|---|---|---|---|
| | Germ | Expelled germ | Spent flake |
| Time constant: seconds | 2 | 2 | 2 |
| R-F level: decibels | 40 | 28 | 28 |
| Sweep time: minutes | 2 | 2 | 2 |
| Sweep amplitude: gauss | 5 | 5 | 5 |
| Sensitivity | 50 | 20 | 5 |
| Modulation amplitude: gauss | 2 | 2 | 2 |
| Integrator | | | |
| Weight setting: | 1600 | 1900 | 2300 |
| Readout multiplier | 10 | 10 | 10 |
| Threshold: millivolts | 0.1 | 0.1 | 0.5 |
| Signal multiplier | 5 | 5 | 5 |
| Sample weight, grams | 12 | 12 | 17 |

An "r-f saturation" technique has been applied successfully to the determination of fat in samples containing significant amounts of moisture. This phenomenon, as applied to NMR spectroscopy, has been discussed by Bloembergen et al. [6]. It involves the relationship between the NMR absorption signal, the r-f energy, and the nuclear spin times, $T_1$ and $T_2$. Relaxation times for protons associated with the fat differ from those associated with the water, and it is possible to use this difference as a basis for discrimination. In process samples where these differences are not sufficiently large, the differences can often be increased by high-speed milling the moisture-containing sample in carbon tetrachloride, thus altering the relaxation times of hydrogen nuclei associated with the fat. The fat content is derived from NMR signals obtained at two r-f levels. At a low r-f level, energy absorbed is proportional to the number of hydrogen nuclei associated with both fat and water. At a high r-f level, signals from fat are preferentially saturated. The amount of signal attenuation due to r-f saturation is proportional to the number of protons in the fat; therefore, the phenomenon can be used to determine fat content.

This approach was used to measure fat (1 to 6%) in spent germ flakes in the presence of 4 to 11% moisture. The correlation between NMR signal differentials and fat contents for approximately 40 random process samples is shown in Fig. 7,

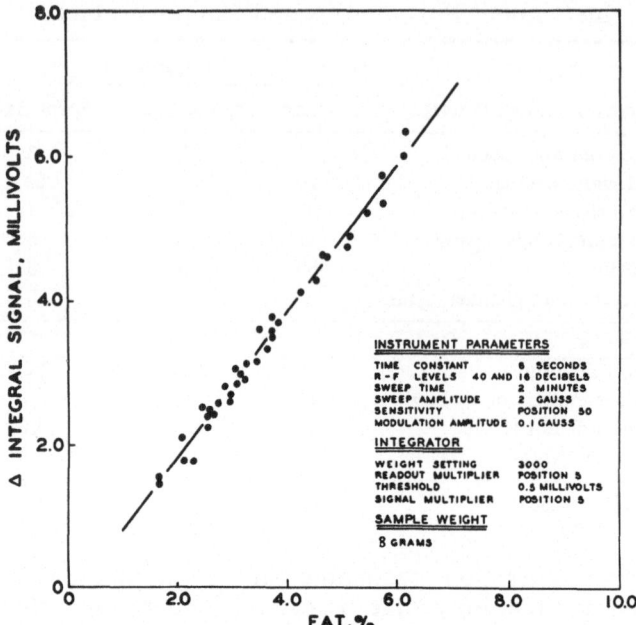

Fig. 7. NMR differential integral signals vs. fat contents for spent corn germ flakes, containing 4 to 11% moisture, dispersed by high-speed milling in carbon tetrachloride.

together with the instrument parameters employed. The method is accurate to within 0.3% fat, absolute (95% confidence limits).

The applicability of NMR absorption techniques to the determination of fat in corn has been demonstrated for several sample types. A typical NMR signal-fat content correlation for a series of samples dried to less than 5% moisture is shown in Fig. 8. Maximum deviations for this correlation were 0.3% fat, absolute. NMR measurements were performed on the whole kernel samples while fat contents were obtained by wet-milling in carbon tetrachloride followed by 16-hour extraction with the same solvent. Samples of commercial hybrid corn, experimental high-oil, hybrid corn, and experimental varieties of high- and low-oil white corn were included in this study.

The utility of the NMR method for measuring fat in 25-gram whole corn samples having been demonstrated, the technique was extended to single kernels. To obtain the requisite instrument sensitivity, a 2-ml "high-gain" r-f coil was used in place of the standard 40-ml coil. A special sample cell and holder were constructed to position each kernel at the optimum loca-

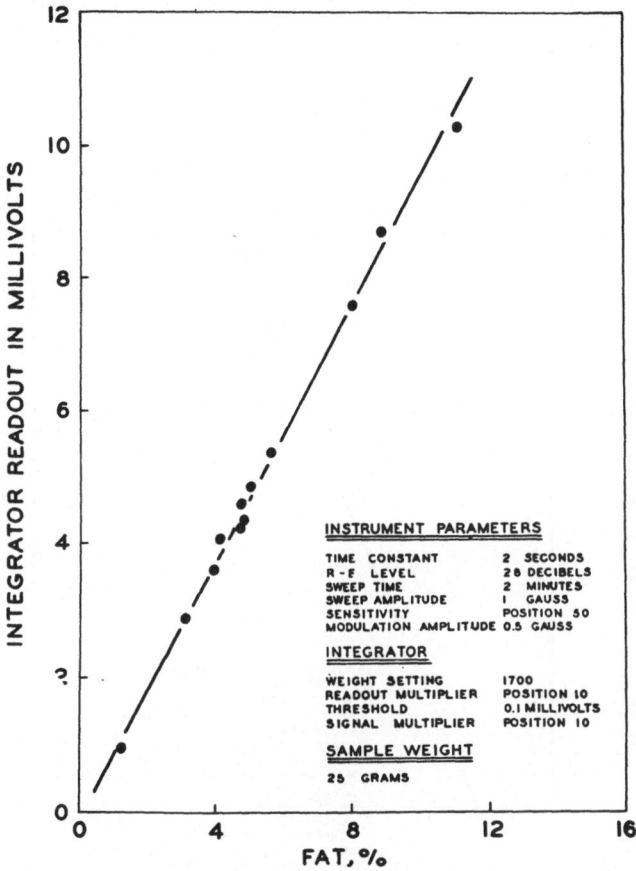

Fig. 8. NMR integral signals vs. fat contents for corn containing less than 5% moisture.

tion within the magnetic field. Instrument parameters selected gave excellent signal-to-noise ratios for kernels weighing 200 to 500 mg.

For calibration purposes, samples of corn containing 4 to 9% fat were dried to approximately 5% moisture, and the viabilities of random kernels were checked before proceeding. Twenty-five kernels from each of four control samples were examined instrumentally (individually), and later the same 25 kernels were ground collectively and analyzed for fat by the referee solvent extraction method. The correlation between fat contents and NMR signal amplitudes (weighted average for 25 kernels) is shown in Fig. 9. Maximum deviations from the

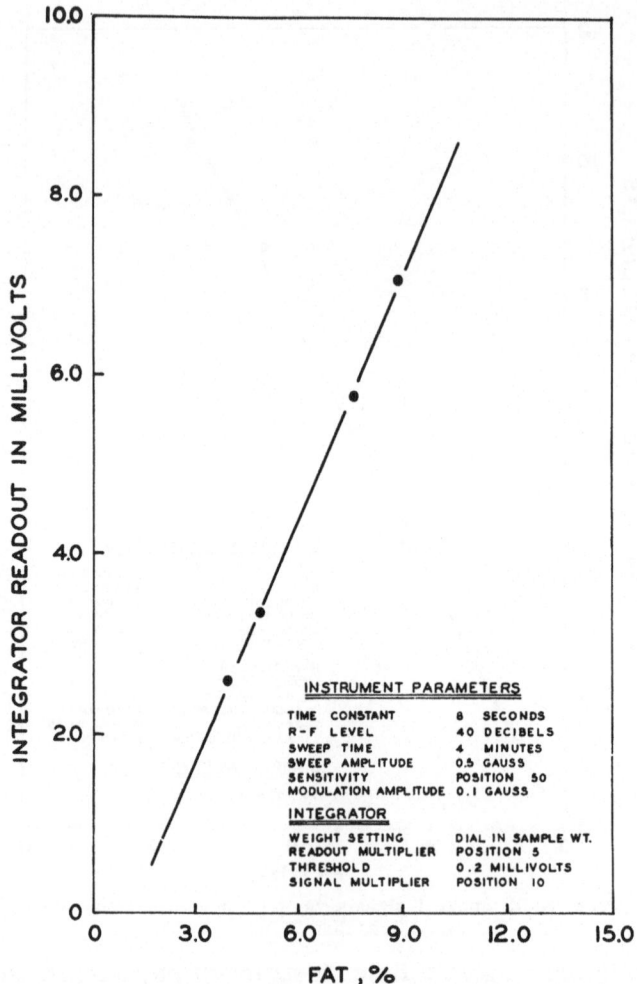

Fig. 9. NMR integral signals vs. fat contents for single kernels
of corn dried to less than 5% moisture.

correlation were 0.1% fat, absolute. The time of measurement
was 4 minutes, although this can be reduced to 2 minutes.

This NMR method is nondestructive; it permits selection
of single kernels for plant breeding without destroying viability.
The ultimate utility of this selection for the geneticist has not
been resolved, and this is the subject of some of our present
NMR research.

## REFERENCES

1. T. F. Conway, R. F. Cohee, and R. J. Smith, Food Eng. (June, 1957).
2. H. Rubin and R. E. Swarbrick, Anal. Chem. 33, 217–220 (1961).
3. Schlumberger Corp., Material Study Report 29.58 (April, 1959).
4. Varian Associates Tech. Info. Bull., Vol. 3, No. 3 (1960).
5. H. Rubin, IRE Trans. Ind. Electronics PGIE 11, 9 (1959).
6. N. Bloembergen, E. M. Purcell, and R. V. Pound, Phys. Rev. 73, 679 (1948).

# ULTRAVIOLET AND
# VACUUM ULTRAVIOLET SPECTROSCOPY

# FAR ULTRAVIOLET ANALYSIS OF STEROIDS
# AND OTHER BIOLOGICALLY ACTIVE SUBSTANCES

## W. F. Ulrich

Beckman Instruments, Inc.
Fullerton, California

## ABSTRACT

Representative steroids, barbiturates, amino acids, and model compounds were studied in the far ultraviolet region to 170 mμ. The absorption of pertinent chromophores is discussed with regard to analytical usefulness. Of these, isolated double bonds and monoaromatic rings offer the most promise. The problems associated with this region are also considered.

## INTRODUCTION

Ultraviolet spectrophotometry provides a simple and rapid means for analyzing many chemical substances of interest in the biosciences. For the most part the near ultraviolet region above 220 mμ has been utilized because of limitations in available instruments. In fact, many absorption maxima reported at shorter wavelengths are actually false maxima caused by stray light effects. Recent improvements in sources, detectors, and optical components have reduced this problem [1] and the practicable range is now extended to 170 mμ or below in several conventional spectrophotometers.

An obvious advantage of the extended range is the improved sensitivity attained for compounds through the stronger absorption bands found at shorter wavelengths. However, benefits may also be found in selectivity and in the additional structural data provided. While these benefits are more striking with the normally transparent simple chromophores, even complex chromophores with near ultraviolet absorptions exhibit characteristic and useful far ultraviolet bands. The analytical significance of these factors is illustrated here by studies on representative steroids, amino acids, barbiturates, and model compounds.

# EXPERIMENTAL

### Apparatus

A Beckman Far Ultraviolet DK-2 Spectrophotometer equipped with 0.01-cm Far Ultraviolet silica cells was used for all measurements. Oxygen and water vapor were purged from the optical system with nitrogen from AR-3 liquid nitrogen tanks (Air Reduction Company, 150 E. 2nd St., New York). Gas chromatographic purifications were achieved with the Beckman Megachrom Gas Chromatograph.

### Reagents

Steroid, amino acid, and barbiturate samples were obtained in varying degrees of purity from several sources. Limited quantities restricted further purification in many instances, which undoubtedly affected absorptivity data. However, comparison with corresponding literature values for several compounds showed reasonable agreement. Cyclohexane, cyclohexene, and cyclohexanol were purified by gas chromatography immediately prior to analysis. Solvents studied included Phillips 99 mole % n-hexane and cyclohexane; distilled water; Eastman Spectrograde methanol, methylene chloride, chloroform, carbon tetrachloride, and reagent grade isopropanol; and Matheson, Coleman & Bell spectro quality acetonitrile, and butyl ether.

### Technique

A major problem in far ultraviolet analysis is the selection of proper solvents. In Fig. 1 transmission curves are given for ten common solvents. The most transparent solvent throughout the range of interest is n-hexane, but it is limited mainly to nonpolar compounds. Cyclohexane has somewhat better solvent properties but has a longer wavelength cut-off. Water and acetonitrile are slightly inferior to n-hexane in transmission, but are more useful for polar compounds. Methanol is perhaps a better general solvent than any of these but has a cut-off at about 187.5 m$\mu$. Isopropanol is similar to methanol in both transmission and solvent properties. The other four solvents shown are too opaque for use in the far ultraviolet region.

In the present study, samples were tested in n-hexane, water, acetonitrile, and methanol, in that order. Each solution was prepared by dissolving a known quantity of solute in 5 ml of solvent. Heating was avoided unless absolutely necessary.

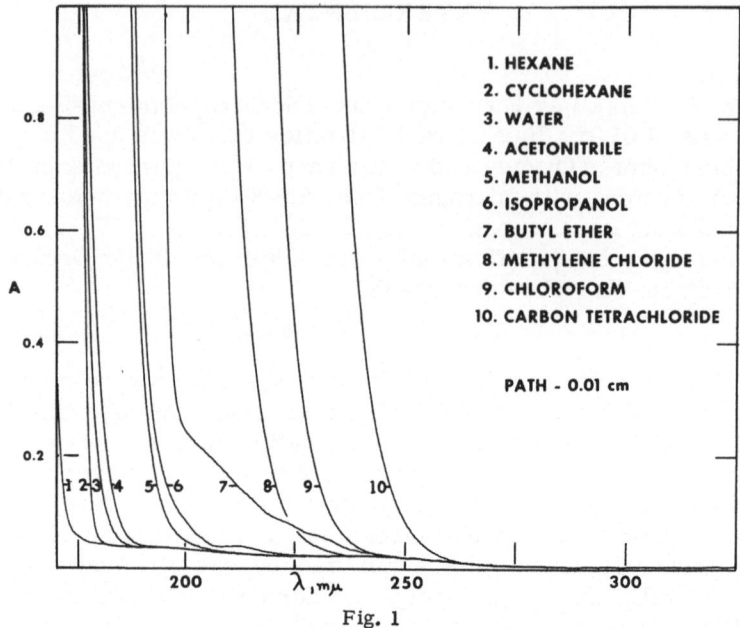

Fig. 1

Before each spectral analysis, the sample cell was examined for cleanliness by measuring its transmittance when filled with solvent. Frequent recleaning of both the internal and external window faces proved necessary. Each solution was then examined against a reference cell filled with solvent. The absorption curves were corrected for cell mismatch and replotted.

## STEROIDS

The near ultraviolet absorption characteristics of steroids have been extensively studied, and a critical review of this work up to 1953 is given by Dorfman [2]. Conjugated dienes, ketoenes, and benzenoid groups are readily measured in this region. Of the simple chromophores, only the ketone group exhibits appreciable absorption in the near ultraviolet and even this is of doubtful analytical importance. Consequently, colorimetric methods have flourished for studies on these groups.

Far ultraviolet studies on steroids have also been made, apparently to a greater extent than for any other group of compounds of biological interest. Henbest and co-workers [3] used a conventional spectrophotometer to measure the absorption of steroids containing isolated ethylenic groups. Stray light problems limited their study to wavelengths above 205 m$\mu$ and gave rise to false maxima. Even so, they found a correlation between the type of monoene and the observed end absorption. Stick et al. [4] obtained more accurate data on similar compounds by correcting for stray light. Their measurements to 190 m$\mu$ supported the earlier conclusions.

Turner [5] used a vacuum spectrograph with fluorite optics to study steroids in the range 160-210 m$\mu$. He found exceptions to the relationship between band position and monoene structure, and concluded that band shape is an important factor. His work also pointed out the need for making corrections for the end absorption of the hydrocarbon skeletal groups. Ellington and Meakins [6] have further emphasized the need for using band width in connection with position and intensity in the spectra-structure correlations. A recent comprehensive study by Micheli and Applewhite [7] provides further information on the isolated double bond structures. Data in the present study also support the correlations between structure and absorption and provide additional information on chromophoric groups other than the simple monoenes. A summary of these correlations is given in the following discussion.

## STEROID BUILDING BLOCKS

Absorption characteristics of single bond chromophores found in steroids are illustrated in Fig. 2. Simple ring systems such as cyclohexane are highly transparent above 170 m$\mu$, although weak end absorption may be observed. Addition of a hydroxyl group gives a pronounced increase in end absorption as does the fusion of rings. Thus, $\epsilon$ at 180 m$\mu$ is less than 0.0 liters mol$^{-1}$ cm$^{-1}$ for cyclohexane and is $5 \times 10^2$ liters mol$^{-1}$ cm$^{-1}$ for cyclohexanol and $24 \times 10^2$ liters mol$^{-1}$ cm$^{-1}$ for cholestanol. None of these systems gives a pronounced maximum above 170 m$\mu$, but both the hydroxyl and the quaternary carbon centers affect the short wavelength absorption and hinder detection of absorption maxima exhibited by other chromophores. However, the absorption does not appear attractive from an analytical standpoint except in rare applications.

The simple keto group exhibits absorption characteristics similar to the hydroxyl group. As shown in Fig. 2, neither stanolone nor androsterone differs appreciably from cholestanol in absorption properties except to some degree in intensity. Again, the analytical importance of this absorption appears marginal.

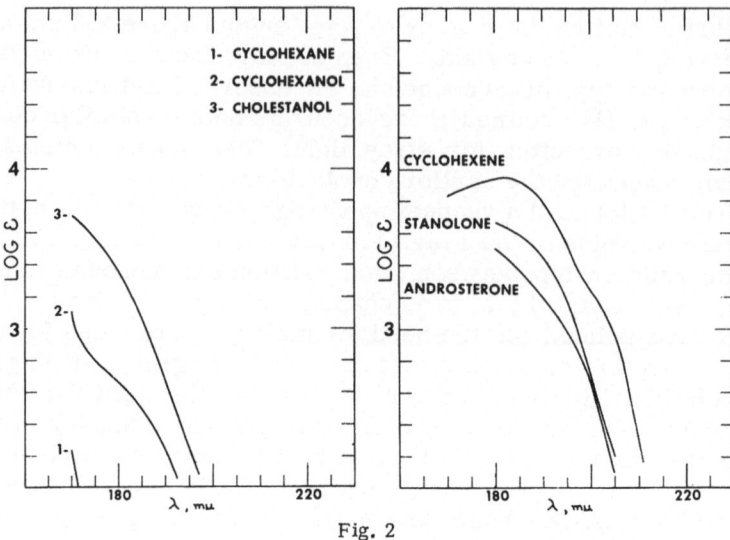

Fig. 2

Absorption by isolated ethylenic groups is more promising and possibly represents the most fruitful application in steroid analysis. These groups are completely transparent in the near ultraviolet unless allowed to react with color-forming reagents. Neither do they give intense infrared absorption in many instances. In the far ultraviolet region, however, the chromophore gives intense absorption with a maximum above 170 m$\mu$.

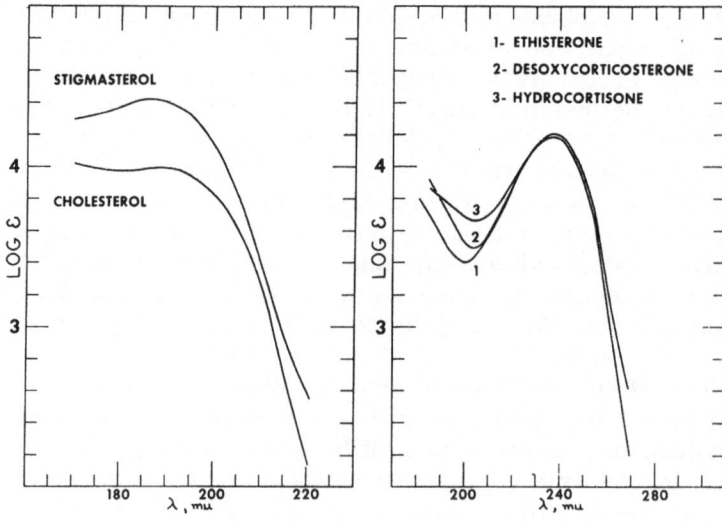

Fig. 3

This is illustrated in the spectra of the model compound, cyclohexene (Fig. 2), and the $\Delta^5$-steroids cholesterol and stigmasterol (Fig. 3). Stigmasterol gives the strongest absorption because of the second ethylenic group in the side chain.

An important feature of the ethylenic absorption is its dependence on substitution. Disubstituted monoenes absorb at the shorter wavelengths, the maximum being found near 180 $m\mu$. Trisubstituted monoenes usually absorb near 190 $m\mu$ and tetrasubstituted monoenes near 200 $m\mu$. Exceptions have been noted; for example, the trisubstituted $\Delta^7$-steroids exhibit a maximum at 200-205 $m\mu$, according to Micheli and Applewhite [7]. For the most part, however, the degree of substitution and band position appear to follow the more general correlations.

There is strong evidence that band intensities may also be correlated to structure, but further work is needed to establish the validity of this premise. If true, the combined use of band position and band intensity would give a strong tool for determining group position. On the other hand, differences in intensity would complicate the determination of total monoenes in mixtures.

Conjugated ketoenes exhibit an intense absorption near 240 $m\mu$, a minimum near 200 $m\mu$, and strong end absorption below 200 $m\mu$. The spectra of ethisterone, deoxycorticosterones, and hydrocortisone shown in Fig. 3 are typical of the $\Delta^5$-3-one steroids. The three differ principally in the substitution on the 17-carbon atom. While the 240-$m\mu$ maxima exhibit only minor differences, the shorter wavelength minima show differences in position and intensity. Possibly these differences can be helpful in characterizing steroids of this type, but more data are required to establish the extent to which this is true.

Conjugated dienes such as ergosterol and 7-dehydrocholesterol give characteristic absorption bands in the 250-300 $m\mu$ region, as shown in Fig. 4. The two steroids studied here have identical structure except for the double bond in the ergosterol side chain. Whereas the near ultraviolet spectra of the two are remarkably similar, ergosterol has a strong far ultraviolet maximum which is easily differentiated from the weaker end absorption in the 7-dehydrocholesterol. The maximum, as would be expected for a disubstituted double bond, is observed at 182 $m\mu$ and provides a simple means for identifying ergosterol.

Aromatic compounds such as estrone and estradiol give intense far ultraviolet absorption. As shown in Fig. 4, the aromatic ring gives moderate absorption near 280 $m\mu$, a strong

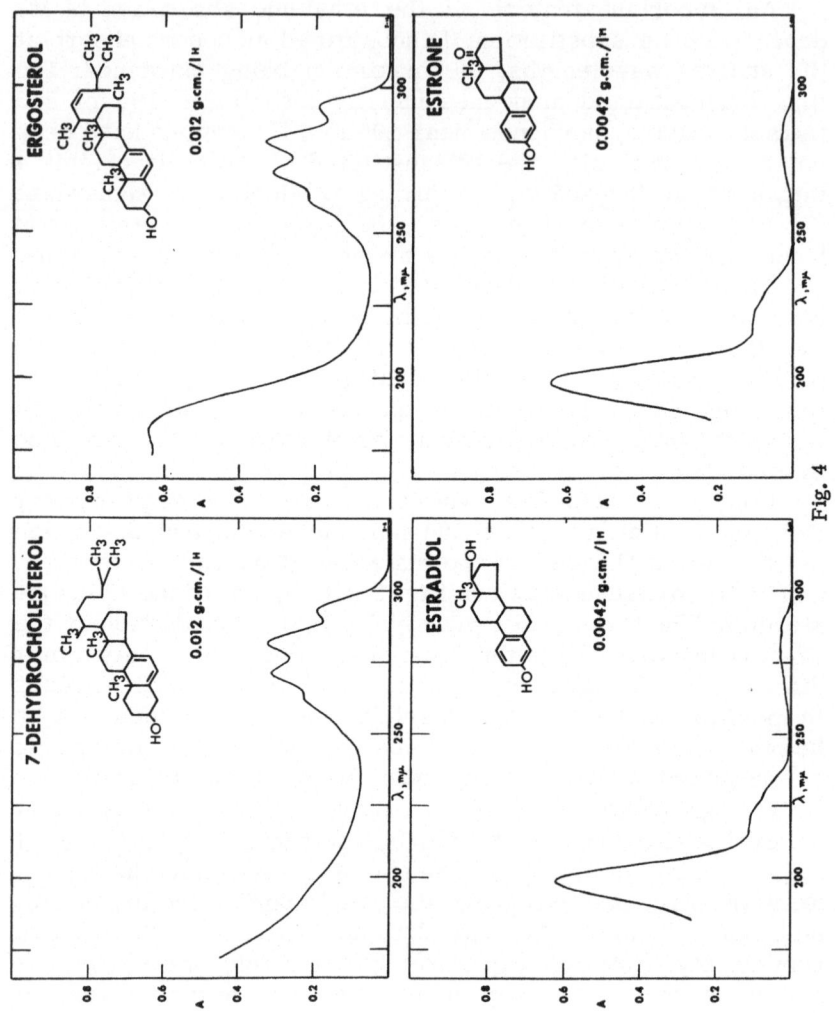

Fig. 4

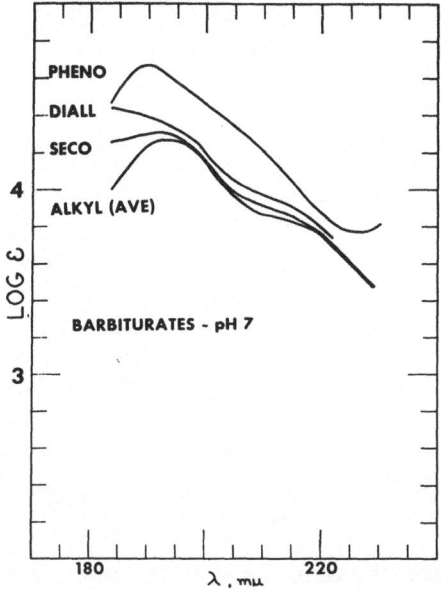

PHENO

DIALL

SECO

4

ALKYL (AVE)

LOG ε

BARBITURATES - pH 7

3

180    λ , mμ    220    Fig. 5

shoulder at 220 mμ, and an intense band near 200 mμ. The molar absorptivity at the maxima is about $40 \times 10^3$ liters mol$^{-1}$ cm$^{-1}$, which is four to five times greater than that of the isolated double bond. The far ultraviolet region provides a highly sensitive means for detecting aromatic compounds and, as discussed later, of indicating the nature of ring substitution.

## BARBITURATES

The far ultraviolet spectra of typical barbiturates are given in Fig. 5. All of these have identical structures except for the type of substitution on the ring. Thus, their spectra should differ only to the extent to which the substituents exhibit spectral differences. Four alkyl compounds (pentobarbital, barbital, metharbital, and amobarbital) give nearly identical spectra, with an average curve shown in Fig. 5.

Secobarbital has a single double bond in one side group which gives a detectable increase in absorption below 200 mμ compared to the saturated models. Diallylbarbital has two unsaturated groups and consequently even stronger absorption. Phenobarbital gives the most striking difference because of the aromatic ring absorption.

The differences in the barbiturate far ultraviolet spectra

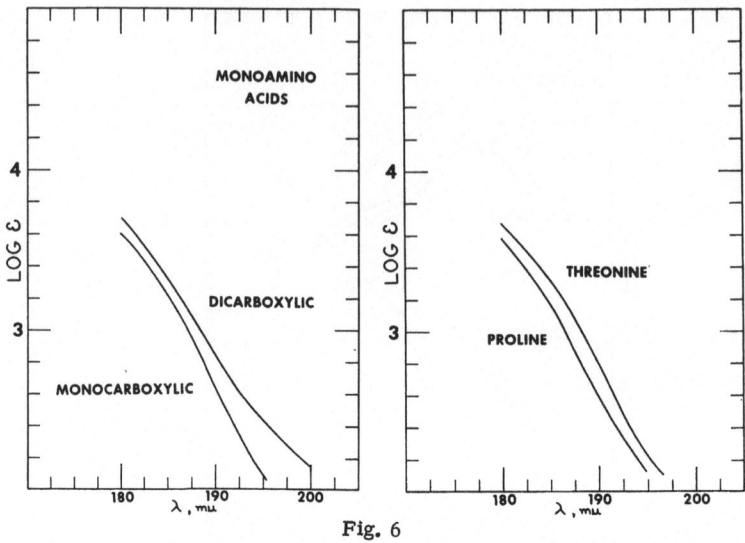

Fig. 6

are considerably greater than those in the near ultraviolet and may prove helpful in characterizing these compounds. However, in practice other factors may limit the reliability of the far ultraviolet absorption. This is particularly true in biological systems in which other components give absorption interferences.

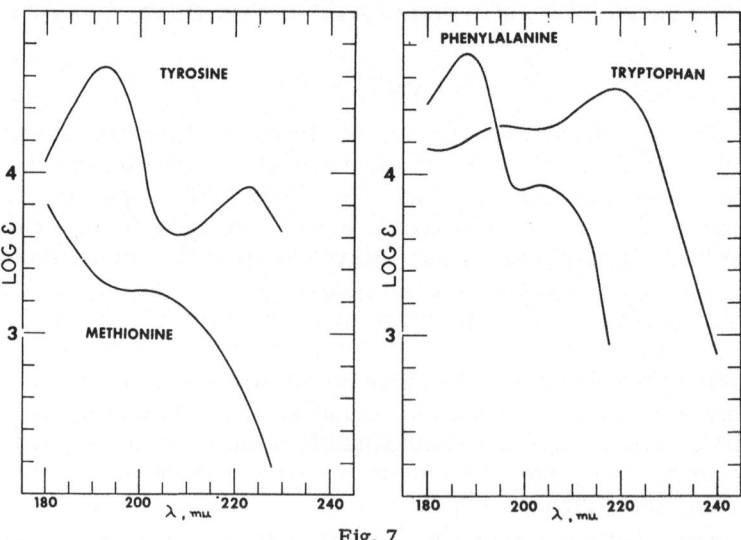

Fig. 7

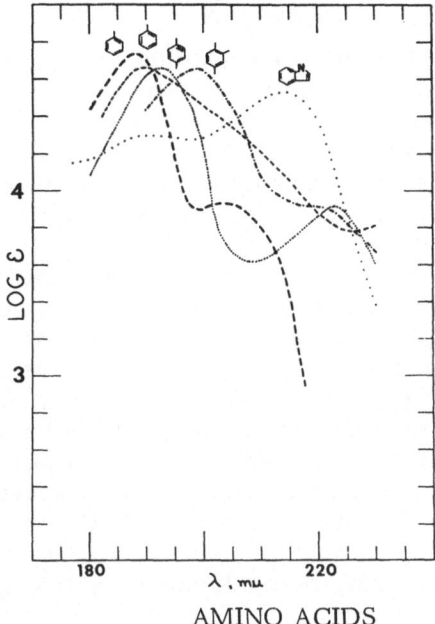

Fig. 8

## AMINO ACIDS

Several far ultraviolet studies on amino acids and proteins have been made [8-10], but not to the same extent as the steroids. In Fig. 6, spectra are shown for mono- and dicarboxylic monoamino acids, proline, and threonine. The monocarboxylic spectrum is the average curve for glycine, alanine, valine, and leucine. Aspartic and glutamic acids were averaged for the dicarboxylic acid curve. Proline is similar to the monocarboxylic acid except for a ring structure. Threonine has a hydroxyl group adjacent to the amino group. None of the chromophores in these compounds gives maxima above 180 mμ; they all exhibit only end absorption.

The spectrum of methionine in Fig. 7 reflects the presence of the alkyl sulfide structure by exhibiting a maximum just above 200 mμ. On this basis methionine is readily distinguishable from the preceding compounds. The compounds tyrosine, phenylalanine, and tryptophan are also readily detectable because of their strong aromatic absorption. As shown in Fig. 7, tryptophan is the most characteristic of the three because of its intense band at 220 mμ, which results from the conjugation of the aromatic ring with an ethylenic group. Tyrosine and phenylalanine give the more normal aromatic spectra.

## DISCUSSION

Even with the limited data now available, practical applications of the far ultraviolet region can be predicted. The detection and characterization of isolated ethylenic groups are relatively simple in the absence of conjugated chromophores, and with corrections can probably often be achieved even in their presence. Aromatic chromophores give the most serious interference in this regard.

Aromatic compounds are commonly determined in the near ultraviolet, but an increase in sensitivity can be achieved at shorter wavelengths. The position of the intense band just below 200 m$\mu$ also provides information in the degree of substitution, as shown in Fig. 8. As the degree of substitution increases, the absorption maximum undergoes a bathochromic shift. Thus, monosubstituted phenylalanine and phenobarbital have maxima at 188 m$\mu$ and 189.5 m$\mu$, respectively; para-substituted tyrosine has its maximum at 192 m$\mu$; and tri-substituted estrone has its maximum at 198 m$\mu$. The conjugated tryptophan, of course, has its maximum at even longer wavelengths.*

Compounds with only C-C, C-H, O-H, C=O, $NH_2$, and COOH groups do not give characteristic maxima in the usable far ultraviolet region, but have end absorptions which might be useful in some instances. However, this region would be quite helpful in determining the purity of such compounds if absorbing impurities are suspected.

In practice, of course, other considerations are involved. The choice of solvents is very limited and one is usually forced to use cell thicknesses of 0.1 cm or less. Cell cleanliness is considerably more important in this region than at longer wavelengths.

Perhaps a more serious problem is the interference from impurities. Most compounds absorb at least weakly in the far ultraviolet region and pose a serious threat to quantitative results. In samples of biological origin, such interferences would normally be expected. This suggests that optimum use of the short wavelength absorptions will depend on the development of suitable clean-up procedures. Fortunately, advances in extractive techniques and in gas, paper, and thin layer chromatography offer considerable promise in this respect. The combination of these separative methods with the high

*In Fig. 8, the tryptophan curve is erroneously displaced 5 m$\mu$ toward shorter wavelengths.

sensitivity of the far ultraviolet region should lead to numerous analytical applications in the biosciences.

## REFERENCES

1. W. I. Kaye, Appl. Spectroscopy 15, 89 (1961).
2. L. Dorfman, Chem. Revs. 53, 47 (1953).
3. P. Bladon, H. B. Henbest, and G. W. Wood, J. Chem. Soc. (1952) 2737.
4. K. Stick, G. Rotzler, and T. Reichstein, Helv. Chim. Acta 42, 1480 (1959).
5. D. W. Turner, J. Chem. Soc. (1959) 30.
6. P. S. Ellington and G. D. Meakins, J. Chem. Soc. (1960) 697.
7. R. A. Micheli and T. H. Applewhite, J. Org. Chem. 27, 345 (1962).
8. L. J. Saidel, A. R. Goldfarb, and S. Waldman, J. Biol. Chem. 197, 285 (1952).
9. M. P. Tombs, F. Souter, and N. F. Maclogen, Biochem. J. 73, 167 (1959).
10. K. Rosenbeck and P. Doty, Proc. Nat. Acad. Sci. US 47, 1775 (1961).

# ABSORPTION SPECTRA OF URANYL HALIDES

## K. V. Narasimham[*]

Physics Department
Andhra University
Waltair, India

## ABSTRACT

The absorption spectra of three different uranyl fluorides and uranyl chloride were investigated in powder form at liquid air temperature. For anhydrous fluoride and uranyl fluoride I, the data of bands are reported for the first time; for uranyl fluoride II, 32 additional bands were obtained. The absorption spectra of the anhydrous fluoride and fluoride I were found to consist of four and three systems, respectively. New analyses of bands are proposed for uranyl fluoride II and uranyl chloride on the basis of five and six systems, respectively. The wavenumbers of the 0,0 bands of the various systems obtained for the four compounds are as follows: anhydrous uranyl fluoride, 20,236, 20,386, 21,183, and 21,277 $cm^{-1}$; uranyl fluoride I, 20,014, 23,160, and 23,429 $cm^{-1}$; uranyl fluoride II, 20,080, 20,753, 21,217, 22,723, and 23,083 $cm^{-1}$; uranyl chloride, 20,530, 21,107, 21,162, 21,234, 22,346, and 22,501 $cm^{-1}$. For uranyl fluoride I and uranyl chloride, the wavenumbers of the fundamentals obtained from the analyses of the absorption spectra were correlated with those obtained in fluorescence, Raman, and infrared spectra by different workers.

## INTRODUCTION

Several detailed studies of the fluorescence and absorption spectra of uranyl salts have been made since they were first investigated by Becquerel [1] and Onnes [2]. Work on these compounds was undertaken in our laboratory to obtain new data wherever possible and give satisfactory analyses of the spectra. The results obtained with some of these compounds were given

[*]Now with Spectroscopy Laboratory; Physics Department; Illinois Institute of Technology; Chicago, Illinois.

in our earlier publications [3-7]. To continue this work, the absorption spectra of three different uranyl fluorides—anhydrous fluoride ($UO_2F_2$) and two hydrated fluorides [$UO_2F_2(n_1H_2O)$ and $UO_2F_2(n_2H_2O)$]—and uranyl chloride ($UO_2Cl_2 \cdot 1H_2O$) in powder form were studied in the present investigation. The spectra of the fluorides provide an interesting study of the different crystal modifications and the effect on the spectra of impurities arising out of differences in the preparation of the samples, as already pointed out in reference [6]. Whereas fluorescence is obtained only for fluoride I [6], the absorption spectra were obtained for all the three fluorides. No previous data on the absorption spectra of anhydrous fluoride and fluoride I are reported in the literature. For fluoride II, the absorption spectrum of which was also studied by Pant [8], a number of additional bands on the short wavelength side were obtained in the present investigation. The absorption spectrum of uranyl chloride was previously studied by Pant [8] and Samoilov [9]. New analyses of the bands for all these absorption spectra have been proposed by the author on the basis of electronic transitions from a common lower state to the different excited states.

## PREPARATION OF THE COMPOUNDS

### Anhydrous Uranyl Fluoride
This compound was prepared by the method of von Unruh [10], i.e., by repeated treatment of uranyl acetate with hydrofluoric acid and evaporating on a water bath. The compound is dried in a vacuum desiccator. The substance formed is extremely hygroscopic.

### Uranyl Fluoride I
A hydrated uranyl fluoride was prepared by the method of Andrews [11]. When a solution of 50 g of uranyl nitrate in 200 ml of water is added to a solution of 75 g of ammonium fluoride in 110 ml of water, a yellow precipitate is obtained. The precipitate is washed and dried in a desiccator. A yellow form of the hydrated fluoride which is quite stable at room temperature is thus obtained. The compound formed by this method was called uranyl fluoride I by Pant [8].

### Uranyl Fluoride II
When the anhydrous fluoride previously described is ex-

posed to the atmosphere for a few days, it takes up water of crystallization and forms a stable grayish compound. The absorption spectrum obtained with this compound is identical to that of uranyl fluoride II, which was studied by Pant [8].

### Uranyl Chloride

This compound was supplied by M/S Johnson and Mathey Co., London.

## EXPERIMENTAL

The absorption spectra of the compounds were recorded at liquid air temperature using the experimental technique described earlier [3]. Using powder layers of different thicknesses in the absorption cell, bands were obtained in the different regions. Spectra were taken on Hilger Fuess and Glass Littrow spectrographs with dispersions of 13 A/mm and 7 A/mm, respectively, at 4000 A.

## RESULTS

For anhydrous uranyl fluoride, about 30 bands in the region 4940-4315 A (Fig. 1), and for uranyl fluoride I, 60 bands in the region 5000-3800 A (Fig. 2), were recorded at the temperature of liquid air. For uranyl fluoride II, a total of 55 bands, compared with 23 bands reported by Pant [8], were obtained in the region 5000-3800 A (Fig. 3). For uranyl chloride, about 60 bands were obtained in the region 4900-3800 A (Fig. 4), which agrees well with the previous data of Pant [8].

## EARLIER ANALYSES

Pant [8] analyzed the absorption spectra of uranyl fluoride II and uranyl chloride on the basis of two electronic transitions from two nearby lower states to a common excited state. The defects of this analysis were discussed in our earlier paper [3].

## PRESENT ANALYSES

### Anhydrous Uranyl Fluoride

The absorption spectrum of anhydrous uranyl fluoride was analyzed as due to four systems with a common lower state.

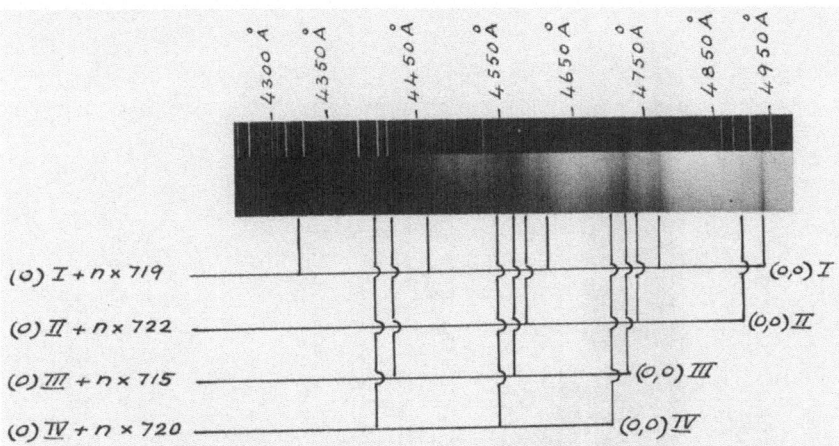

Fig. 1. Absorption spectrum of anhydrous uranyl fluoride at liquid air temperature.

The 0.0 bands of the systems were identified at 20,236, 20,386, 21,183, and 21,277 cm$^{-1}$ (Table I).

In System I, the fundamentals obtained are 719 and 761 cm$^{-1}$, which correspond to the O–U–O symmetric stretching ($\nu_1$) and O–U–O asymmetric stretching ($\nu_3$) frequencies in the excited state. Bands of the $\nu_1$ fundamental and its overtones are the strongest and repeat themselves up to the fourth group. The appearance of the $\nu_3$ fundamental and its combination with considerable intensity may be attributed to the crystalline fields surrounding the uranyl ion. In analyzing the absorption spectrum of cesium uranyl chloride, Dieke and Duncan [12] assigned some bands to this frequency. System II consists of bands belonging to the fundamental 722 cm$^{-1}$ ($\nu_1$) in the excited state and its overtone. System III, the $\nu_1$ frequency, has a wavenumber of 715 cm$^{-1}$ in the upper state. Another fundamental (244 cm$^{-1}$) belonging to the O–U–O bending frequency ($\nu_2$) in the upper state is also obtained, but bands of this frequency are very weak. System IV consists of bands belonging to the $\nu_1$ fundamental (720 cm$^{-1}$) and its overtone.

### Uranyl Fluoride II

If we compare closely the absorption spectra of the anhydrous uranyl fluoride and uranyl fluoride II, it is evident that one series of bands of the anhydrous salt is present as an impurity in the spectrum of uranyl fluoride II (Fig. 3). In fact, the first band of this series is the $a_0$ band of Pant [8] in uranyl

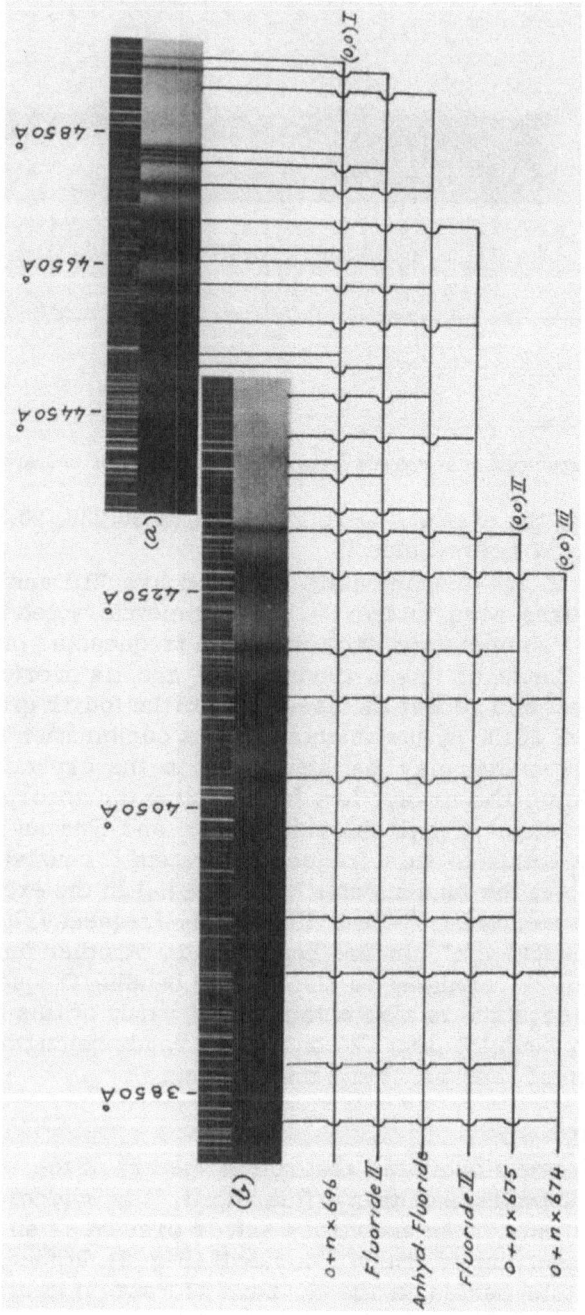

(a) *Long path length*    (b) *Short path length.*

Fig. 2. Absorption spectrum of uranyl fluoride I at liquid air temperature.

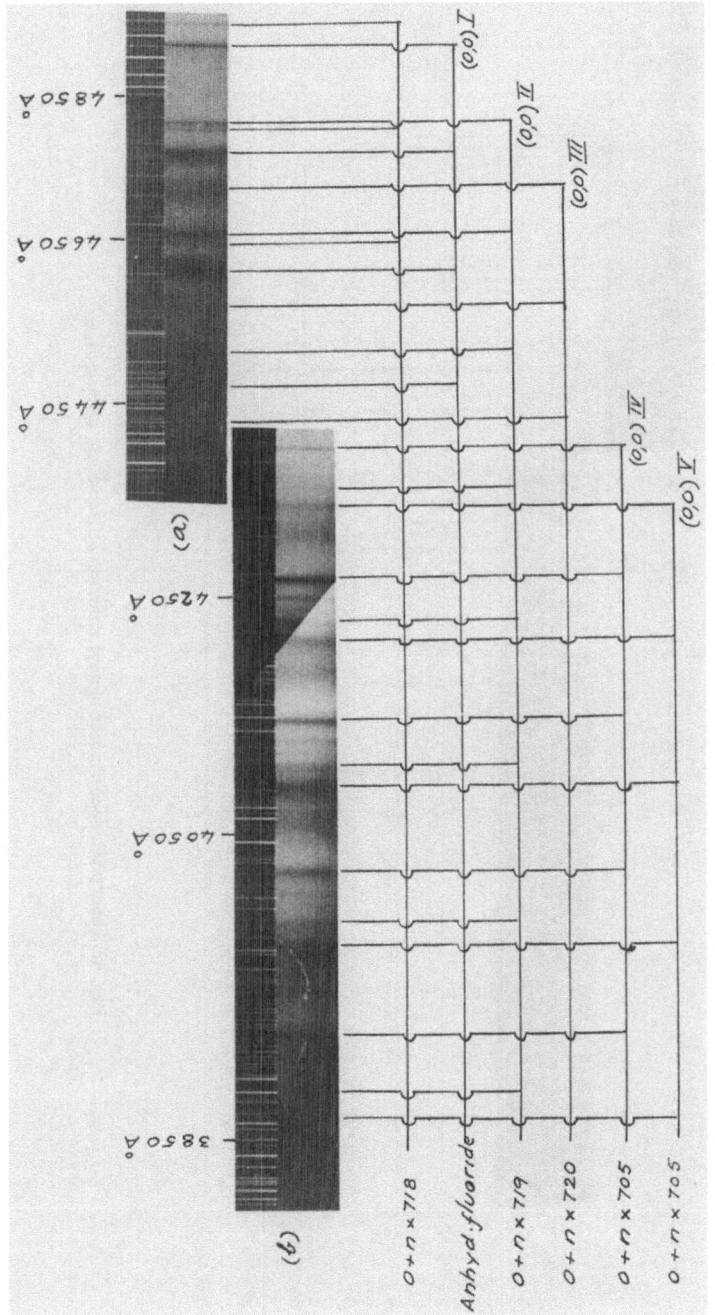

(a) Long path length    (b) Short path length.

Fig. 3. Absorption spectrum of uranyl fluoride II at liquid air temperature.

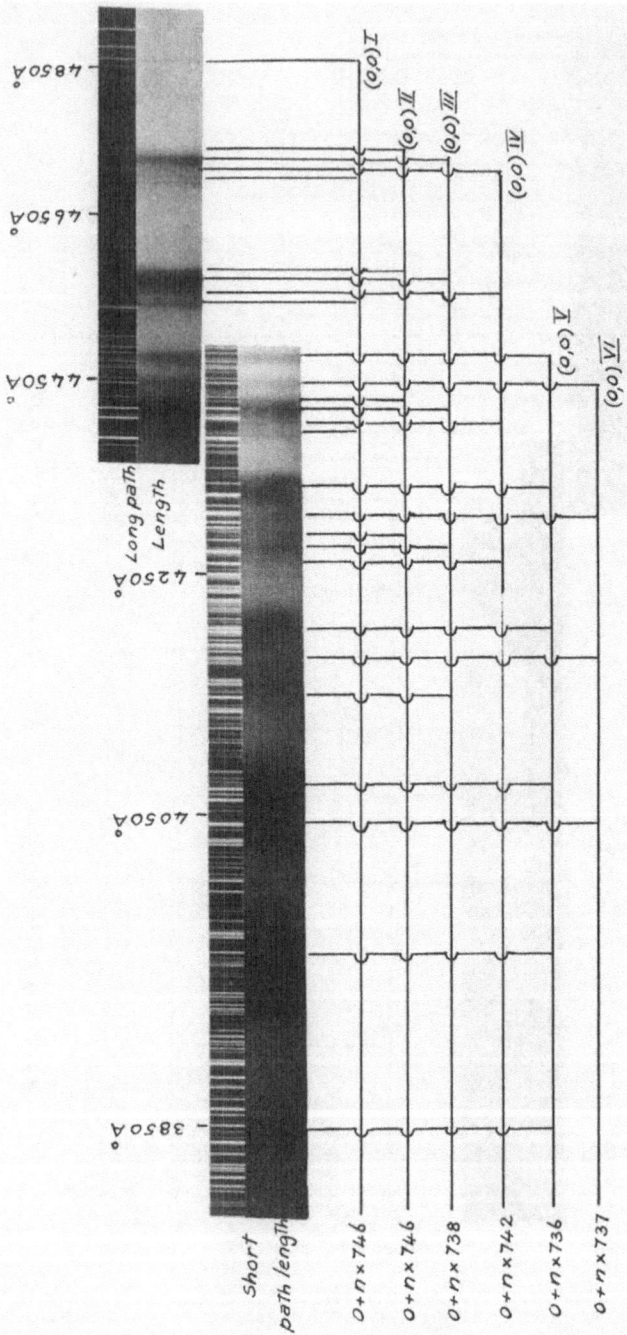

Fig. 4. Absorption spectrum of uranyl chloride at liquid air temperature.

# TABLE I
## Absorption Bands of Anhydrous Uranyl Fluoride

| Wavenumber of band (cm⁻¹) | Intensity | Assignment | | | |
|---|---|---|---|---|---|
| | | System I | System II | System III | System IV |
| 20,236 | 7 vsh * | 0,0 | | | |
| 20,386 | 2 vsh | | 0,0 | | |
| 20,914 | 2 msh * | | | | |
| 20,955 | 7 vsh | 0+719=719 | | | |
| 20,997 | 3 sh * | 0+761=761 | | | |
| 21,043 | 1 d * | | | | |
| 21,108 | 3 sh | | 0+722=722 | | |
| 21,183 | 10 msh | | | 0,0 | |
| 21,277 | 3 d | | | | 0,0 |
| 21,427 | 1 d | | | 0+244=244 | |
| 21,592 | 1 msh | | | | |
| 21,634 | 1 msh | | | | |
| 21,672 | 4 vsh | 0+2×719=1438 | | | |
| 21,725 | 2 sh | 0+761+719=1480 | | | |
| 21,771 | 1 d | 0+2×761=1522 | | | |
| 21,841 | 2 sh | | 0+2×722=1444 | | |
| 21,898 | 7 d | | | 0+715=715 | |
| 21,997 | 4 d | | | | 0+720=720 |
| 22,136 | 2 d | | | 0+244+715=959 | |
| 22,244 | 2 msh | | | | |
| 22,289 | 1 d | | | | |
| 22,316 | 1 d | | | | |
| 22,381 | 2 d | 0+3×719=2157 | | | |
| 22,411 | 1 d | 0+761+2×719=2199 | | | |
| 22,451 | 2 vd * | | | | |
| 22,616 | 3 d | | | 0+2×715=1430 | |
| 22,723 | 2 d | | | | 0+2×720=1440 |
| 22,946 | 1 d | | | | |
| 23.051 | 1 d | | | | |
| 23,168 | 1 d | 0+4×719=2876 | | | |

*The abbreviations for band descriptions used here and in Tables II, III, and V are: sh) sharp; vsh) very sharp; msh) medium sharp; d) diffuse; and vd) very diffuse.

fluoride II. These bands are denoted by asterisks in Table II and the assignments for these bands are not repeated. The remainder of the absorption bands were analyzed on the basis of five systems with a common lower state. The 0,0 bands of the systems are situated at 20,080, 20,753, 21,217, 22,723, and 23,038 cm⁻¹ (Table II).

## TABLE II
### Absorption Bands of Uranyl Fluoride II

| Wavenumber of Band (cm$^{-1}$) | Intensity | System I | System II | System III | System IV | System V |
|---|---|---|---|---|---|---|
| | | | Assignment | | | |
| 20,080 | 5 vsh | 0,0 | | | | |
| 20,232 * | 10 sh | | | | | |
| 20,262 | 1 sh | 0+182=182 | | | | |
| 20,310 | 1 msh | | | | | |
| 20,465 | 1 msh | | | | | |
| 20,753 | 7 sh | | 0,0 | | | |
| 20,773 | 2 sh | | | | | |
| 20,798 | 5 vsh | 0+718=718 | | | | |
| 20,930 * | 2 d | | | | | |
| 20,951 * | 10 sh | | | | | |
| 20,984 * | 3 d | | | | | |
| 21,134 * | 1 vd | | | | | |
| 21,217 | 3 vd | | | 0,0 | | |
| 21,472 | 10 sh | | 0+719=719 | | | |
| 21,524 | 2 vsh | 0+2×718=1436 | | | | |
| 21,619 * | 1 d | | | | | |
| 21,675 * | 7 sh | | | | | |
| 21,716 * | 3 d | | | | | |
| 21,855 * | 1 vd | | | | | |
| 21,937 | 3 vd | | | 0+720=720 | | |
| 22,181 | 5 sh | | 0+2×719=1438 | | | |
| 22,320 * | 1 d | | | | | |
| 22,407 * | 2 d | | | | | |
| 22,590 | 1 vd | | | 0+2×720=1440 | | |
| 22,723 | 7 sh | | | | 0,0 | |
| 22,816 | 2 sh | | | | | |
| 22,885 | 2 sh | | | | 0+162=162 | |
| 22,943 | 3 sh | | 0+3×719=2157 | | | |
| 23,038 | 4 d | | | | | 0,0 |
| 23,198 | 2 vd | | | | | 0+160=160 |
| 23,428 | 10 sh | | | | 0+705=705 | |
| 23,522 | 3 sh | | | | | |
| 23,584 | 2 sh | | | | 0+162+705=867 | |
| 23,648 | 4 sh | | 0+4×719=2876 | | | |
| 23,743 | 5 d | | | | | 0+705=705 |
| 23,806 | 1 d | | | | | |
| 23,917 | 4 vd | | | | | 0+160+705=865 |

## Table II (Continued)

| Wavenumber of Band (cm$^{-1}$) | Intensity | System I | System II | System III | System IV | System V |
|---|---|---|---|---|---|---|
| 24,135 | 10 sh | | | | 0+2×705=1410 | |
| 24,229 | 3 sh | | | | | |
| 24,289 | 2 sh | | | | 0+162+2×705=1572 | |
| 24,355 | 5 sh | | | 0+5×719=3595 | | |
| 24,453 | 7 d | | | | | 0+2×705=1410 |
| 24,634 | 4 vd | | | | | 0+160+2×705=1570 |
| 24,839 | 6 sh | | | | 0+3×705=2115 | |
| 24,932 | 1 sh | | | | | |
| 24,992 | 1 sh | | | | 0+162+3×705=2277 | |
| 25,056 | 2 sh | | | 0+6×719=4314 | | |
| 25,164 | 3 vd | | | | | 0+3×705=2115 |
| 25,314 | 1 vd | | | | | 0+160+3×705=2275 |
| 25,537 | 2 sh | | | | 0+4×705=2820 | |
| 25,636 | 1 d | | | | | |
| 25,779 | 1 d | | | 0+7×719=5033 | | |
| 25,883 | 1 vd | | | | | 0+4×705=2820 |
| 26,256 | 1 vd | | | | 0+5×705=3525 | |

*These bands belong to anhydrous uranyl fluoride.

Two fundamentals corresponding to the $\nu_1$ and $\nu_2$ frequencies of the uranyl ion at 718 and 182 cm$^{-1}$ are obtained in System I. Bands of this system consist of these fundamentals and their overtones and combinations. System II consists of a progression of bands of the $\nu_1$ fundamental (719 cm$^{-1}$) and its overtones. A diffuse series of bands constitutes System III; the value of the $\nu_1$ fundamental obtained in this system is 720 cm$^{-1}$. System IV consists of strong bands with progressions of the $\nu_1$ and $\nu_2$ fundamentals (705 and 162 cm$^{-1}$, respectively) and their overtones and combinations. System V is again a series of diffuse bands. The $\nu_1$ and $\nu_2$ fundamentals obtained for this system have values of 705 and 160 cm$^{-1}$ in the excited state.

### Uranyl Fluoride I

As with uranyl fluoride II, we find in the spectrum of uranyl fluoride I that some bands of the anhydrous salt and uranyl fluoride II are present as impurities. In this case, how-

# TABLE III
## Absorption Bands of Uranyl Fluoride I

| Wavenumber of Band (cm$^{-1}$) | Intensity | Assignment | | |
|---|---|---|---|---|
| | | System I | System II | System III |
| 20,014 | 5 vsh | 0,0 | | |
| 20,060 | 1 sh | | | |
| 20,094 ● ● | 10 vsh | | | |
| 20,127 | 1 msh | | | |
| 20,224 ● | 1 msh | | | |
| 20,443 | 1 msh | | | |
| 20,517 | 1 msh | | | |
| 20,683 | 3 vsh | 0-27+696=669 | | |
| 20,710 | 10 vsh | | 0+696=696 | |
| 20,746 | 7 vsh | | 0+732=732 | |
| 20,792 ● ● | 7 vsh | | | |
| 20,820 | 1 sh | | | |
| 20,960 ● | 6 d | | | |
| 21.008 ● | 3 vd | | | |
| 21,217 ● ● | 5 vd | | | |
| 21,345 | 2 sh | | | |
| 21,377 | 2 msh | 0-27+2×696=1365 | | |
| 21,406 | 8 sh | | 0+2×696=1392 | |
| 21,448 | 5 vsh | | 0+732+696=1428 | |
| 21,486 ● ● | 5 vsh | | | |
| 21,620 ● | 2 d | | | |
| 21,654 ● | 6 d | | | |
| 21,691 ● | 3 vd | | | |
| 21,890 ● ● | 7 vd | | | |
| 22,022 | 2 d | | | |
| 22,056 | 3 d | 0-27+3×696=2061 | | |
| 22,097 | 5 sh | | 0+3×696=2088 | |
| 22,130 | 2 sh | | 0+732+2×696=2124 | |
| 22,166 ● ● | 2 sh | | | |
| 22,272 | 1 d | | | |
| 22,330 ● | 2 d | | | |
| 22,383 ● | 2 d | | | |
| 22,484 | 2 d | | | |
| 22,524 | 3 sh | | | |
| 22,599 ● ● | 2 d | | | |
| 22,682 | 2 msh | | | |
| 22,725 | 3 d | | | |
| 22,765 | 2 msh | | 0+4×696=2784 | |
| 22,826 | 1 msh | | 0+732+3×696=2820 | |

## Table III (Continued)

| Wavenumber of Band (cm$^{-1}$) | Intensity | Assignment System I | System II | System III |
|---|---|---|---|---|
| 22,912** | 2 msh | | | |
| 22,964 | 2 msh | | | |
| 23,041* | 1 msh | | | |
| 23,060* | 1 msh | | | |
| 23,124 | 2 msh | | | |
| 23,160 | 10 sh | 0,0 | | |
| 23,208 | 1 sh | | | |
| 23,358 | 5 msh | | 0+198=198 | |
| 23,429 | 5 vd | | | 0,0 |
| 23,567 | 1 d | | | |
| 23,835 | 10 sh | | 0+675=675 | |
| 24,037 | 5 d | | 0+198+675=873 | |
| 24,103 | 5 vd | | | 0+674=674 |
| 24,315 | 2 vd | | | |
| 24,505 | 7 sh | | 0+2×675=1350 | |
| 24,714 | 3 vd | | 0+198+2×675=1548 | |
| 24,795 | 3 vd | | | 0+2×674=1348 |
| 24,955 | 1 vd | | | |
| 25,177 | 5 sh | | 0+3×675=2025 | |
| 25,436 | 2 vd | | | 0+3×674=2022 |
| 25,662 | 1 vd | | | |
| 25,851 | 1 msh | | 0+4×675=2702 | |
| 26,074 | 1 vd | | | 0+4×674=2596 |
| 26,537 | 1 d | | 0+5×675=3377 | |

*These bands belong to anhydrous fluoride.
**These bands belong to Fluoride II.

ever, only the starting bands (electronic bands) of the common series agree in position. The other members of these series are found to shift to the long wavelength side, indicating a change in the vibrational frequency. This most probably may be explained as being due to interaction occurring in mixed samples. These two series of bands (denoted by two asterisks) belong to fluoride II, and one series (denoted by a single asterisk) belongs to the anhydrous compound. Therefore, assignments for these bands are not given in Table III. The remainder of the bands, which belong to uranyl fluoride I, were analyzed on the basis of three systems with a common lower state with the 0,0 bands at 20,014, 23,160, and 23,429 cm$^{-1}$.

In System I, the $\nu_1$ fundamental of the uranyl ion (696 cm$^{-1}$) in the excited state and its overtones form a long progression of similar bands. A fundamental (732 cm$^{-1}$) corresponding to the asymmetric stretching frequency ($\nu_3$) of the O-U-O ion is also obtained, and the appearance of this fundamental with considerable intensity may be attributed to the crystalline fields surrounding the uranyl ion. This fundamental was also obtained by the author in fluorescence [6], where the value is 891 cm$^{-1}$ in the lower state. A crystal lattice frequency with a value of 27 cm$^{-1}$ in the lower state was also used in the analysis to interpret the weak satellite bands accompanying the $\nu_1$ fundamental and its overtones. System II consists of bands with moderate sharpness, starting at the middle of the spectrum. The $\nu_1$ and $\nu_2$ fundamentals in the upper state have values of 675 and 198 cm$^{-1}$, respectively, for this system. A series of broad and diffuse bands constitutes System III and forms a progression of the $\nu_1$ fundamental (674 cm$^{-1}$) and its overtones.

A correlation of the wavenumbers obtained from fluorescence [6] and from absorption of uranyl fluoride I is presented in Table IV.

TABLE IV

Correlation of the Fundamental Frequencies (cm$^{-1}$) of Uranyl Fluoride I

| Type of Vibration | Fluorescence (author) | | Absorption (author) | | | |
|---|---|---|---|---|---|---|
| | | | System I | | System II | System III |
| | LS * | US * | US | LS | | |
| U-O Symmetric Stretching ($\nu_1$) | 824 | 675 | 696 | | 675 | 674 |
| O-U-O Symmetric Bending ($\nu_2$) | 200 | | | | 198 | |
| U-O Symmetric Stretching ($\nu_3$) | 891 | | 732 | | | |
| Crystal Lattice Frequency | 32 | | | 27 | | |

*LS and US indicate lower and upper state, respectively.

## TABLE V
### Absorption Bands of Uranyl Chloride

| Wave-number of band $(cm^{-1})$ | Intensity | Assignment | | | | | |
|---|---|---|---|---|---|---|---|
| | | System I | System II | System III | System IV | System V | System VI |
| 20,447 | 1 msh | $0-3\times26=-78$ | | | | | |
| 20,469 | 1 msh | $0-2\times26=-52$ | | | | | |
| 20,504 | 1 msh | $0-26=-26$ | | | | | |
| 20,530 | 7 vsh | 0,0 | | | | | |
| 20,555 | 2 sh | $0+25=25$ | | | | | |
| 20,760 | 1 msh | $0+230=230$ | | | | | |
| 21,107 | 3 sh | | 0,0 | | | | |
| 21,162 | 10 d | | | 0,0 | | | |
| 21,193 | 7 sh | | | $0+31=31$ | | | |
| 21,234 | 4 sh | | | | 0,0 | | |
| 21,260 | 2 sh | | | | $0+26=26$ | | |
| 21,276 | 5 sh | $0+746=746$ | | | | | |
| 21,314 | 5 msh | $0+784=784$ | | | | | |
| 21,406 | 1 msh | | | $0+244=244$ | | | |
| 21,506 | 1 msh | $0+230+746=976$ | | | | | |
| 21,536 | 1 msh | | | | | | |
| 21,853 | 5 sh | | $0+746=746$ | | | | |
| 21,900 | 10 d | | | $0+738=738$ | | | |
| 21,938 | 7 sh | | | $0+31+738=769$ | | | |
| 21,976 | 5 d | | | | $0+742=742$ | | |
| 21,994 | 5 sh | | | | $0+26+742=768$ | | |
| 22,021 | 6 sh | $0+2\times746=1492$ | | | | | |
| 22,051 | 8 msh | $0+784+746=1530$ | | | | | |
| 22,078 | 1 msh | | | | | | |
| 22,131 | 2 msh | | | $0+244+738=982$ | | | |
| 22,179 | 1 d | | | | | | |
| 22,243 | 1 d | $0+230+2\times746=1722$ | | | | | |
| 22,288 | 1 d | | | | | | |
| 22,346 | 7 msh | | | | | 0,0 | |
| 22,370 | 7 msh | | | | | $0+24=24$ | |
| 22,396 | 1 d | | | | | | |
| 22,501 | 3 d | | | | | | 0,0 |
| 22,599 | 5 sh | | $0+2\times746=1492$ | | | | |
| 22,641 | 10 d | | | $0+2\times738=1476$ | | | |
| 22,680 | 5 sh | | | $0+31+2\times738=1507$ | | | |
| 22,723 | 3 sh | | | | $0+2\times742=1484$ | | |
| 22,763 | 2 sh | $0+3\times746=2238$ | | | | | |
| 22,789 | 7 sh | $0+784+2\times746=2276$ | | | | | |

## Table V (Continued)

| Wave-number of band (cm⁻¹) | Intensity | System I | System II | System III | System IV | System V | System VI |
|---|---|---|---|---|---|---|---|
| 22,865 | 1 sh | | | $0+244+2\times738=1720$ | | | |
| 23,082 | 10 vd | | | | | $0+736=736$ | |
| 23,210 | 2 msh | | | | | | |
| 23,238 | 3 d | | | | | | $0+737=737$ |
| 23,308 | 4 msh | | | | | $0+226+736=962$ | |
| 23,345 | 1 msh | | $0+3\times746=2238$ | | | | |
| 23,375 | 6 msh | | | $0+3\times738=2214$ | | | |
| 23,423 | 2 msh | | | $0+31+3\times738=2245$ | | | |
| 23,465 | 1 msh | | | | $0+3\times742=2226$ | | |
| 23,537 | 2 msh | $0+784+3\times746=3022$ | | | | | |
| 23,810 | 10 vd | | | | | $0+2\times736=1472$ | |
| 23,949 | 3 d | | | | | | $0+2\times737=1474$ |
| 24,050 | 5 d | | | | | $0+226+2\times736=1698$ | |
| 24,114 | 2 msh | | | $0+4\times738=2952$ | | | |
| 24,262 | 1 d | $0+784+4\times746=3768$ | | | | | |
| 24,551 | 10 vd | | | | | $0+3\times736=2208$ | |
| 24,690 | 2 d | | | | | | $0+3\times737=2211$ |
| 24,786 | 5 d | | | | | $0+226+3\times736=2434$ | |
| 25,288 | 5 vd | | | | | $0+4\times736=2944$ | |
| 25,499 | 3 d | | | | | $0+226+4\times736=3170$ | |
| 26,027 | 3 vd | | | | | $0+5\times736=3680$ | |

### Uranyl Chloride

The absorption spectrum of uranyl chloride was analyzed on the basis of six systems designated I-VI, with the possibility that Systems II, III, and IV might be multiplet components of a triplet state. The 0,0 bands of these six systems were chosen at 20,530, 21,107, 21,162, 21,234, 22,346, and 22,501 cm⁻¹, respectively (Table V).

System I corresponds to the fluorescence series, as some bands of the first group are also obtained in fluorescence [6]. The $\nu_1$ fundamental (746 cm⁻¹) and its overtones form a long progression of bands. The $\nu_3$ fundamental at 784 cm⁻¹ in the excited state is also obtained; this corresponds to the value of 964 cm⁻¹ in the lower state observed in fluorescence [6]. Combinations of this frequency with the $\nu_1$ frequency are also obtained in the higher groups. Other fundamentals obtained in this system are 230 cm⁻¹ of the $\nu_2$ frequency in the upper state

## TABLE VI
### Correlation of Fundamental Frequencies (cm$^{-1}$) of Uranyl Chloride

| Type of Variation | Fluorescence (author) | | Raman | | Infra-red | Absorption (author) | | | | | | |
|---|---|---|---|---|---|---|---|---|---|---|---|---|
| | LS* | US* | [13] | [14] | [13] | System I US / LS | | System II (US) | System III (US) | System IV (US) | System V (US) | System VI (US) |
| U-O Symmetric Stretching ($\nu_1$) | 880 | | 860 | 865 | 860 | 746 | | 746 | 738 | 742 | 736 | 737 |
| O-U-O Symmetric Bending ($\nu_2$) | 246 | | 210 | 197 226 | | 230 | | | 244 | | 226 | |
| U-O Symmetric Stretching ($\nu_3$) | 964 | | | 906 | 930 | 784 | | | | | | |
| Crystal Lattice Frequency | | 26 | | | | 25 | 26 | | 31 | 26 | | |

*LS and US indicate lower and upper state, respectively.

and 25 and 26 $cm^{-1}$ of the crystal lattice frequency in the upper and lower states. System II consists of a progression of bands corresponding to the $\nu_1$ fundamental (746 $cm^{-1}$) and its overtones. System III consists of bands belonging to the fundamentals 738, 244, and 31 $cm^{-1}$ of the $\nu_1$ and $\nu_2$ frequencies of the uranyl ion and crystal lattice frequency, respectively. A series of sharp bands forms System IV, in which the $\nu_1$ fundamental has the value 742 $cm^{-1}$ in the excited state. A fundamental at 26 $cm^{-1}$ of the crystal lattice in the excited state is also obtained. In the middle of the spectrum, two series of diffuse bands begin to form Systems V and VI. In System V, the fundamentals obtained have values of 736 and 226 $cm^{-1}$, which correspond to the $\nu_1$ and $\nu_2$ frequencies. For System VI, the $\nu_1$ fundamental has a value of 737 $cm^{-1}$.

In view of the small and nearly equal separations of the electronic levels of Systems II, III, and IV, and of the similarity of the bands, these three systems might be considered to be due to multiplet components of a triplet excited state. This reduces to four the number of distinct systems in the absorption spectrum of the chloride.

Table VI presents a correlation of the wavenumbers of the fundamental frequencies obtained from fluorescence and absorption spectra by the author and from Raman and infrared spectra obtained by Satyanarayana [14] and Conn and Wu [13].

## ELECTRONIC TRANSITIONS

For all the uranyl halides studied in this investigation, from three to five systems were observed in the absorption spectrum of each compound. This compares with the four systems in uranyl nitrate and even more in cesium uranyl nitrate and cesium uranyl chloride observed by Dieke and Duncan [12]. From the electronic configuration of the uranyl ion [15], it is obvious that the lower state is a totally symmetric singlet state $'\Sigma_g$. As observed in the absorption spectrum of uranyl chloride, triplet states are also possible in the excited states; however, the nature of these excited states can only be established by further experiments.

## ACKNOWLEDGMENTS

The author wishes to thank Dr. V. Ramakrishna Rao for his valuable guidance throughout the course of the work, and Mr. K. Srinivasulu for preparing the fluoride compounds.

## REFERENCES

1. E. Becquerel, Compt. rend. 75, 296 (1872).
2. E. Becquerel and K. Onnes, Leiden Communications 110 (1909).
3. V. Ramakrishna Rao and K. V. Narasimham, Indian J. Phys. 30, 334 (1956).
4. K. V. Narasimham and V. Ramakrishna Rao, J. Sci. Ind. Research (India) 19B, 285 (1960).
5. K. V. Narasimham, Indian J. Phys. 34, 321 (1960).
6. K. V. Narasimham, Ibid. 35, 282 (1961).
7. K. V. Narasimham and V. Ramakrishna Rao, Spectrochim. Acta 18, 1055 (1962).
8. D. D. Pant, Proc. Indian Acad. Sci. 31A, 35 (1950).
9. B. N. Samoilov, Zh. eksp. i teor. fiz. 18, 1030 (1948).
10. A. von Unruh, Dissertation, University of Rostok (1919).
11. W. S. Andrews, Amer. Mineralogist 7, 19 (1922).
12. G. H. Dieke and A. B. F. Duncan, Spectroscopic Properties of Uranium Compounds, McGraw-Hill, New York (1949).
13. G. K. T. Conn and C. K. Wu, Trans. Faraday Soc. 34, 1483 (1938).
14. B. S. Satyanarayana, Proc. Indian. Acad. Sci. 15A, 414 (1942).
15. S. P. McGlynn and J. K. Smith, J. Mol. Spectroscopy 6, 164 (1961).

# EMISSION SPECTROSCOPY

# TIME-RESOLUTION SPECTROSCOPY

## Francis D. Harrington

U. S. Naval Research Laboratory
Washington, D. C.

## INTRODUCTION

Time-resolution spectroscopy is concerned with the evaluation of temporal spectral variations in a radiating source, which is frequently a rapidly occurring optical phenomenon. The observational information required for this type of spectrum analysis is recorded either by film, using a time-resolving spectrograph, or by electronic sensing devices adapted to a standard spectrograph. The characteristics of these instruments and their applications to studies of exploding wires and nuclear explosions will be discussed in terms of the information desired and results obtained.

The Radiometry Branch, Optics Division of the U. S. Naval Research Laboratory (NRL) has been engaged for a number of years in a variety of researches on rapidly occurring, high-velocity, luminescent plasmas. Most of the projects were supported by the Los Alamos Scientific Laboratory (LASL) of the University of California.

The time-resolving spectrographs which evolved as part of the work performed on these projects are capable of recording spectral time histories of these fast optical events. The many unusual high-speed spectrographs, i.e., dynamic spectrographs, developed during this period possess a wide variety of time-resolutions, wavelength resolutions, and wavelength coverages.

The spectrographs will be discussed in three general categories: the moving-film type, the rotating-drum type, and the rotating-mirror type. A method of externally time-resolving a static spectrograph, as well as photoelectric methods adaptable to conventional spectrographs, will also be described.

## GENERAL FEATURES OF TIME-RESOLVING SPECTROGRAPHS

A typical image-streaking time-resolving spectrograph is shown schematically in Fig. 1. The optical phenomenon to be

162

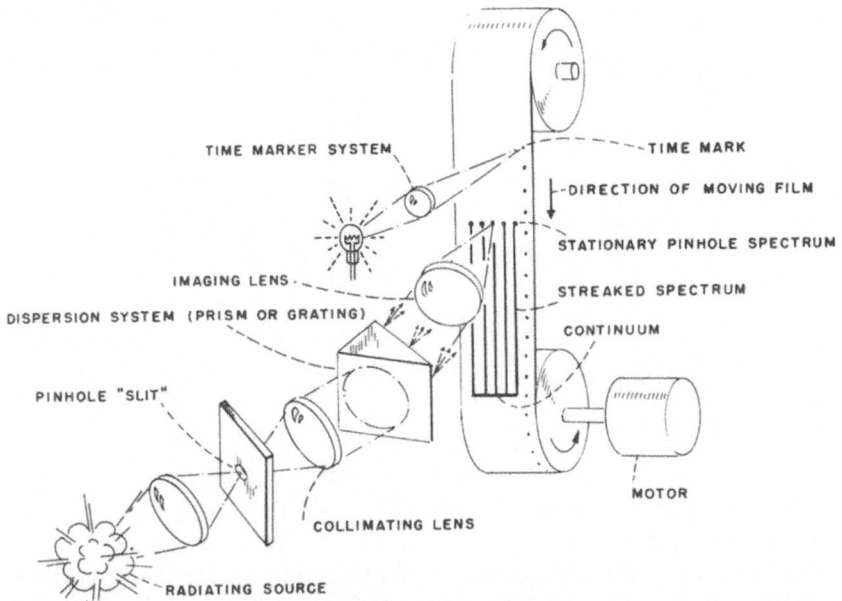

Fig. 1. Moving-film spectrograph.

observed is focused on the entrance plane of the spectrograph. The "slit" of the spectrograph in this instrument is a pinhole aperture. The light passing through the pinhole entrance is collimated, dispersed, and then focused into a spectrum on the photographic emulsion. The optical system of the spectrograph uses essentially a Wadsworth mounting for the dispersing element [1, 2], which produces stigmatic images in the focal

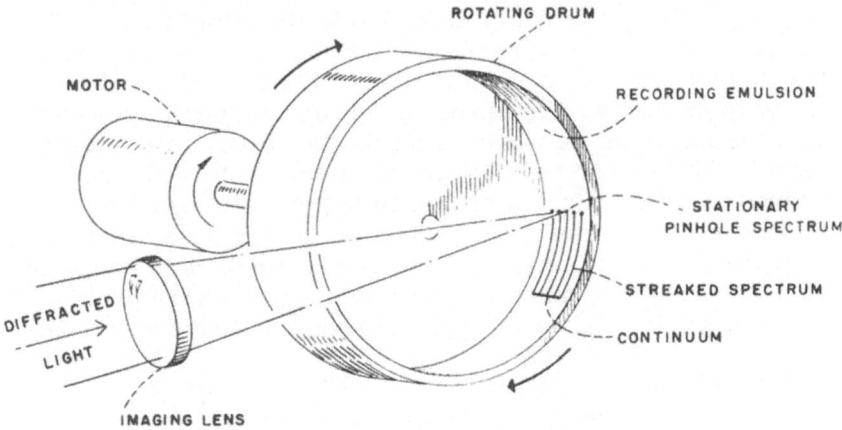

Fig. 2. Rotating-drum spectrograph.

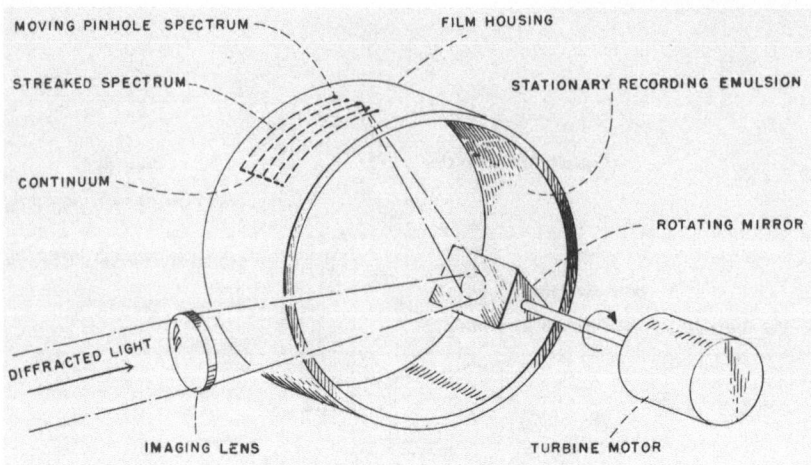

Fig. 3. Rotating-mirror spectrograph.

plane. For continuous radiation the static image is a line of dispersed continuum oriented in the direction of spectral dispersion. For monochromatic radiation the spectral image is a "point" on the film. A time-streaked spectrum results if there is relative motion between the spectral image and the recording emulsion at right angles to the spectral dispersion in the focal plane. In Fig. 1, the image-streaking is produced by moving the film.

In a rotating-drum spectrograph as shown in Fig. 2, it is the film drum with the film inside which is rotated to streak the spectrum. In a turbine-driven rotating-mirror spectrograph such as illustrated in Fig. 3, the rotating mirror [3, 4] streaks the image along stationary film inside the film housing. The distance from the mirror face to the film is the "writing arm." The velocity of the image relative to the film is usually referred to as the "writing speed" and is determined in this case from the length of the writing arm and the speed of the rotating mirror. The image velocity and the image size of the pinhole determine the exposure time, which is generally considered to be the time resolution of the spectrograph. The time resolution is defined as the time required to displace a pinhole image on the film one full diameter. The wavelength resolution of the spectrograph is the wavelength interval between two just-resolved streaked spectrum lines.

The instruments illustrated in Figs. 2 and 3 are continuous-writing spectrographs and will continue to record or rewrite over the initial spectrum with each revolution of the

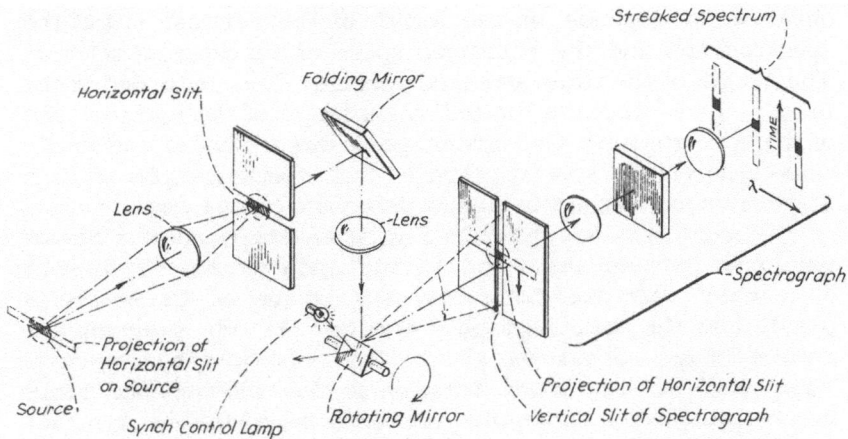

Fig. 4. Method of externally time-resolving a static spectrograph.

film drum, or mirror-rotor, unless a capping shutter is used. An explosive activated shutter [5] has been found to be the most successful capping device. Detonation of an explosive attached to a block of quartz or glass of good optical quality placed in the line of sight pulverizes the transparent optical block and makes it opaque. An electronic system actuated by the light from the phenomenon triggers the explosive shutter. This type of spectrograph is activated to the desired operating conditions shortly before zero time of the optical event. Because of the continuous-writing feature of the instrument, synchronization of zero time with spectrograph operation is not necessary.

A method of externally time-resolving a static or conventional stigmatic spectrograph is illustrated in Fig. 4. The method of time-scanning is similar, for example, to that used at LASL [6] and at the Atomic Weapons Research Establishment, England [7, 8]. The horizontal slit in the optical train, which is shown near the source, is used for spatial studies of a desired region of the phenomenon. The confined area defined by the horizontal slit is imaged on the slit plane of the spectrograph. A high-speed rotating mirror is located near the slit and is used to sweep the source image along the entire length of the vertical slit of the spectrograph. The rotating mirror is synchronized with the optical source so that the mirror is in the right position to scan the slit. Time resolution of the order of 1 $\mu$sec is obtainable during total time observation periods lasting 20 $\mu$sec to 50 $\mu$sec. The total time of the observation depends on the length of the vertical slit of the spectrograph and the rotational speed of the external mirror. The length of the time-streaked spectral lines recorded in the focal plane is likewise limited by the length of the entrance slit of the spectrograph. One advantage of this method of externally time-resolving a spectrograph is that spectrographs of high dispersion and large wavelength coverage can be used.

Although it is not shown in Fig. 4, an image-rotator can be employed between the rotor-mirror and entrance slit in such a manner that the horizontal slit image of the source is parallel to the spectrograph's vertical slit. By sweeping the source image across the slit of the spectrograph, a spectral "snapshot" of very short duration in the microsecond range can be obtained. If the optical source is reproducible, a series of these "snapshot" spectra can be obtained by progressively varying the time at which the image of the source sweeps the slit. This process will result in a representative time history of the source given by a series of individual spectra.

If spatial resolution is not required at the source, a simpler system of externally scanning the slit of the spectrograph is obtained by using a rotating disc with a very narrow single radial slit cut in its face. The rotating disc is placed as close to the slit as possible. An optical system is placed between the source and the instrumentation so that the optical event will be focused on the slit of the spectrograph. By rotating the disc at relatively high speed, the radial slot will traverse the slit and produce a time-resolved spectrum as described before. As in the case of the external rotating mirror, the position of the radial slit in the disc must be correlated with the source excitation. For very large sources, a degree of spatial resolution at the source is maintained by this method of externally time-scanning the spectrograph with a rotating disc.

A photoelectric method of time-resolving a standard spectrograph at selected wavelengths is shown schematically in Fig. 5. By placing a number of exit slits of predetermined widths and selected wavelength positions in the focal plane, the light variations as seen on the entrance slit of the spectrograph can be displayed on oscilloscope screens by means of photomultipliers placed behind the exit slits. Each oscilloscope displays intensity with respect to time for the spectrum line at whose position in the focal plane the exit slit is placed. Time-intensity of a band of continuum radiation can be measured in the same way. Since there is a space requirement for

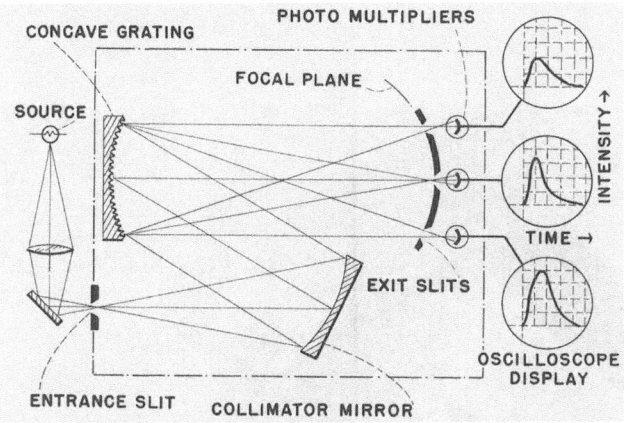

Fig. 5. Photoelectric method of time-resolving a static spectrograph.

the photomultiplier and exit slit, only a limited number of these sensing devices can be located in the focal plane of the spectrograph. Time resolution of $10^{-8}$ sec can be obtained with good electronic circuitry. Techniques and applications of this form of time-resolution spectroscopy are well described in the literature [9–13].

Another electro-optical system for obtaining time-resolved information in the microsecond range from low light level optical phenomena is shown schematically in Fig. 6. The

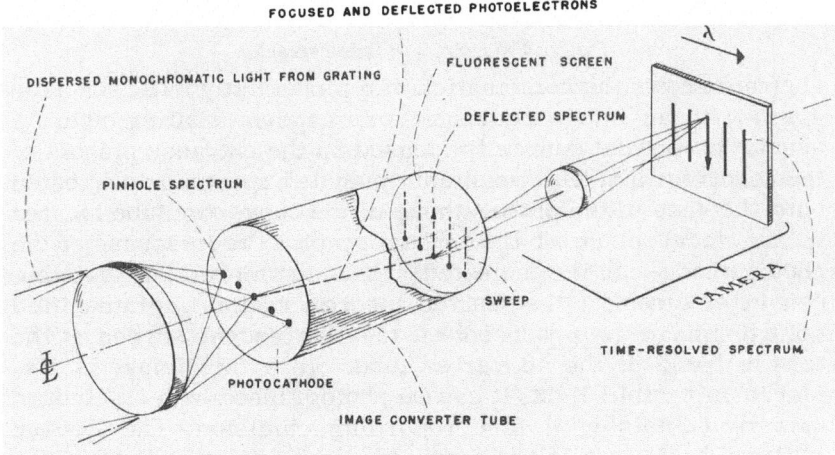

Fig. 6. Method of time-resolving a static spectrograph by placing an image-converter tube in the focal plane.

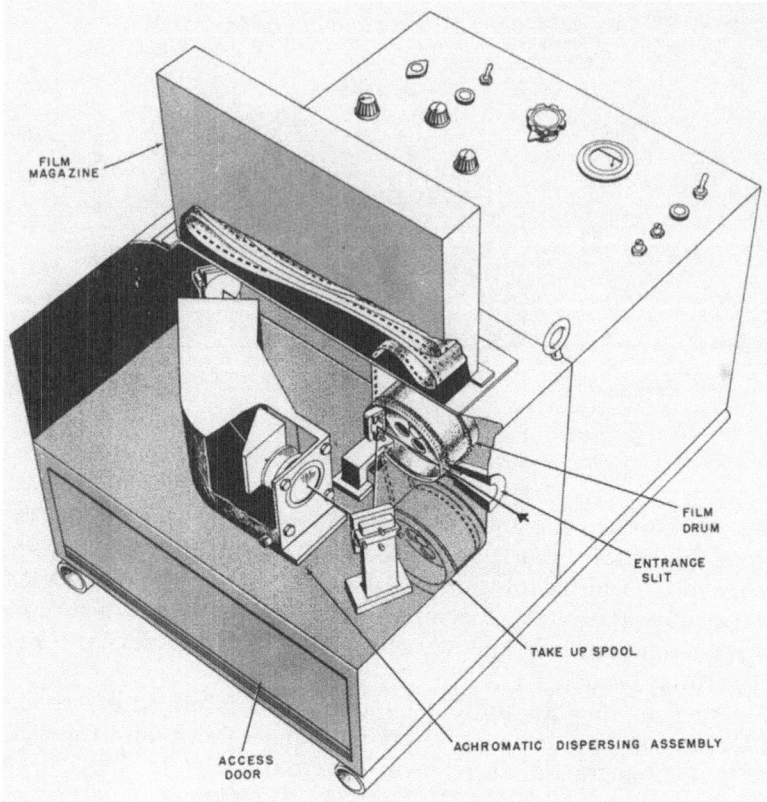

Fig. 7. The f/6.6 Cine Spectrograph.

diagram shows the combination of a pinhole stigmatic spectrograph with an image-converter or image-intensifier tube. As
usual, the optical source is focused on the entrance pinhole of
the spectrograph. The resultant "pinhole" spectrum is directed
onto the face of the photocathode of the converter tube located
in the focal plane of the spectrograph. The response of the
photocathode can be selected from commercially available
converter tubes for the desired spectral region. An intensified
optical image is produced on the fluorescent screen at the
terminal end of the converter tube. Since this image is displayed in visible light, it can be photographed with a standard
camera containing a fast recording emulsion. The focused
electron beam, within the converter tube, can be made to sweep
the fluorescent image screen to produce a time-resolved
spectrum which in turn is recorded by the camera. This

process combined with pulsed shuttering can produce a few individual spectra; and in this case, a conventional spectrograph entrance slit could be used. Principles and operation of image-converter tubes are discussed in [14] and [15]. An application of electro-optical time-resolution spectroscopy to high-intensity discharges in air and argon has been described by D. P. C. Thackeray [16].

EXAMPLES OF TIME-RESOLVING
SPECTROGRAPHS AND APPLICATIONS

### The f/6.6 Cine High-Speed Spectrograph

The f/6.6 Cine High-Speed Spectrograph [17] is a prismatic instrument designed for use at low-dispersion in the near ultraviolet and the visible wavelength regions. This spectrograph (Fig. 7) is a moving-film instrument and has two film speeds: 9 ft/sec and 90 ft/sec. The spectrograph can record 460 frames/sec at low film speed and 4600 frames/sec at high film speed. The spectra which this spectrograph records are not streaked images but groups of individual spectra, one to five per frame, on a strip of 70-mm film. The cine feature of the spectrograph repetitively exposes a moving image on moving film in such a manner that no relative motion exists between the image and the film. Unit magnification is maintained between the slit and image on the film plane during the

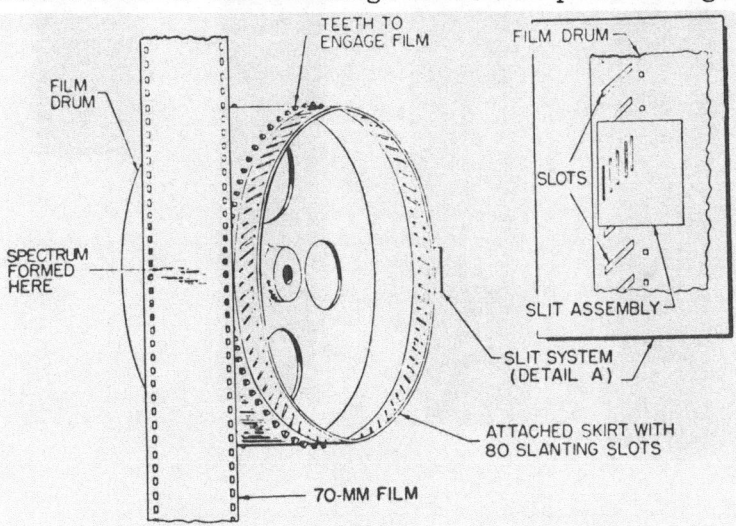

Fig. 8. Details of the film drum and slit assembly in Fig. 7.

FRANCIS D. HARRINGTON

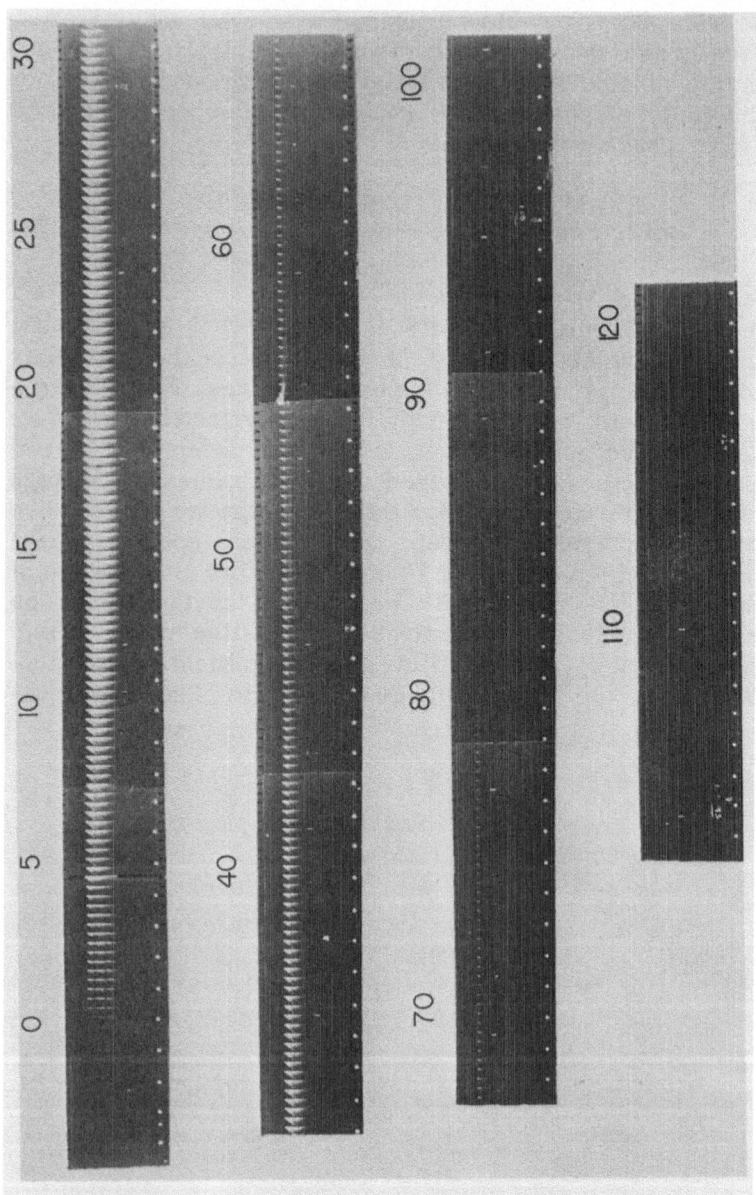

Fig. 9. Cine Spectrograph recording of a G. E. No. 22 flashbulb explosion.

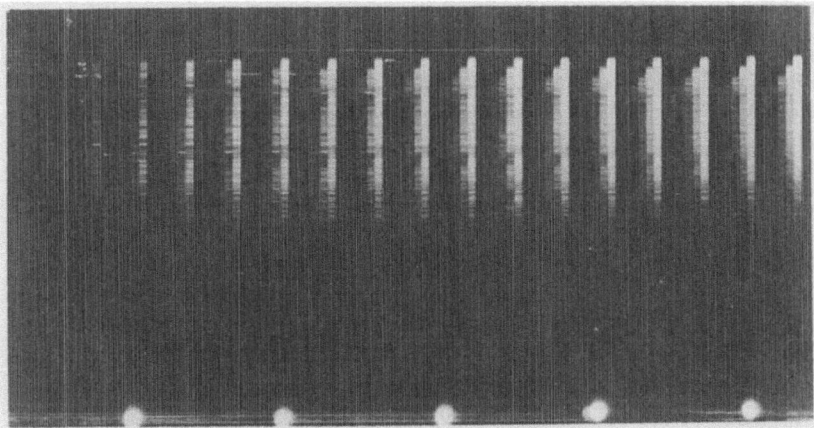

Fig. 10. Enlargement of the first 5 msec in Fig. 9.

exposure. An illustration of the framing process and the multiple slit arrangement is shown in Fig. 8. Four data-taking slits and one wavelength calibration slit compose the slit assembly. The calibration slit is sometimes used to obtain an additional data spectrum. The film drum contains a number of slanting slots in the periphery of its skirt. The downward motion of film rotates the film drum and causes the slanting slots in the skirt of the film drum opposite the film contact to move upward and travel across the assembly of data slits, thus causing the image on the film to move downward (see Fig. 8). One frame on the film corresponds to one transit of a slanting slot across the slits. Each spectrum in the frame is displaced wavelengthwise due to the displacement in the multiple slits. The first data slit in the arrangement is unattenuated. The next three data slits are attenuated with neutral density filters of known calibration. The wavelength calibration slit will also be attenuated when used to obtain an extra data spectrum. If the optical event should overexpose the unattenuated spectrum, the desired spectrum can be found among one of the attenuated spectra. The optical system of the Cine Spectrograph consists of quartz and lithium fluoride lenses and a Young-Thollan prism arrangement using two crystalline quartz prisms. The spectral range of the spectrograph is from 2200 to 7000 A and at 4000 A the wavelength resolution is approximately 5 A. The wavelength dispersion is 30 A/mm at 3000 A, 125 A/mm at 4000 A, and of the order of 400 A at 6000 A. By substituting a pinhole or a slit of small height for the multiple-slit assembly and eliminating the cine feature, the framing spectrograph can be converted to a streak spectrograph. For this modification of the spectrograph, time resolutions of $10^{-5}$ and $10^{-6}$ sec are obtainable at low and high film speeds, respectively. An optical event can be observed for both the Cine Spectrograph and its modification as a streak spectrograph for 9.75 sec (low film speed) and 0.75 sec (high film speed). The film cassette located on top of the spectrograph contains a maximum of 100 ft of 70-mm film which is folded over in layers so that the film is not creased. In this manner, the conventional loading spool and its consequent inertia are eliminated. The film is freely drawn from the cassette and collected on a motor driven take-up spool in the bottom compartment of the spectrograph.

The spectral time-history of an exploding G. E. No. 22 flashbulb as recorded by a Cine Spectrograph is presented in Fig. 9. Timing marks are recorded on the edge of the spectrogram as a row of dots spaced 1 msec apart. It can be observed that the peak brightness appears between 15 and 20 msec and the duration of the flash is about 80 msec. The spectral distribution and the intensity of the light can also be seen with respect to time, chiefly in the form of an increasing and decreasing continuum. Figure 10 is an enlargement of the first 5 msec of the same flashbulb spectrogram. The origin of the explosion is recorded as an emission spectrum which merges and then fades into a growing continuum. The first spectrum in each frame is unattenuated, and adjacent to it are the four attenuated spectra.

Another application of the Cine Spectrograph by L. F. Drummeter, J. A. Curcio, and C. H. Duncan of NRL was to determine fireball "color" temperatures (spectral distribution temperatures) of nuclear explosions as they developed with respect to time. It was assumed that these fireballs would behave as gray bodies and that the shape of the spectral radiance curve in the selected wavelength region could therefore be matched to a Planckian curve for the determination of color temperature.

Wien's radiation law, which is an approximation to Planck's law, was used in the determination of color temperature from the spectrograms. The approximation is very good for values of $\lambda T \leq 0.3$ cm-deg and the data generally fell below this limit. Wien's law was used in the form

$$N_\lambda = C_1 \lambda^{-5} \exp(-C_2/\lambda T) \qquad (1)$$

where $N_\lambda$ is the measured spectral radiance in $w/cm^2 \cdot sr \cdot A$; $C_1$ is the first radiation constant; $C_2$ is the second radiation constant; $\lambda$ is the wavelength; and $T$ is the temperature in °K.

In general it is not necessary to know the absolute spectral radiance and frequently only the relative spectral radiance $N_\lambda' = KN_\lambda$ ($K$ = constant) was determined. Equation (1) may thus be written

$$N_\lambda' = B\lambda^{-5} \exp(-C_2/\lambda T) \tag{1a}$$

where $B$ is a lumped and perhaps unknown constant.

By rearranging equation (1a) and taking its logarithm the relation becomes

$$\ln(N_\lambda'\lambda^5) = \ln B - C_2/\lambda T \tag{2}$$

The slope which ultimately provides the color temperature $T$ at wavelength $\lambda$ is obtained by differentiating equation (2) as follows:

$$d(\ln N_\lambda' \lambda^5) \, / \, d(1/\lambda) = -C_2/T \tag{3}$$

Figure 11 is part of a sample fireball spectrum obtained with the Cine Spectrograph. Each group of separated individual spectra is a separate frame and represents a finite exposure established by the framing rate which is provided by the chosen film speed. Film densities vs. wavelength were obtained from microdensitometer traces of the envelope of each spectrum. At selected wavelengths, the densities were converted into spectral radiances relative to a standard tungsten calibration of the spectrograph by means of film H & D curves (density vs. log exposure). Corrections for the transmission of the external optics used in front of the spectrograph slit and for the attenuation of the air path were applied if absolute values of spectral radiance were computed. These absolute values would represent the spectral radiances, $N_\lambda$, which would be measured very close to the source for the time intervals of the successive frames.

On a plot of the logarithm of the product of $N_\lambda'\lambda^5$ as the ordinate against $1/\lambda$ as the abscissa for the selected wavelength interval, the slope of the best straight line drawn through the points represents the average color temperature for the particular time at which an individual spectrum was exposed.

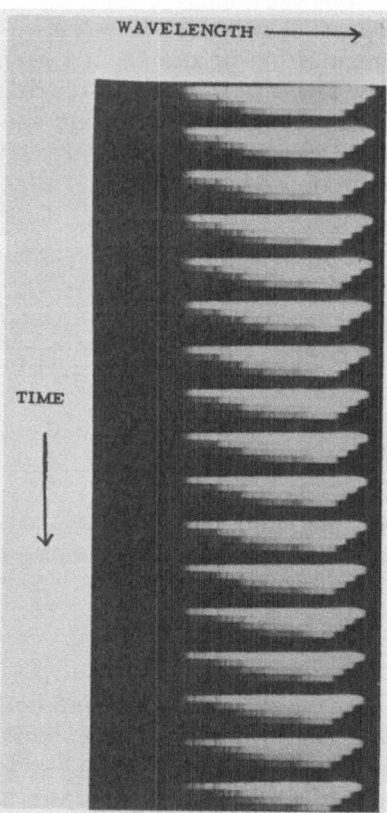

WAVELENGTH ⟶

TIME

Fig. 11. Sample spectra of nuclear fireball light obtained with the Cine Spectrograph.

Figure 12 presents a sample plot of $\ln(N_\lambda' \lambda^5)$ vs $1/\lambda$. The scatter of points about the straight line shows the combined effect of experimental errors, sensitometric errors introduced in the reduction of the data, and the departure of the source from a gray body for a specific observation. Figures 13-15 show graphically the time histories of fireball color temperatures obtained with the Cine Spectrographs. Figure 13 shows the first 10 msec plotted with more observational data than that for the same region of Figs. 14 and 15. The reason for this is that a faster framing rate was used by the Cine Spectrograph to obtain data for Fig. 13.

Another use of the Cine Spectrographs by Drummeter, Curcio, and Duncan was in the measurement of the effective thickness of nitric oxide, $NO_2$, in the line of sight with respect to time when observing a nuclear explosion. Since the effects of temperature at the source and similarly the temperature of

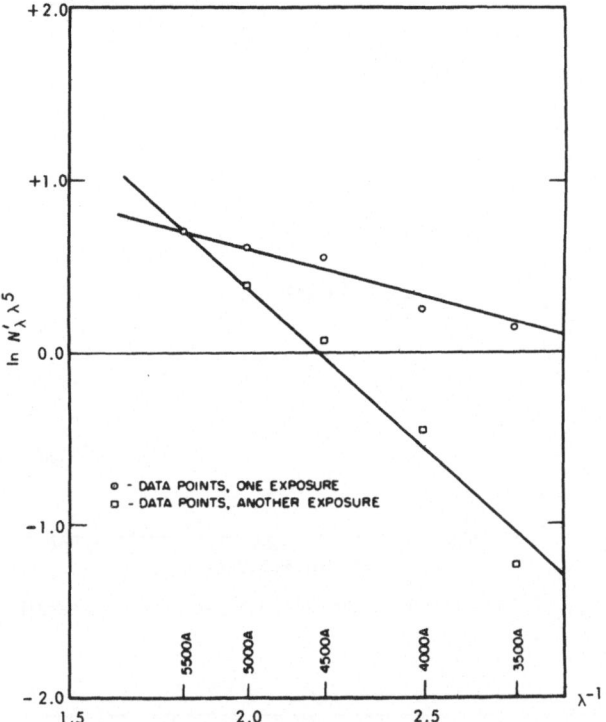

Fig. 12. Sample plot of $\ln N_\lambda' \lambda^5$ vs $\lambda^{-1}$

$NO_2$ were unknown, published values of Hall and Blacet's [18] absorption coefficients for $NO_2$ at 25°C and standard pressure were utilized in the reduction of the observational data. For this measurement, the spectrograph was modified by replacing the multiple slit assembly with a single vertical slit of approximately the same height, and replacing the film drum by another drum from which the skirt containing the frame slots had been eliminated. An external rotating disc which contained a single radial slot was placed ahead of the slit of the spectrograph and the slot was imaged on the instrument's slit. Further external optics were used to focus the light from the source onto the face of the rotating disc. With each rotation of the disc, the radial slot traversed the single slit and produced a frame with minimum image blurring on the moving film. The dead time which was provided by each rotation of the disc was sufficient to properly space the frames on the film.

The envelope of each frame was determined from a microphotometer trace, and the density was converted to relative

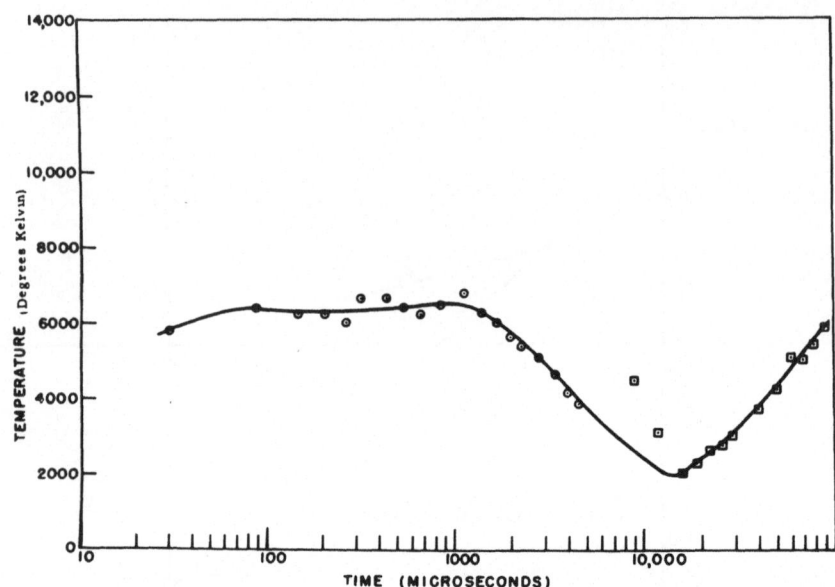

Fig. 13. Time history of the color temperature of a fireball.

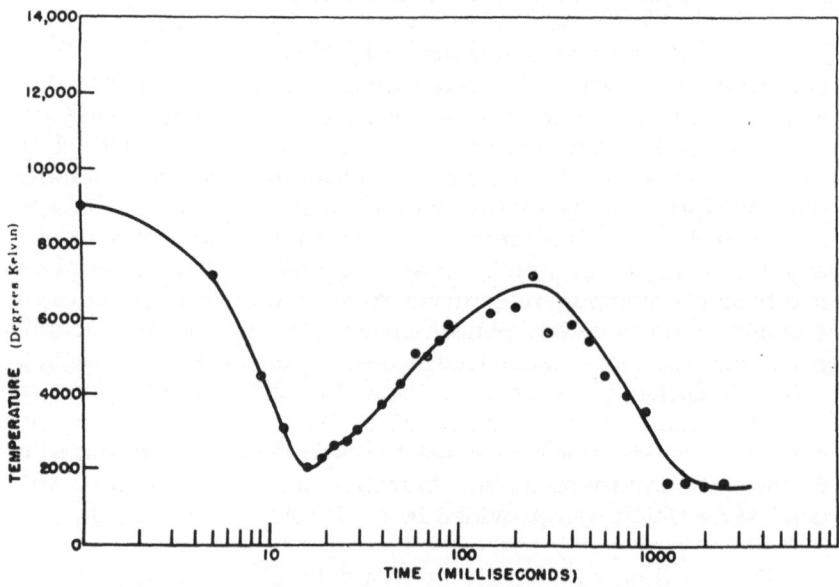

Fig. 14. Time history of the color temperature of a fireball.

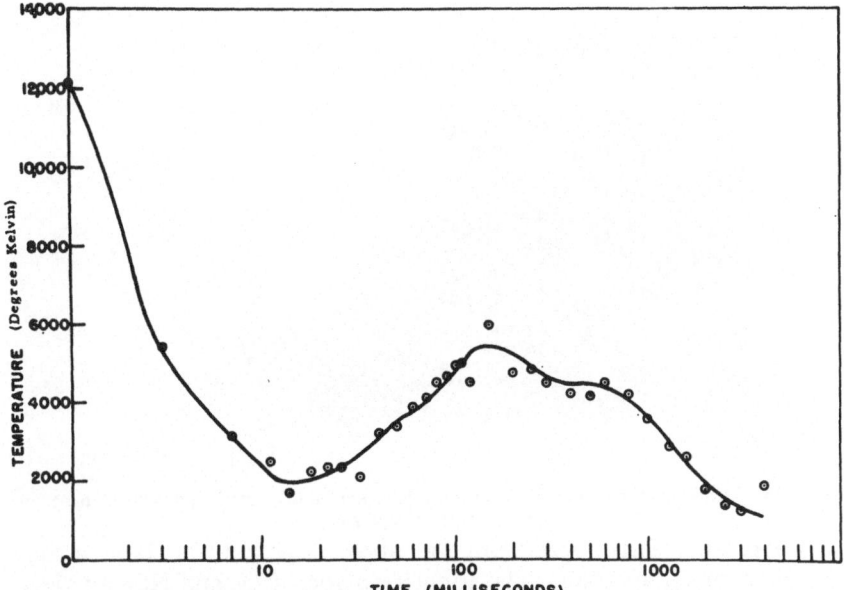

Fig. 15. Time history of the color temperature of a fireball.

exposure by H & D curves. The transmission of a $NO_2$ band was obtained from the ratio of the relative exposure at the center of the band to the relative exposure of the envelope at the point from which the absorption band originated. Differential attenuation coefficients from Hall and Blacet's curves were used for the bands selected from the observational data. By subtracting the absorption coefficients for the envelope of the continuum in the referenced curves from the coefficients at the center of the desired bands on the same envelope, the differential attenuation coefficients for the whole series of $NO_2$ bands observed in the observational data were obtained. The effective thickness of $NO_2$ was then calculated by means of the relation

$$T = \exp(-\sigma t) \tag{4}$$

where $T$ is the transmission of a band determined by the ratio of relative exposures; $\sigma$ is the differential attenuation coefficient in $cm^{-1} \cdot (mm\ Hg)^{-1} \cdot 10^3$; and $t$ is the effective thickness of $NO_2$ in centimeters at 25°C and 760 mm Hg pressure.

Figure 16 presents a photograph of a time-resolved spectrum of a nuclear explosion showing the development of $NO_2$

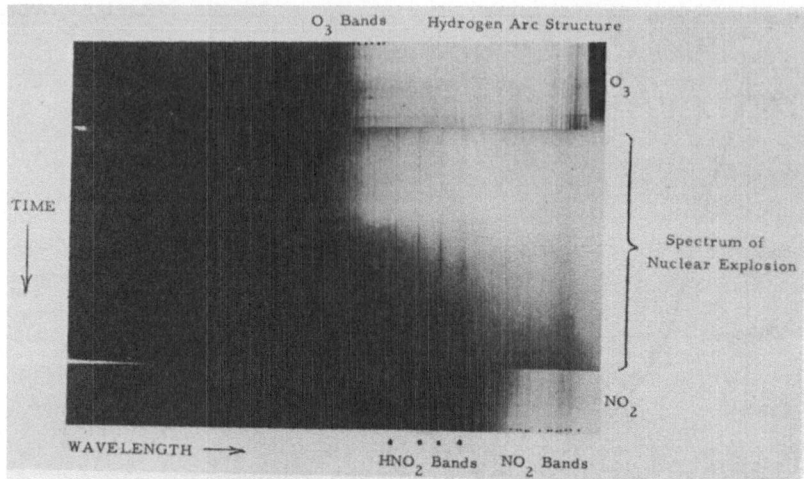

Fig. 16. Time-resolved spectrum of a nuclear explosion compared with laboratory spectra of $O_3$ and $NO_2$.

absorption. This spectrum is superimposed on laboratory spectra of $O_3$ and $NO_2$. The identification of $O_3$ and $NO_2$ as well as $HNO_3$ are seen in the time-resolved spectrum. From chronologic microphotometer traces and the methods of data reduction described above, a typical time history of the $NO_2$ observable in the spectrograph's line of sight is graphically shown in Fig. 17.

### The f/2.8 N4GS Spectrograph

An outgrowth of the Cine Spectrograph was a companion

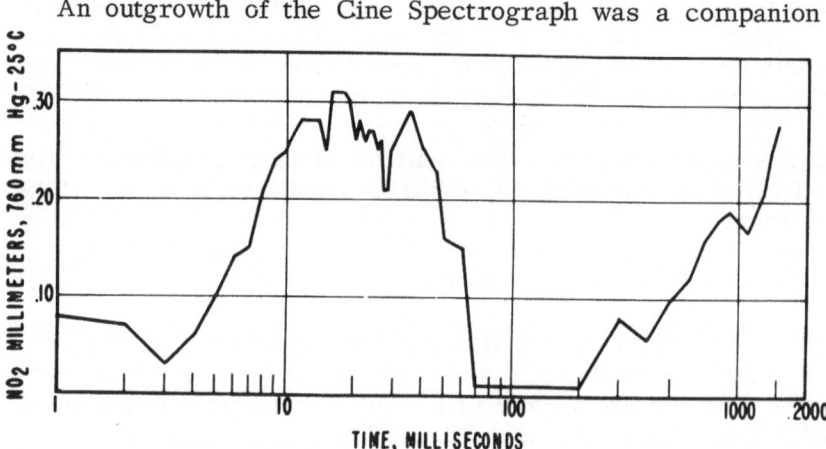

Fig. 17. Typical time history of the effective thickness of $NO_2$ in the line of sight to a nuclear explosion.

Fig. 18. The f/2.8 N4GS Spectrograph.

grating spectrograph with better light-gathering power. It was designed to extend the wavelength coverage available with the Cine Spectrographs from the visible region into the near infrared region. This f/2.8 grating instrument is known as the N4GS Spectrograph [19] and is shown in Fig. 18. The spectrograph uses the housing, mechanical film drive, and electronic features of the Cine Spectrograph. The framing device was eliminated and the N4GS operates as a streak spectrograph with a pinhole entrance aperture. The optical system consists of glass optics with a 600-grooves/mm plane replica grating. The wavelength dispersion of the N4GS approximates that of the Cine Spectrograph at 3500 A, which is near the low wavelength cutoff of glass optics. The N4GS Spectrograph has more uniform and better dispersion than a comparable prism spectrograph for wavelengths beyond 3500 A. Wavelength coverage, in

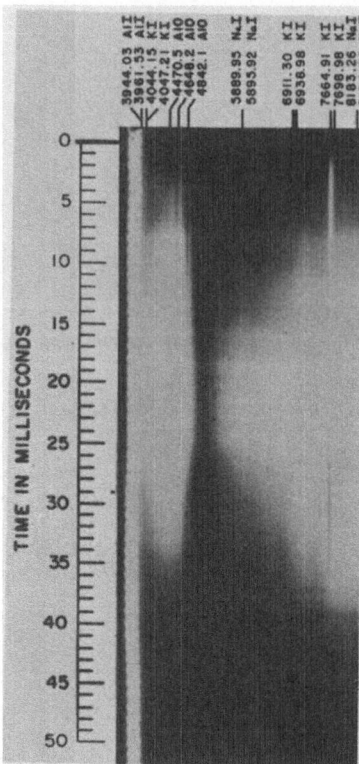

Fig. 19. Time-resolved spectrogram of an exploding flashbulb obtained with the N4GS Spectrograph.

the first-order spectrum, is from 3537 A to 8367 A on 70-mm film. The reciprocal linear wavelength dispersion at 5893 A in the first order is 89.3 A/mm and the average wavelength resolution in the first order spectrum is approximately 3 A. The time resolutions for the two nominal film speeds of 9 and 90 ft/sec are $1.3 \times 10^{-5}$ sec and $1.3 \times 10^{-6}$ sec, respectively.

A typical time-resolved spectrogram of an exploding flashbulb obtained with the N4GS Spectrograph is shown in Fig. 19. It presents visually the spectral time history of the flash, the total duration of which is approximately 45 msec. Atomic lines in emission and absorption, molecular bands in emission, and varying degrees of the continuum are recorded as they develop and change with respect to time. The absorption noted in the region of 5000 A is due to the insensitivity of the spectroscopic emulsion in that region.

### The f/2.0 Mark 55 Spectrograph

The f/2.0 high-speed ultraviolet spectrograph [20] is shown

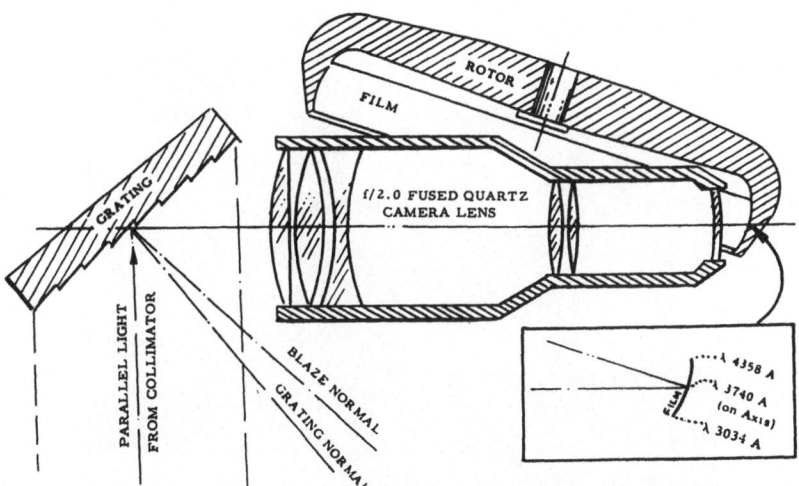

Fig. 20. The f/2.0 Mark 55 Spectrograph.

in Fig. 20. It is known as the Mark 55 Spectrograph. The Mark 55 was designed and built for NRL by Dr. Gordon Milne of the University of Rochester. It is a 35-mm rotating-drum spectrograph. The film drum is 12 in. in diameter and is driven by a 10,000-rpm motor. Its recycling, or rewrite, time is 6 msec and the time resolution is $10^{-7}$ sec. Figure 20 shows only the grating and camera portion of the instrument; it does not show the entrance pinhole aperture or the collimator mirror. The optical system consists of reflection and fused quartz refracting optics together with a 600-grooves/mm plane grating. The wavelength coverage, in the first order spectrum, is from 3034 to 4358 A. The average reciprocal linear dispersion is 45 A/mm in this order and the wavelength resolution is 1.3 A at 3700 A. The Mark 55 belongs to the class of continuous-writing spectrographs.

### The f/6.6 Model 102 Spectrograph

The f/6.6 Mod. 102 prism streak spectrograph is another continuous-writing spectrograph in which the image streaking is accomplished by a turbine-driven three-faced mirror. The turbine which rotates the mirror is driven by helium. The maximum rotor speed is 5000 rps, and the spectrograph's rewrite time is 67 $\mu$sec. The time resolution available at maximum rotor speed is $10^{-8}$ sec. Figure 21 shows two views of the spectrograph. The optical system consists of quartz and lithium fluoride lenses, and a double crystalline quartz prism

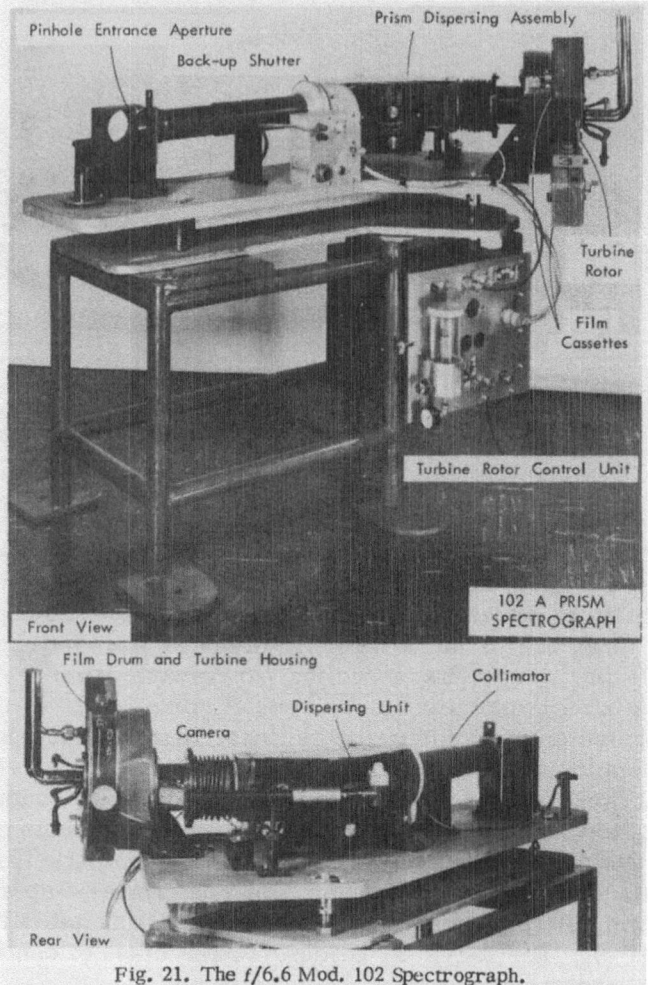

Fig. 21. The f/6.6 Mod. 102 Spectrograph.

system arranged in a Young-Thollan dispersion mounting. The dispersion unit allows the adjustment of two spectral regions in low dispersion, one beginning in the near ultraviolet from 2553 to 4046 A and the other originating in the upper wavelengths of the ultraviolet at 2850 A and extending to 5790 A in the visible region. Both wavelength intervals are recorded on 35-mm film located in the stationary film drum. The wavelength dispersion is 125 A/mm at 4000 A with a wavelength resolution of 12.5 A.

A Mod. 102 Spectrograph time-resolved spectrogram is

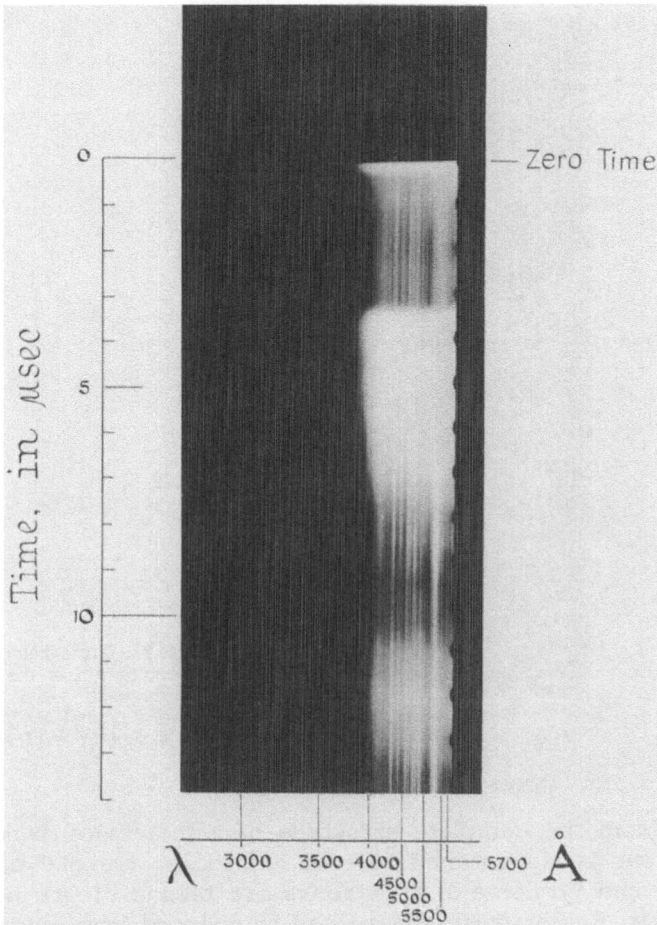

Fig. 22. Time-resolved spectrogram of an exploding wire obtained
with the Mod. 102 Spectrograph.

shown in Fig. 22. It represents the spectral history of an exploding 2-mil copper wire, 5 cm in length, which was exploded
in an evacuated chamber. The turbine mirror speed of 3394
rps produced the image streaking in the spectrogram. The
emission of excited neutral copper and changes in the continuum
with respect to time are observed in the figure.

Figure 23 illustrates a Mod. 102 time-resolved spectrogram of the air fluorescent emission from a typical air burst
of a nuclear explosion. Due to the low dispersion of the spectrograph and the size of the pinhole entrance aperture, a molecular
band appears as a streaked line and represents the integrated

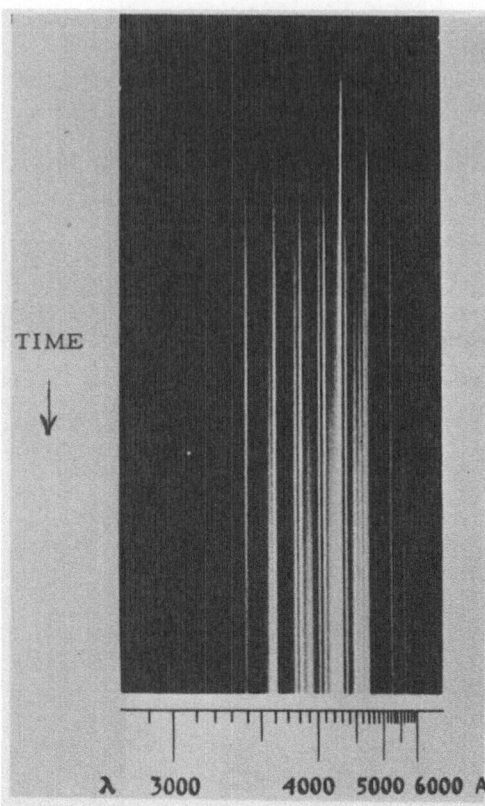

TIME

↓

Fig. 23. Time-resolved spectrogram of the air fluorescent emission from a typical air burst of a nuclear explosion.

λ   3000          4000  5000 6000 A

light from the complete structure of a molecular band. The spectrum is an intermixture of excited $N_2$ (second positive group) and $N_2^+$. Some of the streaks are blends of one or both systems. Radiometric values can be deduced from a streaked spectrogram of this nature by the usual methods of calibration and data reduction.

### The f/6.6 N9GS Spectrograph

Another continuous-writing image-streaking spectrograph is the proposed N9GS grating spectrograph [21]. It is a high-dispersion instrument with an f/6.6 aperture. Figure 24 schematically illustrates the optical system, which is composed of off-axis parabolic mirrors and a 1200-grooves/mm plane replica grating. The time-axis of the spectrum is produced by means of an air-driven high-speed three-faced rotating mirror, with a maximum safe speed of 2200 rps. The rewrite time at maximum speed is approximately 150 $\mu$sec, and the time reso-

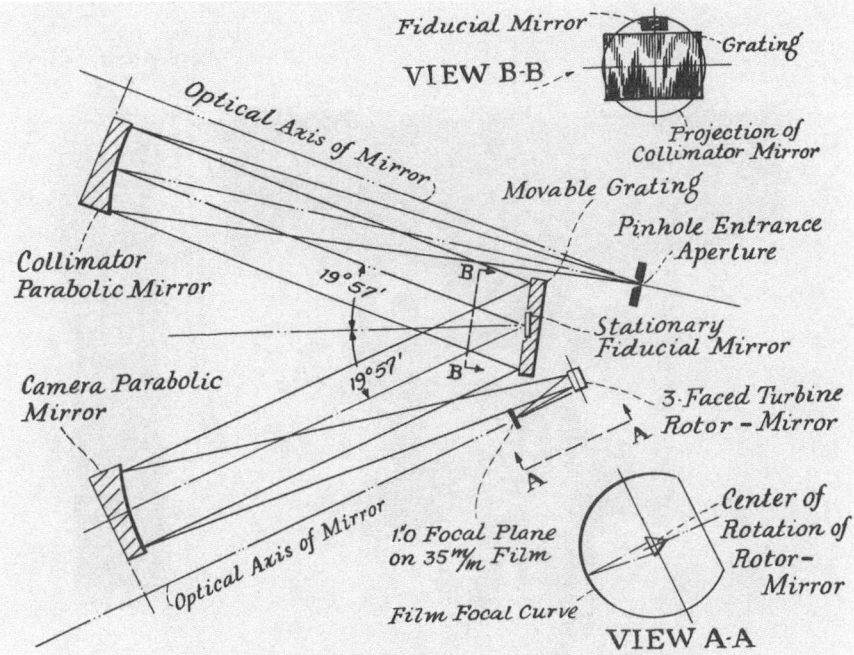

Fig. 24. Optical system of the f/6.6 N9GS Spectrograph.

lution of the spectrograph is $10^{-8}$ sec. In the first-order spectrum the total wavelength coverage is from 2400 to 7000 A with a reciprocal linear dispersion of 5.4 A/mm at 4700 A. The wavelength resolution is approximately 0.2 A. Due to the high–dispersion of the instrument, the spectrograph can only observe the spectrum in wavelength intervals of 137 A on 35-mm film. Rotation of the grating provides the means by which the desired wavelength intervals are adjusted to the focal plane.

A prototype N9GS was built from available components to check the performance of the proposed design. The collimating and camera elements were identical 10° off-axis parabolic mirrors which were cut from the same f/1.9, 48-in.-focal-length parabola. These off-axis elements are the mirrors around which the optical system was designed. The turbine and film drum housing which was incorporated into the proto-type was from a Mod. 102 Spectrograph. Figure 25 shows the prototype spectrograph. This type of spectrograph is applicable to the measurements of instantaneous spectrum line profiles and Doppler line shifts. Several exploding wire spec-

Film Drum and Turbine Housing

Turbine Rotor

Film Cassettes

Pinhole
Entrance
Aperture

Turbine Rotor Control Unit

Air (or helium) Supply

Fig. 25. A prototype N9GS Spectrograph.

trograms obtained with the prototype instrument are shown in the following figures.

Figure 26 is a time-resolved exploding wire spectrogram of three 1-mil twisted aluminum wires dusted with chalk. The wires were 5 cm long and exploded in an evacuated chamber at an ambient pressure of about $5 \times 10^{-5}$ mm Hg. The initial stored energy of the discharge was of the order of 300 joules and the initial capacitor voltage was approximately 350 kv. The Ca II lines are seen to appear along with the excited neutral Al lines.

The wavelength fiducial marker, from which any wavelength shift in the spectrum can be measured, is the long streak seen throughout the spectrogram and is produced by the total light from the phenomena. Since the grating aperture is smaller than that of the collimator mirror, the unused light from the collimator is reflected by the small mirror above the ruled area of the grating to the camera mirror. The camera mirror in turn focuses this undiffracted light onto the focal plane as an image of the entrance pinhole aperture. The fiduciary light is reflected from the same face of the rotating mirror as that

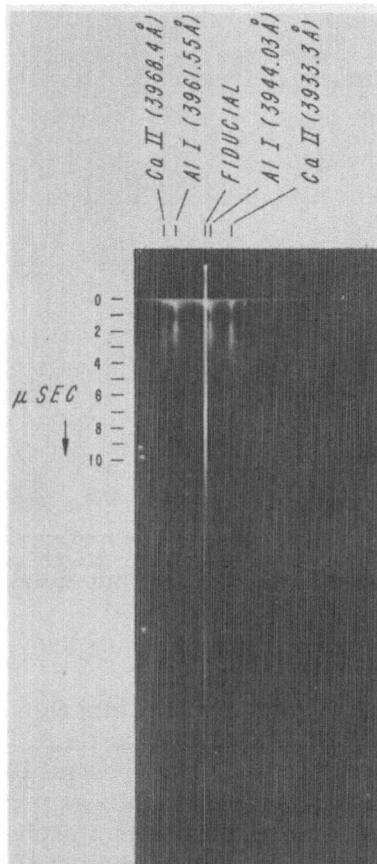

Fig. 26. Time-resolved exploding wire spectrogram of three 1-mil twisted aluminum wires dusted with chalk obtained with the prototype N9GS Spectrograph.

for the spectrum. By slightly tipping the plane mirror above the grating, the fiducial image can be made to appear just ahead of the origin of the spectrum. In this way the fiducial streak can be easily distinguished from a spectrum line. The fiducial streak is a measure of the total duration of the phenomena, since it represents the total integrated light of the event as it develops in time. The length of the streak in Fig. 26 indicates that light was still being produced in other regions of the spectrum, whereas the spectral lines show that in the region observed the duration was much shorter. The spectrogram in Fig. 26 was obtained at a rotor-mirror speed of 1100 rps.

Another N9GS time-resolved spectrogram is that shown in Fig. 27 of a copper foil exploded in a small spherical

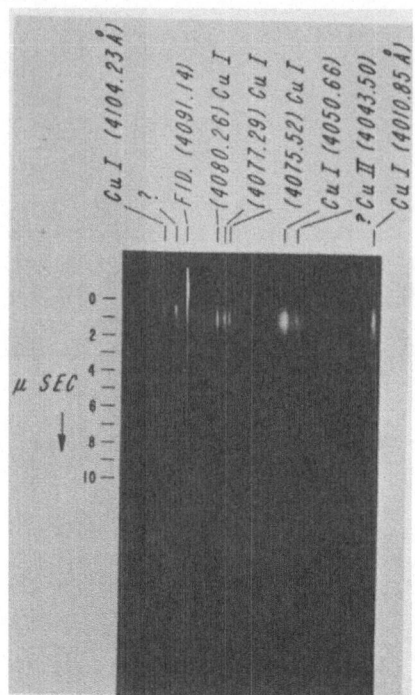

Fig. 27. Time-resolved spectrogram of an exploding copper foil obtained with the prototype N9GS Spectrograph.

chamber at a pressure of approximately $4 \times 10^{-5}$ mm Hg. The foil was 1 cm long by 0.3 cm wide and 0.0002 in. thick. The identified spectrum is that for excited Cu I. The spectral lines appear slightly shifted to the blue during their brightness period from 0.4 to 1.5 sec, which indicates that there is a probable Doppler line shift during this period. The major portion of the energy of the explosion was dissipated during the first 2 $\mu$sec, as shown by the appearance of the fiducial line. It was found from the calibrated zero position of the fiducial image that in the first 0.4 $\mu$sec the phenomenon was obscured in the spectrum. The rotor speed for this spectrogram was 1220 rps.

The exploding wire group at NRL is now studying time-resolved spectra emitted by hollow exploding-imploding aluminum cylinders in a vacuum. The purpose is to determine the temperature and density of electrons and also the densities of neutral and ionized atoms. These determinations are presently being investigated in the optical phenomenon when conditions of thermal equilibrium can be assumed. It is expected that the determination of these parameters will lead to a better understanding of the processes prevailing in the aluminum plasma with respect to time.

The procedure is as follows: A static, time-integrated spectrogram obtained with a survey prismatic spectrograph showed that Al I, Al II, and Al III atomic systems were present in the near ultraviolet and visible regions during the cylinder explosion. Since the N9GS time-resolving spectrograph can record only a small spectral region of 137 A at one time, regions of interest were selected from the survey spectrogram to be time-resolved.

In the preliminary determination of the above parameters of the discharge, it was assumed that the plasma generated was optically thin. It was further assumed, on the first approach to the problem, that thermal equilibrium existed. If this is the case, the temperature of the electrons and the neutral and ionized atoms will be the temperature which the Boltzmann distribution prescribes. Further experimentation will confirm the validity of, or deviations from, these assumptions.

One spectral region which was selected to be time-resolved was one in which two Al III multiplets displayed strong lines. Another criterion in this selection was that the excitation potential of the upper energy levels of one multiplet was several electron volts greater than the corresponding upper levels of the other multiplet. The ratio of the line intensities of selected lines of the two multiplets will be temperature sensitive as indicated by the relation

$$\frac{I_1}{I_2} = \frac{g_1 A_1 \lambda_2}{g_2 A_2 \lambda_1} \exp\left(-\frac{E_1 - E_2}{kT}\right) \tag{5}$$

where $I_1$ and $I_2$ are relative intensities of lines 1 and 2; $g_1$ and $g_2$ are statistical weights $(2J+1)$ for the upper energy levels; $A_1$ and $A_2$ are calculated spontaneous transition probabilities; $\lambda_1$ and $\lambda_2$ are wavelengths of lines 1 and 2; $E_1$ and $E_2$ are excitation potentials of upper energy levels; $k$ is Boltzmann's constant; and $T$ is absolute temperature. The "$A$" coefficients of spontaneous transition probabilities were computed according to the Bates and Damgaard method as outlined in [22]. A series of line intensity ratios was calculated by equation (5) for four combinations of lines from the multiplets for increasing values of $kT$. These ratios were computed from the strongest lines of the Al III multiplets of $(3^2D - 4^2P^\circ)$ and $(4^2P^\circ - 5^2S)$. The computed values of the ratios were plotted against temperature. From microphotometer traces of the selected Al III lines and appropriate H & D curves of the photographic emulsion the

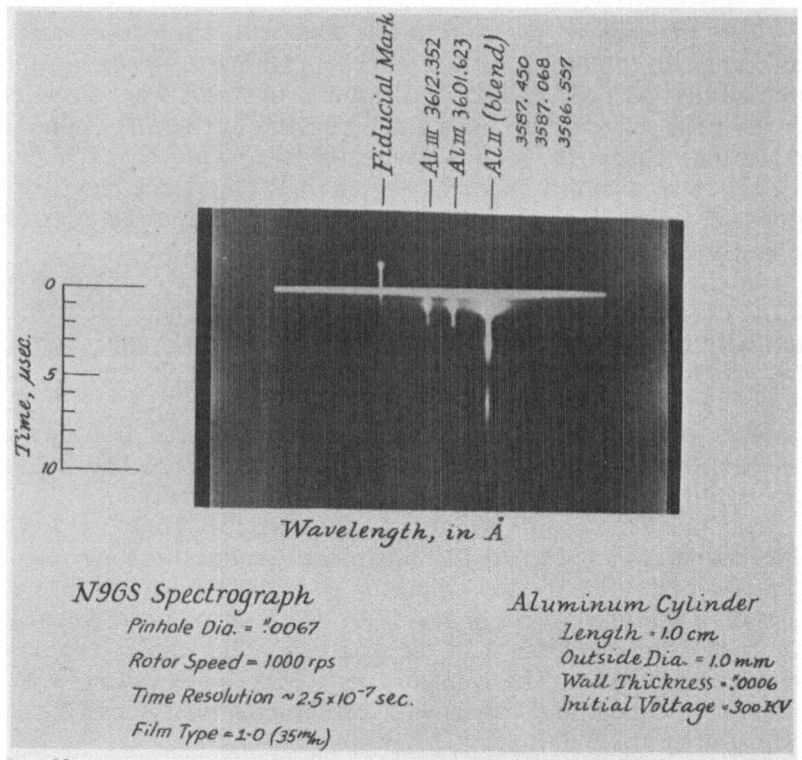

Fig. 28. Time-resolved spectrogram of an exploding-imploding aluminum cylinder obtained with the prototype N9GS Spectrograph.

relative line intensities were determined. The ratios determined from the measured relative intensities can be entered into the theoretical ratio vs. temperature graph and the corresponding temperatures read directly.

The time-resolved spectrogram of another region which contained Al II and Al III spectral lines is shown in Fig. 28. Relative line intensities were obtained as before from microdensitometer traces at the same position on the phenomenon time scale at which the intensities of Al III lines were previously made. The ratio of the ion densities of $N^{++}$ to $N^+$ was determined from the equation

$$\frac{I^{++}}{I^+} = \frac{g^{++} U^+ \lambda^+ A^{++} N^{++}}{g^+ U^{++} \lambda^{++} A^+ N^+} \exp\left(-\frac{E^{++} - E^+}{kT}\right) \tag{6}$$

By substituting the known quantities (such as the measured relative line intensity ratio of Al III to Al II, the temperature which was determined from line intensity ratios of Al III lines,

and the other quantities which have been previously identified),
the ratio of doubly ionized atoms to singly ionized atoms can
be obtained. The superscripts (++) and (+) indicate doubly ionized and singly ionized atoms, respectively. The quantity $U$ is
the partition function and in many cases is approximately equal
to the statistical weight $g$ for the ground state of the atom or
ion. The statistical weight of an energy level or state is equal
to $(2J + 1)$. $J$ is the designation for the inner quantum number of
the energy level.

Since the temperature $T$ and the ratio of the ion densities
of $N^{++}/N^+$ have been determined from relative line intensities,
the next step is to substitute these values into the Saha equation
and solve for the electron density $N_e$. One arrangement of
Saha's equation is

$$\frac{N^{++}N_e}{N^+} = \frac{2U^{++}}{U^+} \left(\frac{2\pi mkT}{h^2}\right)^{3/2} \exp\left(-\frac{\chi^+}{kT}\right) \tag{7}$$

where all the notations have been previously identified except
for $m$, the mass of an electron, and $\chi^+$, the ionization potential
of the singly ionized system. It is obvious for this degree of
ionization that

$$N_e = N^+ + 2N^{++} \tag{8}$$

and

$$N^{++} = N^+ \times (\text{ratio of ion densities})$$

from these two equations the density of singly and doubly ionized atoms of aluminum, i.e., the number of atoms per cubic
centimeter for each species, can be obtained.

The procedure described is for only one point on the time
scale of the phenomenon. The method can be progressively
repeated many times along the time axis. Thus, the variations
and time-history of the parameters of $T, N^+, N^{++}$, and $N_e$ can be
recorded. The spectroscopic phase of the above work has been
a combined effort of Ihor Vitkovitsky, William B. Buchanan,
and the author.

Another interesting time-resolved spectrogram obtained
from an exploding-imploding aluminum cylinder is shown in
Fig. 29. It shows the *raie ultime* lines of Al I together with the
Ca II lines. The Ca trace element was provided by dusting the
aluminum cylinder with chalk. The Al I lines show a shift to

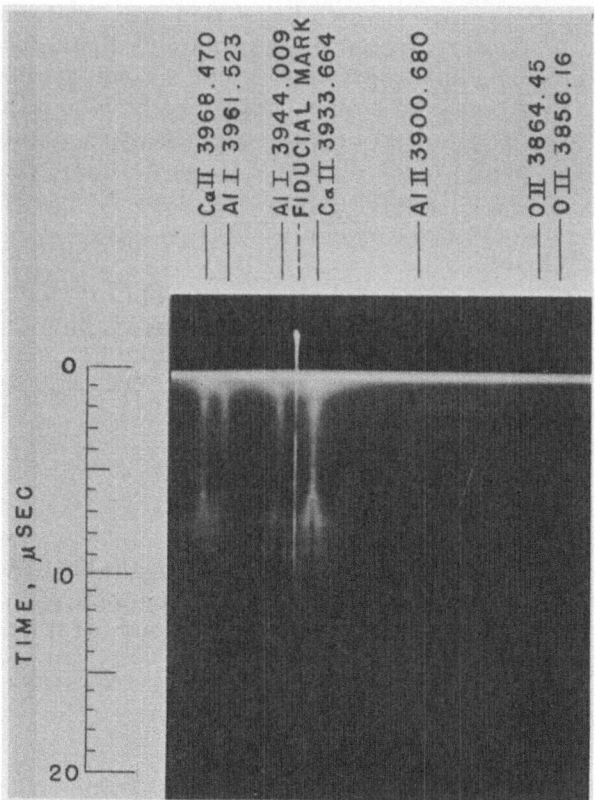

Fig. 29. Time-resolved spectrogram of an exploding-imploding aluminum cylinder obtained with the prototype N9GS
Spectrograph.

the red at their origin just below the initial continuum, whereas the Ca II lines appear to come directly out of the continuum unshifted. The red shift of Al I is apparently due to
pressure broadening. This red shift is not apparent in other
spectrograms of Al II and Al III. Another interesting feature
is the self-reversal exhibited near the end of the spectral lines
of Al I and Ca II. The self-reversal or absorption is symmetrical for Ca II, whereas that for Al I is unsymmetrical.

### The f/3.5 Spectrograph

The optical system of an f/3.5 static spectrograph [23] is
shown in Fig. 30. It is a stigmatic medium-dispersion grating
instrument which uses spherical mirrors with a Schmidt-type
corrector mirror. A 1200-grooves/mm plain replica grating
is used with this system. The average dispersion is 20.6

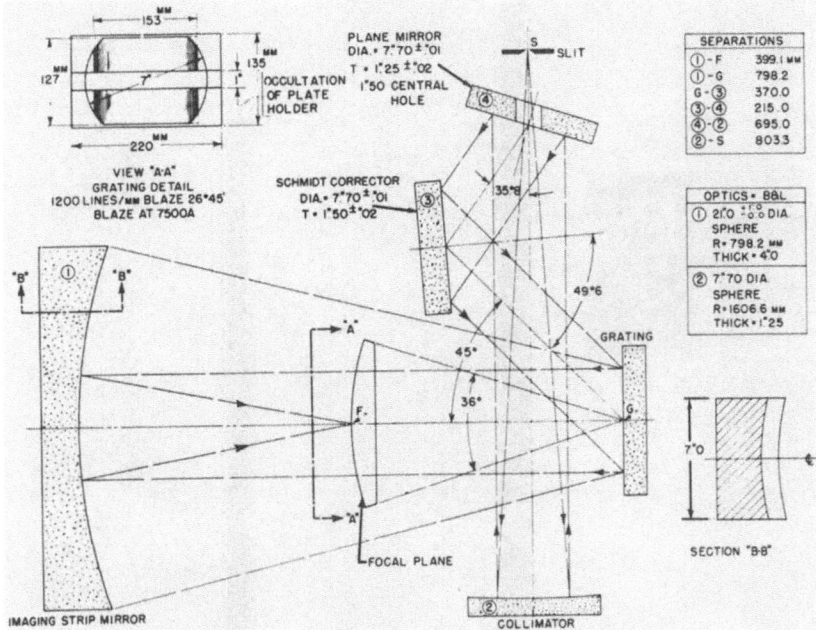

Fig. 30. Optical system of the f/3.5 static spectrograph.

Fig. 31. The f/3.5 static spectrograph.

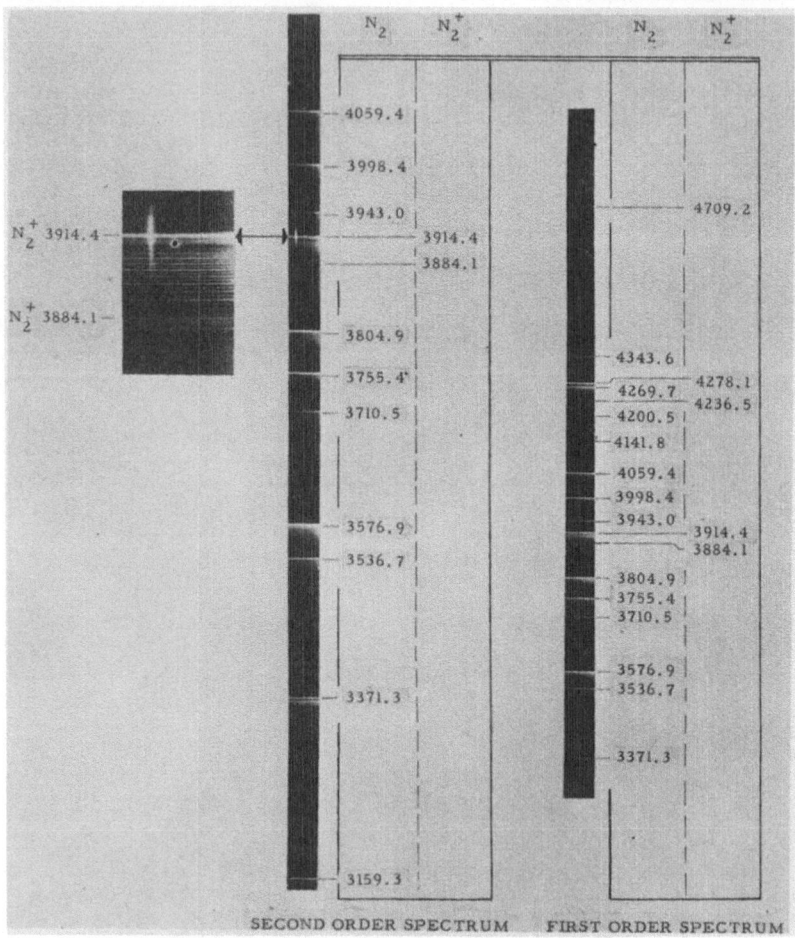

Fig. 32. Spectrogram of air fluorescent emission from a typical sea-level nuclear explosion obtained with the f/3.5 static spectrograph.

A/mm and 10.3 A/mm in the first- and second-order spectra, respectively. The first-order wavelength coverage is from 3317 to 8468 A with a wavelength resolution of 0.5 A. Figure 31 is a photograph of the spectrograph with the outer cover removed. Because of its excellent f/number, the spectrograph has found many applications where either light levels were low or exposures in the microsecond time region were required. Figure 32 is a reproduction of a sea-level nuclear explosion obtained in first- and second-order spectrum with the f/3.5 spectrograph. The spectrogram is an example of the instrument's capabilities when used for very short duration exposures. A discussion of the air fluorescent emission observed in this spectrogram is given in [24].

The $f/3.5$ spectrograph is now being used in the 2-megajoule magnetic compression experiment by the Plasma Physics Branch at NRL. The instrument has been modified, using an external rotating-mirror to provide two types of data. First, the image of the radiating plasma may be streaked along the slit of the spectrograph to produce the time-history of impurity line radiation. The impurity lines are due to foreign gases added to the deuterium discharge. Second, the plasma image may be streaked across the slit as previously described and thereby record space-resolved spectral "snapshots" of very short duration. These show the distribution of the radiating plasma at a particular time. The space-resolved spectro-grams will be used for probing the radial distribution of electron temperature at early stages of the discharge when the temperature is not too high. The temperature determina-tions wi'l be obtained by observing the relative intensities of spectrum lines of different ionization species.

## CONCLUSION

It can be seen that time-resolution spectroscopy has many uses in the study of rapidly radiating and varying sources. This review of the subject includes material presented in two previous papers with additional material included. The first paper was presented at the 5th International Congress on High-Speed Photography held in Washington, D. C., October 19, 1960, and subsequently published in the Proceedings of that Congress [25]. The second paper was presented on June 16, 1961, at the symposium on Spectroscopic Studies Using Time-Resolving Techniques held in Manchester, England. The symposium papers were abstracted in Nature [26] and in the British Journal of Applied Physics [27].

## ACKNOWLEDGMENTS

This paper would be incomplete without an acknowledgment of the significant advice and encouragement of Dr. Harold S. Stewart, who directed most of the projects under which the NRL instrumentation was developed.

I wish to take this opportunity to thank Dr. Herman Hoerlin of LASL for his many contributions on photographic techniques and theory which greatly benefited the whole spectroscopic project.

I wish to thank Dr. L. F. Drummeter of NRL for his encouragement and advice in the preparation of this paper, and for permission to present a few of the hitherto unpublished results of an extensive NRL program of spectroscopic observations on nuclear explosions.

Also, I wish to thank the Society of Motion Picture and Television Engineers (SMPTE) for their permission to include herein the essence of the earlier paper [25] concerning the spectrograph instrumentation and to reproduce Figs. 7-10, 20, 22, 26, and 27.

## REFERENCES

1. F. L. O. Wadsworth, Astrophys. J. 3, 54 (1896).
2. W. F. Meggers and K. Burns, Bureau of Standards Scientific Papers, No. 18, 185 (1922).
3. W. E. Buck, High-speed turbine-driven rotating mirrors, Rev. Sci. Instr. 25, 115 (1954).
4. B. Brixner, High-speed turbine-driven rotating mirrors: Notes on design, construction, and performance, Rev. Sci. Instr. 30, 1041 (1959).
5. B. Brixner, A high-speed rotating mirror frame camera, SMPTE 59, 502 (1952).
6. R. J. Reithel et al., The current pause in an exploding wire, in W. G. Chace and H. K. Moore (eds.): Exploding Wires, Vol. 1, Plenum Press, Inc., New York (1959), p. 26.
7. K. R. Coleman, The photography of high temperature events, in H. Schardin and O. Helwich (eds.): Kurzzeitphotographie, IV Internationaler Kongress, Köln, 1958, Verlag Dr. Othmar Helwich, Darmstadt (1959), pp. 32-39.
8. A. E. Huston, Some developments in rotating mirror cameras, Ibid., pp. 163-166.
9. R. W. Engstrom, J. Opt. Soc. Am. 37, 420 (1947).
10. E. A. McLean et al., Spectroscopic study of helium plasmas produced by magnetically driven shock waves, Physics of Fluids 3, 843 (1960).
11. E. A. McLean, A. C. Kolb, and H. R. Griem, Visible precursor radiation in an electromagnetic shock tube, Physics of Fluids 4, 1055 (1961).
12. B. R. Bronfin, E. A. McLean, and H. R. Griem, Absolute intensity calibration of a time-resolved spectrogram, J. Opt. Soc. Am. 52, 224 (1962).
13. E. A. McLean, The measurement of transition probabilities using an electromagnetic shock tube, Proceedings of the Fourth Symposium on Temperature, Its Measurement and Control in Science and Industry, Reinhold Publishing Company, New York (1962).
14. Image Intensifier Symposium, U. S. Army Engineer Research and Development Laboratories, Corps of Engineers, Oct. 6-7 (1958).
15. M. M. Butslov et al., Electron-optical method for studying short-duration phenomena, in H. Schardin and O. Helwich (eds.): Kurzzeitphotographie, IV Internationaler Kongress, Köln, 1958, Verlag Dr. Othmar Helwich, Darmstadt (1959), pp. 230-242.
16. D. P. C. Thackeray, Current developments in the production and assessment of high intensity discharges, Ibid., pp. 123-129.
17. D. J. Lovell, H. S. Stewart, and S. Rosin, J. Opt. Soc. Am. 44, 799 (1954).
18. T. C. Hall, Jr., and W. Blacet, Separation of absorption spectra of $NO_2$ and $N_2O_4$, J. Chem. Phys. 20, 1745 (1952).
19. F. D. Harrington, An $f/2.8$ low-dispersion time-resolving grating spectrograph, NRL Report 5576, Feb. 7 (1961).

20. G. G. Milne, T. E. Putnam, and W. Staundenmaier, Interim report: High-speed streak spectrograph designs for the visual and ultraviolet regions, The University of Rochester, July 31 (1956).
21. F. D. Harrington, An $f/6.6$ high-dispersion time-resolving grating spectrograph, NRL Report 5533, Sept. 23 (1960).
22. D. R. Bates and A. Damgaard, Phil. Trans. A242, 101 (1949).
23. F. D. Harrington, An $f/3.5$ medium-dispersion grating spectrograph, NRL Report 5446, March 22 (1960).
24. D. R. Westervelt, E. W. Bennett, and A. Skumanich, Air fluorescence excited by gamma rays and X-rays, LASL Report J-10-546, Aug. 18 (1959).
25. F. D. Harrington, High-speed time-resolved spectroscopic instruments, in J. S. Courtney-Pratt, SMPTE (ed.): Proceedings of the 5th International Congress on High-Speed Photography, New York (1962), pp. 277-282.
26. J. D. Craggs, Time-resolved spectroscopy, Nature 192, 4807, 1032 (1961).
27. Brit. J. Appl. Phys. 13, 3, 98 (1962).

# INTRODUCTORY REMARKS AT THE
# OPTICAL EMISSION PROBLEM CLINIC

## J. F. Woodruff
Armco Steel Corp.
Middletown, Ohio

Last year at this Conference, I presented a paper which described the accuracy of photoelectric spectrometers used in Armco's production control laboratories as compared to the classical chemical procedures formerly used in these same plants. To determine the accuracies of the two methods of analysis a study was made of the analyses reported in a monthly check analysis program. This program was initiated at Armco more than 25 years ago.

In each production laboratory, the works chemist in charge makes out a daily report and certifies the ladle analyses as determined by the "turn" analysts on all heats and cast produced for a 24-hr period. Each month the chemical research department reviews these daily reports and selects one day's analytical production for checking. The day selected is varied from month to month. A wire is sent to each production laboratory requesting ladle test samples of certain heats and casts produced on the selected day.

With this check analysis program we have a basis for determining whether or not the photoelectric spectrometers are giving reliable analyses to the melt shops.

## TABLE I
Comparison of Research and Production Control Carbon Values Using Data from Monthly Check Analysis Reports

| Range, % C | Standard Deviation from Research, ± % C | |
| | P. C. Combustion Method | P. C. Quantovac Method |
| --- | --- | --- |
| 0.010-0.049 | 0.0029 | 0.0027 |
| 0.050-0.099 | 0.0038 | 0.0024 |
| 0.10 -0.49 | 0.0072 | 0.0057 |
| 0.50 -0.99 | 0.0137 | 0.0086 |

## TABLE II

Comparison of Research and Production Control Sulfur Values Using Data from Monthly Check Analysis Reports

| Range, % S | Standard Deviation from Research, ± % C | |
| --- | --- | --- |
| | P. C. Combustion Method | P. C. Quantovac Method |
| 0.004-0.019 | 0.0016 | 0.0027 |
| 0.020-0.029 | 0.0020 | 0.0020 |
| 0.030-0.099 | 0.0023 | 0.0020 |

## TABLE III

Comparison of Research and Production Control Phosphorus Values Using Data from Monthly Check Analysis Reports

| Range, % P | Standard Deviation from Research, ± % C | |
| --- | --- | --- |
| | P. C. Chemical Method | P. C. Quantovac Method |
| 0.003-0.019 | 0.0018 | 0.0022 |
| 0.020-0.049 | 0.0036 | 0.0046 |

Using the data from the check analysis reports, a similar study has been made of the results obtained with vacuum photoelectric spectrometers, namely, the ARL Quantovacs. We now have five of these instruments in use in our control laboratories. I felt that you might be interested in the accuracy of the carbon, sulfur, and phosphorus results by these units. Tables I-III show comparisons between the production control combustion and chemical methods and the Quantovac methods.

With the exception of sulfur values below 0.019% and phosphorus values from 0.020% to 0.049%, the Quantovac results are equal or better than the former chemical methods.

# QUANTOVAC ANALYSIS OF COPPER AND COPPER-BASE ALLOYS

## H. T. Dryer and B. R. Boyd

Applied Research Laboratories, Inc.
Dearborn, Michigan

The Quantovac, introduced by Applied Research Laboratories in this country in 1957, was developed specifically for the complete analysis of iron and steel. Rapid, accurate, and simultaneous analyses were needed for carbon, phosphorus, and sulfur in addition to the usual alloying and residual elements, which had been handled successfully by the ARL Production Control Quantometer (PCQ) for many years. Most of the sensitive lines for the metalloids are in the vacuum ultraviolet region and require that the instrument for this application be usable in both the vacuum ultraviolet and ultraviolet regions. Methods evaluation with the Quantovac for the analysis of iron and steel indicated that this instrument offered a number of advantages over conventional direct-reading instruments and presented great possibilities for other metals and alloys, including nonferrous metals.

The Quantovac (Fig. 1) is a vacuum Quantometer consisting of a vacuum spectrometer with an enclosed excitation stand, a Multisource excitation unit and a recording console. A cutaway view of the vacuum spectrometer (Fig. 2) shows the enclosed spark stand, the location of the primary slit, grating, secondary slits, secondary mirrors, and two photomultiplier tube sections. The excitation stand is flushed with argon to insure transmission of the vacuum ultraviolet wavelengths and to provide a highly reproducible and high intensity discharge. The spectrometer is evacuated—including the section from the primary slit to the quartz windows or photomultiplier tubes. The EMI tubes, used for the shortest wavelengths, are end-window tubes and provide their own vacuum seal, while quartz

The analytical data presented in this paper represent the results from independent investigations at two ARL Application Laboratories in Lausanne, Switzerland, and Dearborn, Michigan.

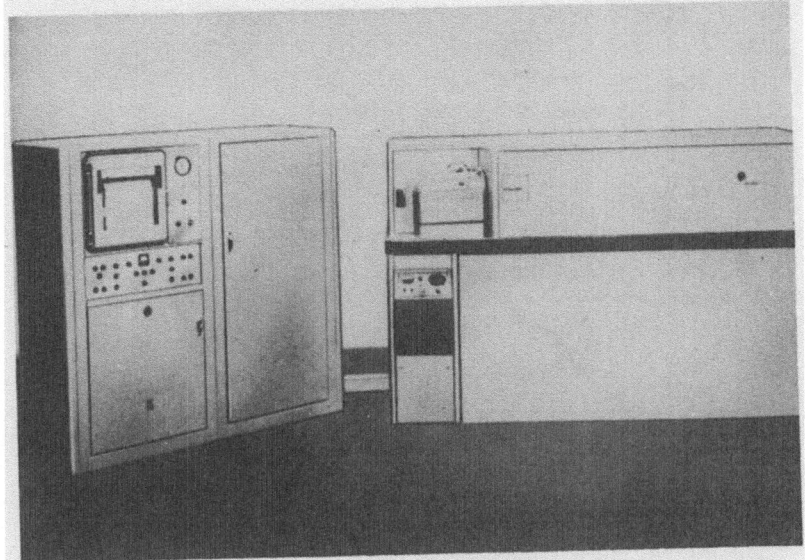

Fig. 1. The Quantovac.

windows are used for vacuum seals on the RCA or ultraviolet portion.

The investigations on the Quantovac soon established that the critically damped Multisource discharge case provided the

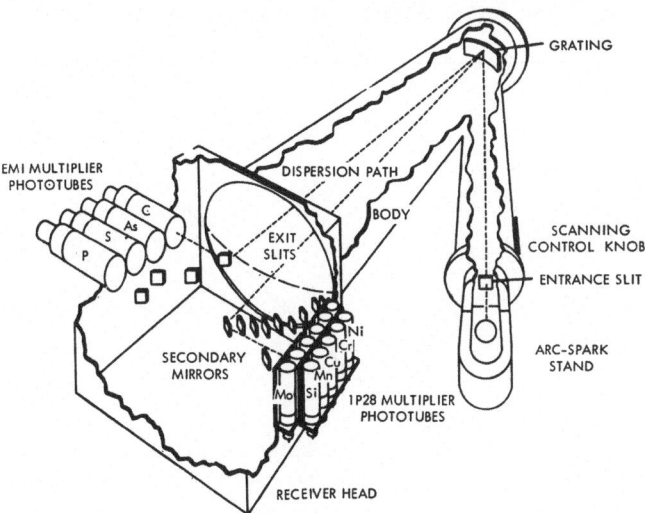

Fig. 2. Schematic drawing of the Quantovac direct-reading spectrometer.

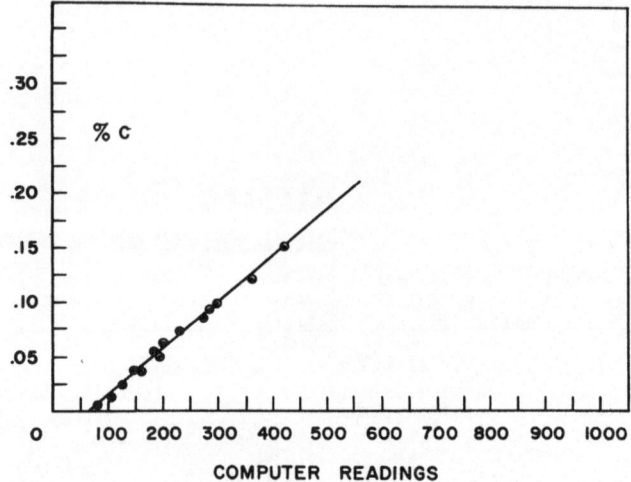

Fig. 3. Quantovac analysis of C in 18-8 type stainless steel.

highest intensity and the best line-to-background ratio and, consequently, extremely good precision and sensitivity. This discharge is used with the sample negative, a silver counter electrode, and an argon flush.

Quantovac (Qvac) performance data for the analysis of low-alloy steels, cast irons, stainless steels, and tool steels have been evaluated for a large variety of samples from a number

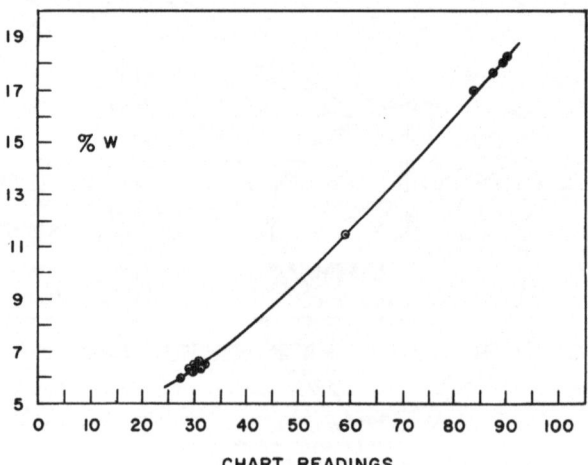

Fig. 4. Quantovac analysis of W in tool steel.

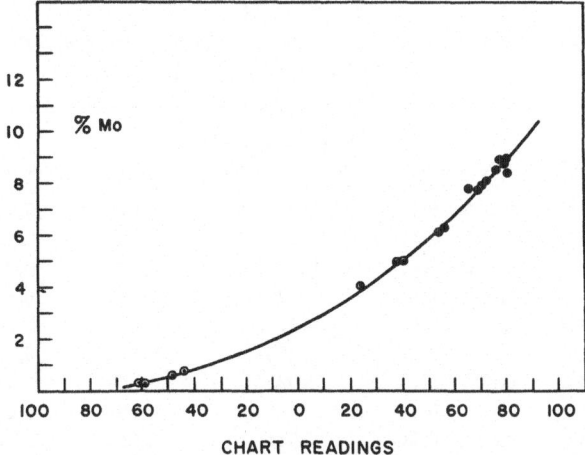

Fig. 5. Quantovac analysis of Mo in tool steel.

of laboratories. Typical curves for these analyses are shown in Figs. 3 to 5. Several of these curves serve to demonstrate the superiority of the Quantovac over conventional direct-reading instruments, namely, tungsten in tool steels and molybdenum in tool steels. A summary of the accuracy and precision data for Cr-stainless is presented in Table I, for 18-8 type stainless steel in Table II, and for tool steels in Table III. A comparison of data obtained with a conventional

TABLE I

Precision and Accuracy Data for Qvac Analysis of Cr-Stainless Steel

| Element | Conc. Range | Av. Dev. from Chem. Anal.* | Std. Dev. | At Conc. |
|---|---|---|---|---|
| C | 0.008- 0.30% | 0.003% | ± 0.003 % | 0.114% |
| P | 0.008- 0.11 | 0.0013 | 0.0007 | 0.018 |
| S | 0.12 - 0.55 | 0.005 | 0.0067 | 0.317 |
| Ni | 0.05 - 0.50 | 0.012 | 0.007 | 0.45 |
| Cr | 11.5 -14.5 | 0.073 | 0.029 | 12.97 |
| Si | 0.10 - 0.90 | 0.026 | 0.013 | 0.67 |
| Mn | 0.20 - 0.70 | 0.020 | 0.008 | 0.40 |
| Cu | 0.06 - 0.30 | 0.005 | 0.0025 | 0.12 |
| Mo | 0.01 - 0.60 | 0.01 | 0.008 | 0.13 |

*Duplicate analyses.

TABLE II

Precision and Accuracy Data for Qvac Analysis of 18-8
Type Stainless Steel

| Element | Conc. Range | Av. Dev. from Chem. Anal.* | Std. Dev. | At Conc. |
|---------|-------------|----------------------------|-----------|----------|
| C | 0.008- 0.15% | 0.0026% | ±0.0036% | 0.100% |
| P | 0.008- 0.15 | 0.002 | 0.001 | 0.018 |
| S | 0.005- 0.35 | 0.0022 | 0.008 | 0.205 |
| Ni | 7.5 -11.5 | 0.073 | 0.089 | 9.86 |
| Cr | 17.0 -22.0 | 0.097 | 0.042 | 17.50 |
| Si | 0.30 - 0.90 | 0.025 | 0.012 | 0.61 |
| Mn | 0.50 - 2.0 | 0.023 | 0.009 | 0.79 |
| Cu | 0.04 - 0.30 | 0.006 | 0.009 | 0.24 |
| Mo | 0.01 - 0.40 | 0.014 | 0.012 | 0.30 |

*Duplicate analyses.

unit (PCQ) to data obtained with the Quantovac is presented in
Table IV.

The nonferrous copper base metals were investigated for
possible application. A number of elements are extremely
difficult to determine by the usual direct-reading or spectrographic methods; large variations in time are experienced,
depending on impurities and levels. The determination of

TABLE III

Precision and Accuracy Data for Qvac Analysis
of Tool Steels

| Element | Conc. Range | Av. Dev. | Std. Dev. | At Conc. |
|---------|-------------|----------|-----------|----------|
| C | 0.1 - 1.5% | 0.015% | 0.009% | 1.16% |
| P | 0.01 - 0.05 | 0.002 | 0.0015 | 0.019 |
| S | 0.005- 0.14 | 0.001 | 0.0006 | 0.013 |
| M | 0.1 - 0.5 | 0.017 | 0.003 | 0.26 |
| Si | 0.1 - 0.5 | 0.015 | 0.008 | 0.31 |
| W | 0.5 - 5.0 | 0.063 | 0.042 | 1.50 |
| W | 5.0 -19.0 | 0.13 | 0.15 | 6.38 |
| Mo | 0.1 - 2.0 | 0.040 | 0.009 | 0.29 |
| Mo | 2.0 -10.0 | 0.087 | 0.025 | 6.20 |
| V | 0.5 - 4.5 | 0.060 | 0.012 | 2.87 |
| N | 0.05 - 0.4 | 0.018 | 0.002 | 0.16 |
| C | 2.5 - 5.0 | 0.085 | 0.018 | 4.00 |
| Cu | 0.05 - 0.25 | 0.010 | 0.005 | 0.12 |

## TABLE IV
### Comparison of Accuracy of PCQ and Qvac Analysis
### of Carbon and Low Alloy Steels

| Element | Conc. Range | No. of Samples | Average Deviation from Chemical Analysis | |
| :---: | :---: | :---: | :---: | :---: |
| | | | Qvac | PCQ |
| C | 0.04 -0.09 % | 12 | 0.004 % | |
| C | 0.01 -0.21 % | 11 | 0.008 % | 0.010% |
| C | 0.20 -1.50 % | 19 | 0.024 % | |
| S | 0.015-0.055% | 20 | 0.0021% | |
| P | 0.010-0.050% | 24 | 0.0011% | |
| P | 0.050-0.10 % | 6 | 0.0027% | 0.016% |
| Mn | 0.10 -0.50 % | 31 | 0.012 % | 0.022% |
| Si | 0.03 -0.60 % | 19 | 0.012 % | 0.037% |
| Ni | 0.02 -0.20 % | 12 | 0.008 % | 0.020% |
| Cr | 0.02 -0.20 % | 4 | 0.0075% | 0.005% |
| Cu | 0.02 -0.20 % | 12 | 0.006 % | 0.018% |
| Al | 0.04 -0.50 % | 4 | 0.005 % | |
| Mo | 0.03 -0.33 % | 8 | 0.012 % | |

phosphorus in copper and copper-base alloys could not be performed with the speed and accuracy currently required. Improved speed, accuracy, and uniformity of method were needed for the analysis of pure copper.

The analysis of pure copper samples presents a special problem for the conventional spectrochemical laboratory. Because of the tremendous changes in electrical and thermal conductivity characteristics with impurities and concentrations, the excitation and control of pure copper samples will usually exhibit rather unique phenomena. For example, using a direct-reading instrument and integrating to a constant amount of radiation from a copper internal standard line, one sample of pure copper may integrate in 20 seconds while another pure copper sample may integrate in 60 seconds.

Although a number of discharges were studied for pure copper using the Multisource unit, optimum conditions were identical to those used for iron and steel samples, i.e., the critically damped Multisource discharge, a silver counter electrode, sample negative, and an argon flush.

The copper samples were ground prior to excitation on a 60X alundum disc grinder and excited by the standard Quantovac

TABLE V
Program for Qvac Analysis of Cu-Base Alloys

| Element | Conc. Range | Element | Conc. Range |
|---------|-------------|---------|-------------|
| Zn | 0.0005-0.020% | S | 0.001-0.014% |
| Pb | 0.002 -0.22 % | As | 0.001-0.012% |
| Sb | 0.025 -0.35 % | Bi | 0.001-0.010% |
| Sn | 0.002 -0.020% | | |
| Ag | 0.0002-0.020% | Al | 0.02 -3.3 % |
| Mn | 0.001 -0.020% | Zn | 0.05 -1.0 % |
| Cd | 0.0004-0.024% | Pb | 0.5 -3.5 % |
| P. | 0.0005-0.07 % | Sn | 0.1 -12 % |
| Se | 0.0001-0.01 % | P | 0.005-0.40 % |
| Fe | 0.002 -0.015% | Fe | 0.002-0.22 % |
| Ni | 0.0002-0.020% | As | 0.035-0.50 % |

discharge. The analytical cycle includes a 10-second flush, 10-second prespark, and a 20-second integration controlled by the internal standard line. The first phenomenon which was observed was the reproducibility of the integration time between samples as opposed to the large variations experienced with the air discharge. In addition to eliminating the variations

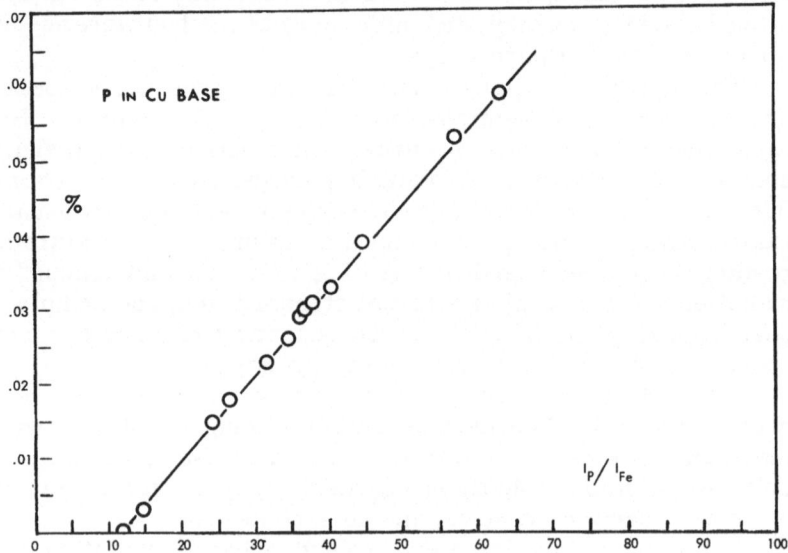

Fig. 6. ARL (Qvac) analysis of P in Cu.

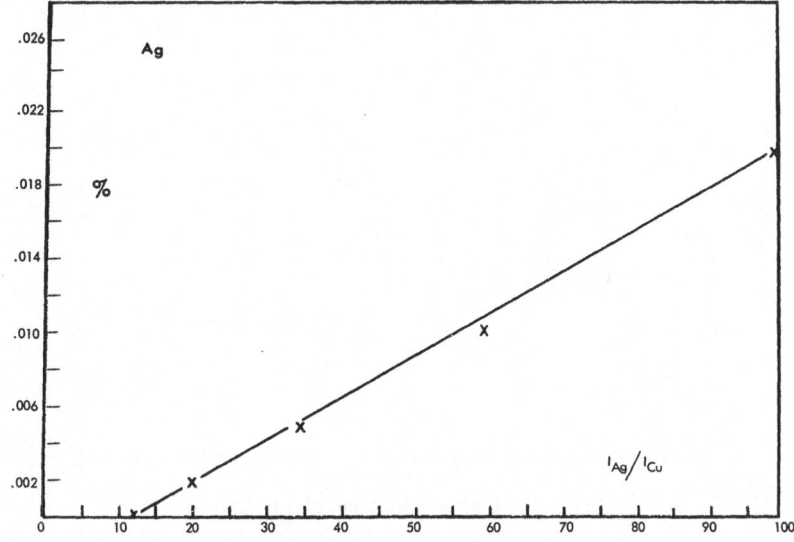

Fig. 7. ARL (Qvac) analysis of Ag in Cu.

in time between samples, the reproducibility of the analyses was far superior to that obtained by other methods.

The analytical program for this study is shown in Table V. The high and low limits of the concentration ranges are not

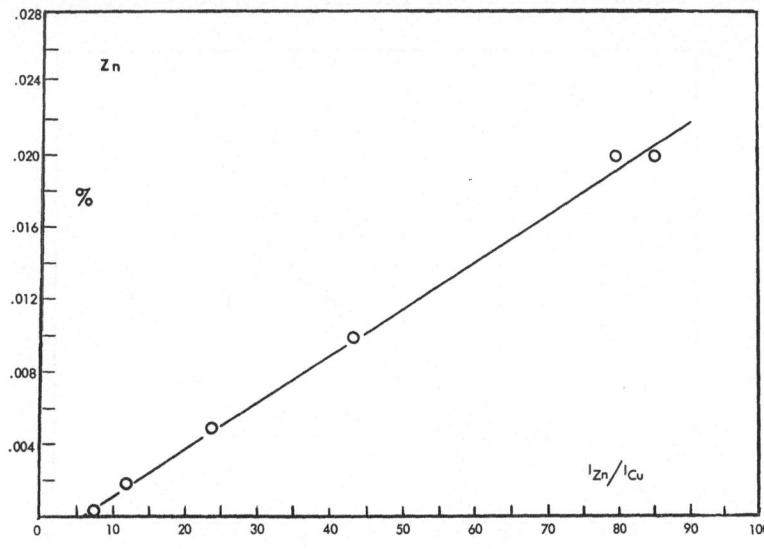

Fig. 8. ARL (Qvac) analysis of Zn in Cu.

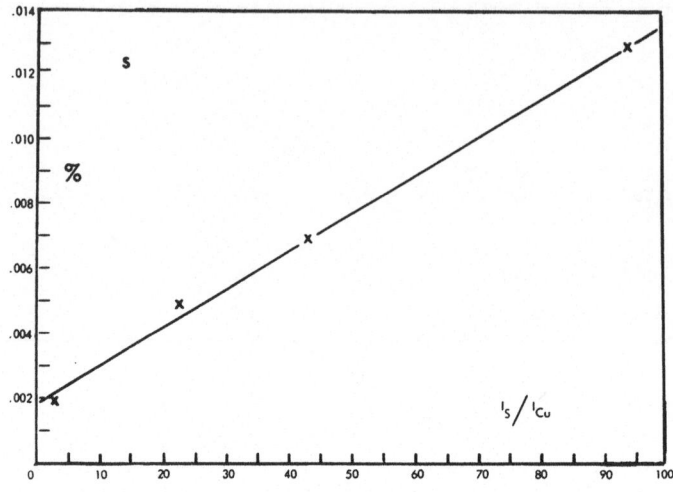

Fig. 9. ARL (Qvac) analysis of S in Cu.

imposed by the analytical method but are the ranges covered by the copper and copper base standards.

The analytical curves for the various elements in "pure" copper are shown in Figs. 6-10. As shown in Fig. 6, the sensitivity for phosphorus is excellent (0.0005%) and the speed of analysis is about 1.5 minutes for duplicate analyses not including sample preparation. The analytical curve for silver is

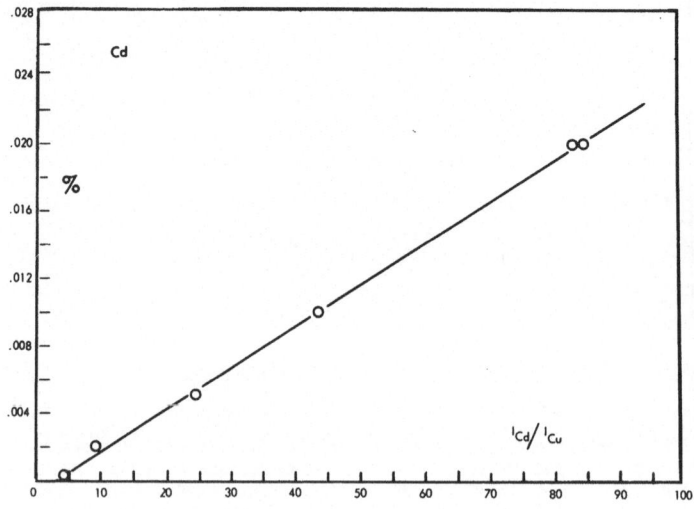

Fig. 10. ARL (Qvac) analysis of Cd in Cu.

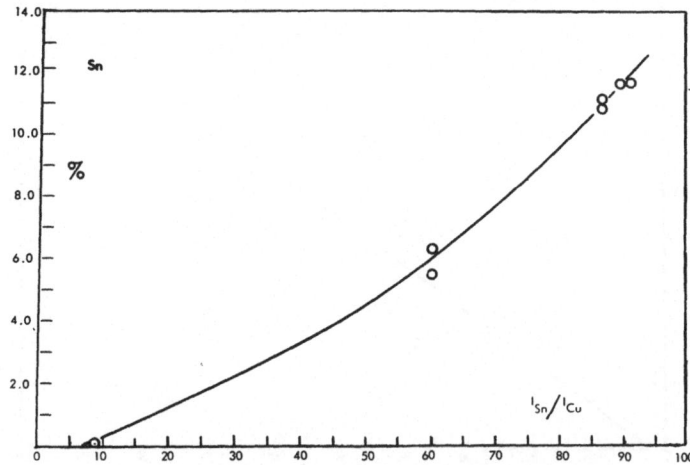

Fig. 11. ARL (Qvac) analysis of Sn in Cu–base alloy.

of special interest (Fig. 7). The silver used for the counter electrode contributes little if any radiation to the analytical discharge. High purity copper rods have been substituted for silver as the counter electrode with no change in integration times or changes in analytical curves.

Additional studies were made on the analysis for several of the alloying elements in copper-base alloys. The analytical data was excellent for the elements studied and indicated that no difficulties would be experienced providing the alloying

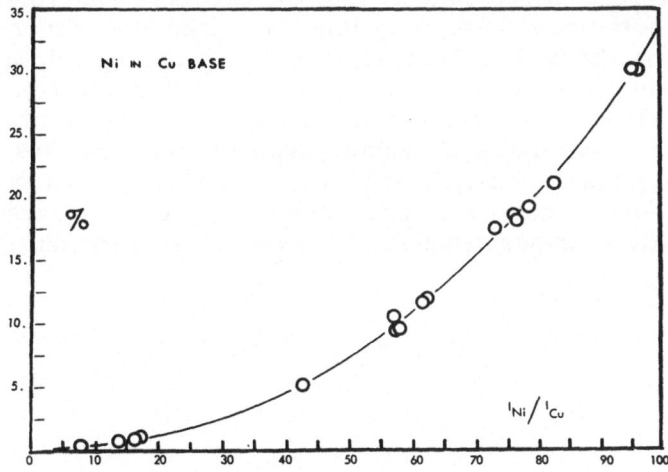

Fig. 12. ARL (Qvac) analysis of Ni in Cu–base alloy.

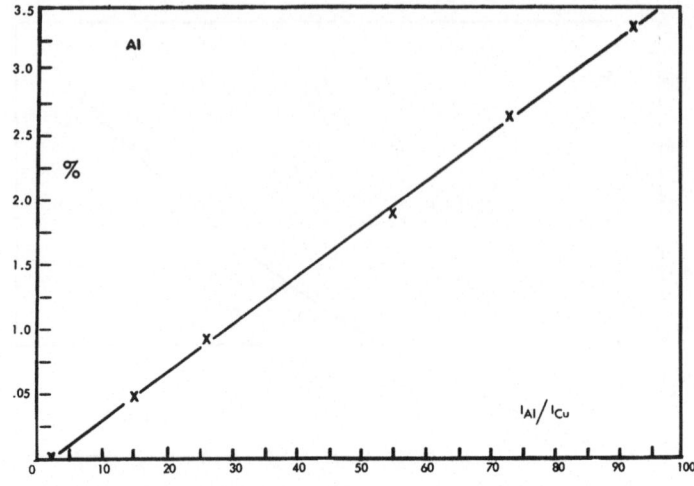

Fig. 13. ARL (Qvac) analysis of Al in Cu-base alloy.

constituents do not exhibit macrosegregation. The analyses of
the copper alloys were limited to some extent by the available
standards and concentration ranges covered by these standards.
The analytical curves for these studies are shown in Figs.
11-13.

## SUMMARY

The analytical data from the analyses of the copper and
copper-base alloys demonstrate the excellent performance of
the Quantovac on these materials. The difficulties experienced
with conventional instruments have been eliminated, and the
accuracy and precision data are excellent. The performance of
this method on both trace and alloy analyses makes it attractive
for the convenience of handling both types of analyses with a
single method. Analytical data on both alloyed steels and
copper-base materials have been shown to demonstrate the
versatility and performance of the Quantovac methods.

# USE OF A VACUUM CUP ELECTRODE FOR SPARK EMISSION SPECTROGRAPHIC ANALYSIS OF PLANT TISSUE*

## S. T. Bass and J. Soulati
Department of Biochemistry
Michigan State University
East Lansing, Michigan

Emission spectrographers have used many different methods to get a homogeneous sample into the analytical gap [1-9]. Many other works could be cited, but these should suffice to show the great number of techniques that are or have been used for this purpose.

Recognizing that some of these techniques might be adapted to the analysis of plant tissue, we conducted an investigation to select one that could be used successfully with a method previously described for analyzing plant materials [10]. The systems investigated were the porous cup electrode, the spray-in spark technique, and the vacuum cup electrode [2].

In preliminary work, poor reproducibility of results with the porous cup electrode caused it to be discarded and the lack of sensitivity caused the elimination of the spray-in technique. The vacuum cup electrode, however, showed promise of functioning successfully.

The circuit parameters used with the vacuum cup electrodes were investigated because of the pumping effect on the sample caused by the number of interruptions per half cycle of current. It would appear from Zink [2] that when more breaks occur in the current, more sample will be pumped into the analytical gap. The best circuit parameters may not be the same as were shown for salt-capped electrodes [11].

Unfortunately, these electrodes are very expensive and their use would be prohibitive if each spectrum required a new electrode. An investigation of multiexcitation on such

*Published with the approval of the Director of the Michigan Agricultural Experiment Station as Journal Article No. 2885. This paper was published in the Journal of the Association of Official Agriculture

## TABLE I
Summary of Equipment and Operating Conditions Used

| | |
|---|---|
| Spectrograph | Large glass-quartz Littrow prism |
| slit width | 20 μ |
| electrodes | |
| sample | United Carbon Co. No. 6010, 6011 |
| counter | 3/16 in. regular grade graphite, 120° cone |
| A.O.A.C. [14] | 3/16 in. regular grade graphite, flat top |
| Excitation unit | National Spectrographic Lab. air interrupted—spark |
| Analytical gap | 2 mm |
| Auxiliary gap | 2-4 mm |
| Spark primary voltage | 240 v |
| Secondary RF amps | varied |
| Inductance | 50-500 μh |
| Capacitance | 0.01 μf |
| Resistance ohms | residual |
| Breaks per half cycle | varied |
| Preburn | 15 sec |
| Exposure | 60 sec |
| Film | Kodak S. A. 1 |
| Processor | N. S. L. plate processor |
| Microphotometer | Jarrell-Ash Recording |
| Calculating board | Jarrell-Ash Siedel |

electrodes to reduce the cost of electrodes is in press [12]; this investigation demonstrated that multiexcitation on the same electrode was possible, and that multiple samples could be analyzed with the same electrode if the same elements were present in similar concentration ranges.

The purpose of this paper is to present a comparison of the results obtained with the vacuum cup electrode and those obtained with the A.O.A.C. systems [10, 13, 14].

## APPARATUS AND MATERIALS

The operating conditions and the nature of the equipment used in this experiment are presented in Table I. The samples had known past histories, since all of the fruit tree leaves had been analyzed by a number of investigators and the synthetic samples were similar to the A.O.A.C. standards [13], except that a controlled amount of potassium [10] was added to each standard.

## TABLE II

Effect of Varying Inductance on Precision of Intensity Ratios (expressed as % standard deviation of the mean)

| Inductance* | B | P | Fe | Mg | Mn | Ca | Cu |
|---|---|---|---|---|---|---|---|
| 500 μh | 9.96 | 8.21 | 7.18 | 6.87 | 12.03 | 3.20 | 7.95 |
| 300 μh | 16.21 | 14.27 | 5.72 | 11.53 | 13.12 | 4.65 | 14.45 |
| 50 μh | 8.56 | 8.83 | 7.39 | 7.73 | 8.10 | 6.28 | 6.73 |

*Eight exposures, one electrode for each inductance.

## PROCEDURE

The ash from 1 g of plant tissue (for A.O.A.C. standards the amount was measured and evaporated to dryness) was dissolved in 4 ml of 1:1 HCl and 1 ml of internal standard ($CoCl_2 \cdot 6H_2O$, 25 g/liter). In most of these investigations more sample was required, so a 5 or 10 g sample was used with the above ratios of acids and internal standard.

The source conditions of the flat top electrode system [14] were described by Connor and Bass [11]. The known parameters were varied with the vacuum cup electrode in two ways: by changing the inductance, and by varying the secondary air gap. The source inductances of 50, 300, and 500 μh were tested for precision and sensitivity. The secondary air gap was examined for its effect on the precision of results with the vacuum cup electrode.

An additional experiment was conducted to determine whether any one level of potassium was superior to any other, since previous work had showed that potassium concentration affected the results obtained in the analysis of plant tissue.

After the best conditions were found, a comparison was

## TABLE III

Effect of Inductance Changes on Precision of Intensity Ratio of the Element Manganese

| Inductance* | 50 μh | 300 μh | 500 μh |
|---|---|---|---|
| Intensity ratio | 1.705 | 1.28 | 1.33 |
| % Stand. dev. of the mean | 6.04 | 7.42 | 6.75 |

*Twenty exposures per inductance.

## TABLE IV
Effect of Varying Auxiliary Gaps on the Precision of Intensity Ratios (expressed as % standard deviation of the mean)

| Auxiliary gap* | B | P | Fe | Mg | Mn | Ca | Cu |
|---|---|---|---|---|---|---|---|
| 4 mm | 6.49 | 11.27 | 6.84 | 6.92 | 3.76 | 4.97 | 3.66 |
| 2 mm | 9.81 | 11.69 | 6.30 | 13.81 | 7.90 | 5.43 | 5.97 |

*Two electrodes, nine exposures per electrode per width.

## TABLE V
Effect of Varying Potassium Levels on Precision of Intensity Ratios (expressed as % standard deviation of the mean)

| Percent K* | B | P | Fe | Mg | Mn | Ca | Cu |
|---|---|---|---|---|---|---|---|
| 0 | 10.92 | 8.77 | 7.87 | 7.86 | 14.81 | 4.28 | 4.17 |
| 2 | 9.90 | 10.84 | 11.44 | 8.66 | 7.40 | 5.06 | 7.63 |
| 4 | 4.69 | 7.78 | 5.50 | 6.49 | 8.17 | 4.44 | 7.04 |
| 6 | 10.97 | 15.46 | 7.35 | 15.85 | 5.56 | 11.08 | 12.77 |
| 8 | 12.42 | 19.02 | 6.73 | 11.20 | 7.15 | 2.91 | 8.90 |

*Nine exposures per electrode per level.

## TABLE VI
Comparison of Results with Two Types of Vacuum Cup Electrodes with Those Obtained with Flat Top Graphite Electrodes (expressed as % standard deviation of the mean of intensity ratios)

| Electrode | B | P | Fe | Mg | Mn | Ca | Cu |
|---|---|---|---|---|---|---|---|
| Flat top* | 4.47 | 5.77 | 2.09 | 4.58 | 4.21 | 3.08 | 6.24 |
| U.C.P. No. 6010[†] | 6.80 | 5.47 | 5.76 | 5.01 | 4.74 | 3.99 | 4.09 |
| U.C.P. No. 6011[†] | 4.65 | 6.38 | 4.47 | 4.12 | 7.17 | 2.69 | 3.80 |

*Twenty pairs of electrodes used.
†Two electrodes, ten exposures per electrode.

TABLE VII

Comparison of Accuracy and Precision of Several Analytical Methods for Determining the Concentration of Several Constituents of Peach Tree Leaves

| Method | %<br>Ca | %<br>Mg | %<br>P | ppm<br>B | ppm<br>Cu | ppm<br>Fe | ppm<br>Mn |
|---|---|---|---|---|---|---|---|
| Kenworthy et al. [15] | 1.98 ±.17‡ | 0.53 ±.047 | 0.218 ±.023 | 37.5 ±12.1 | 21.0 ± 5.69 | 230 ±44.9 | 80.0 ±19.40 |
| Kenworthy [9] | 1.93 ±.10 | 0.47 ±.030 | 0.195 ±.014 | 40.5 ± 4.8 | 17.3 ±1.10 | 257 ±56.9 | 71.0 ± 3.90 |
| U.C.P. No. 6010* | 2.53 ±.22 | 0.57 ±.078 | 0.218 ±.022 | 46.6 ± 4.0 | 21.1 ±1.74 | 228 ±20.2 | 82.6 ± 7.03 |
| Flat top graphite† | 1.66 ±.15 | 0.60 ±.064 | 0.164 ±.010 | Not det. | 20.6 ±1.02 | 268 ±34.9 | 90.8 ± 5.67 |

*Two electrodes, seven exposures on one and eight on the other.
†Eighteen sets of electrodes.
‡Standard deviation of the mean.

made between two different vacuum cup electrodes and the flat top graphite electrode.

## RESULTS

The effect of varying inductance on precision of intensity ratios is shown in Tables II and III. Table II shows these effects on results for a group of elements, while Table III shows the effect on a sample with a single element.

Table IV illustrates the effects of two widths of auxiliary air gap on the precision of intensity ratios of the vacuum cup electrodes.

The effect on precision caused by changing the potassium content is shown in Table V.

In Table VI a comparison of two types of vacuum cup electrodes and flat top electrodes (14) is made.

The results from three different analytical procedures are compared with vacuum cup electrode procedure in Table VII. Peach tree leaves which were used in this experiment, have been used for some time as a standard for fruit tree leaves [9, 15].

## DISCUSSION

The vacuum cup electrode system appears to give the spectrographer an extra technique for introducing the sample into the analytical gap. Some of the difficulties that might be encountered with these electrodes are discussed by Bass and Soulati [12], but the most objectionable problems can be eliminated to make the system operational in most laboratories.

In this report an effort is made to establish the best source unit parameters. The one basic difference in source parameters between the vacuum cup system and parameters of Connor and Bass [11] is that the new system causes the solution to be pumped into the analytical gap. Therefore, the authors attempted to adjust the number of interruptions for each half cycle of the current.

The best ways to change the number of interruptions of current per half cycle of electricity are to increase the inductance of the source or to decrease the width of the auxiliary gap of the source unit. An effective change in the number of impulses per half cycle of electricity was obtained, with either method, but both resulted in a decrease in total energy. In many instances, this decrease in energy cannot be tolerated. The effects on precision of making these changes (Tables II-IV) indicate that neither of these changes offers any advantage over the conditions set forth by Connor and Bass [11], which are $50\,\mu h$ and 4 mm auxiliary gap in these experiments.

Bass and Soulati [10] reported that the potassium concentration affects spectrographic analytical results. In our experiments, potassium levels of 0, 2, 4, 6, and 8% were run with the new type electrodes; the results showed that high levels of potassium decrease the precision of intensity ratios. This is believed to be caused by the salting-out of compounds at these high concentrations. The precision with 4% potassium was best, that with 2% was next best, and third best was in the absence of potassium. Since biological materials contain the element potassium, the zero potassium level has no place in an analytical procedure for samples of this type.

A comparison was made of results with two types of vacuum cup electrodes and with the flat polished top electrodes. The data in Table VI indicate that the precision of each system is good; it would be difficult to find any one superior to the other. However, if the preparation time is considered, the use of the vacuum cup electrode saves a great deal of time, compared to the time required to put solutions on the flat top electrodes, the time to dry the solution, and the time required to make the many flat top electrodes. Vacuum cup electrodes can be purchased preformed for approximately the same price per sample as other electrodes, which reduces the time required in preparation.

In comparison with other methods of analysis (Table VII), the vacuum cup gave good results with all elements except calcium. The reason for this is not apparent, since all other data for vacuum cup systems show excellent precision for the element calcium.

Another solution method that should be mentioned is that of Kenworthy [9]. He used the rotating disk with a photoelectric spectrometer and obtained excellent results. In Table VII his results are compared with those of the authors and those from several other laboratories that originally standardized these samples [15].

This investigation indicates that other methods can be used to introduce a sample into the analytical gap. This experiment was carried out using the solution method of preparation used in the first action method A.O.A.C. [13] with certain modifications [10].

The data reveal that the vacuum cup electrode may be an added tool with which to analyze plant tissue. The use of this electrode should allow the operator to increase over-all output without increasing the cost per sample.

REFERENCES

1. B. L. Vallee and Ruth W. Peattie, Anal. Chem. 24, 434 (1952).
2. T. H. Zink, Appl. Spectroscopy 13, 94 (1959).
3. A. Strasheim and E. J. Tappere, Ibid 13, 12 (1959).
4. E. F. Runge and F. R. Bryan, Ibid 13, 74 (1959).
5. M. P. Brash, Ibid 14, 43 (1960).
6. A. Strasheim and D. J. Eve, Ibid 14, 97 (1960).
7. W. K. Baer and E. S. Hodges, Ibid 14, 141 (1960).
8. D. M. Shaw, O. Wichremashinghe, and C. Yip, Spectrochim. Acta 13, 197 (1958).
9. A. L. Kenworthy, Proc. Thirty-Sixth Annual Meeting Council on Fertilizer Application, pp. 39-50 (1960).
10. S. T. Bass and H. Soulati, J. Ass. Offic. Agr. Chemists 44, 517 (1961).
11. Jane Connor and S. T. Bass, Ibid 42, 382 (1959).
12. S. T. Bass and J. Soulati, Paper No. 178, Pittsburgh Conference on Analytical Chemistry and Applied Spectroscopy (1962).
13. Official Methods of Analysis, 9th ed., The Association of Official Agricultural Chemists, Washington, D. C. (1960), Sec. 41.008.
14. S. T. Bass and Jane Connor, J. Ass. Offic. Agr. Chemists 43, 113 (1960).
15. A. L. Kenworthy, E. J. Miller, and W. T. Mathis, Amer. Soc. Hort. Sci. 67, 16 (1956).

# ANALYSIS OF HIGH-PURITY CHROMIUM*

## R. E. Heffelfinger, E. R. Blosser, O. E. Perkins, and W. M. Henry

Analytical Spectroscopy Division
Battelle Memorial Institute
Columbus, Ohio

## ABSTRACT

A combination chemical-spectrographic method for determining metallic impurities in high-purity chromium is described. Chromium is removed from the metallic impurities by volatilization as chromyl chloride. The remaining solution is examined spectrographically for elements such as iron, nickel, aluminum, manganese, titanium, vanadium, magnesium, and copper in the 0.1- to 30-ppm range. Direct arcing of chromic sulfate is used for the determination of silicon.

## INTRODUCTION

Chromium, a metal used for modern high-temperature applications, has good oxidation resistance, cold workability, high-temperature toughness, and availability. Brittleness, one of its less desirable properties, appears to be related to the impurities present.

When the importance of the purity of chromium was first learned, the only satisfactory method for the determination of metallic impurities was the spectrographic analysis by DC arc excitation of chromic sulfate, which permitted detections in the order of 10 to 50 ppm. However, it soon became necessary to improve both the detectability and the accuracy of the determination of the impurity elements. To accomplish this, we selected a chemical separation step—volatilization of chromium as the chromyl chloride—prior to spectrographic analysis. The impurities were thus concentrated from a large amount of sample into a small volume of solution free of the matrix material. Then the highly reproducible spark excitation could

*Published in Anal. Chem. 34, 621 (1962).

be used with solution standards to determine these concentrated impurities. Others have added the separated impurity elements to a powder matrix such as calcium carbonate to handle the very small amount of impurity and then completed the analysis by DC arc spectrographic technique [2]. The solution spark offers greater precision and accuracy because of the inherently higher precision of spark analysis and because of less handling of the impurities.

## PROCEDURE

Dissolve 5.1-5.5 g of chromium in 80 ml of 6N hydrochloric acid and dilute in a 100-ml graduated cylinder to give a chromium concentration of 0.1 g per ml.

The excess above 50 ml, i.e., 1-5 ml, may then be tapped off and put into a crucible with 2 ml of concentrated sulfuric acid, dried, and ignited at 800°C to obtain chromic sulfate, which is analyzed by ordinary DC arc techniques to determine silicon and other impurities present in detectable amounts.

To the 50 ml of chromium solution which contains 5 g of chromium, add 40 ml of 60% perchloric acid and 0.15 mg of cobalt (the cobalt is added as a 5-ml tap from 3 mg of pure cobalt metal dissolved in 5 ml of concentrated $HNO_3$ and diluted to 100 ml). Heat the solution gently to fumes of perchloric and until the chromium is oxidized to the sexivalent oxidation state (deep red color). Continue heating the solution strongly and add concentrated hydrochloric acid, specific gravity 1.19, a few milliliters at a time, until the chromium is evolved. This will require about 40 to 60 ml of hydrochloric acid. Fume off any perchloric acid remaining.

Remaining in the beaker from which the chromium was removed are the impurities and a trace of chromium. To the cooled beaker, add 1 ml each of concentrated sulfuric acid, hydrochloric acid, and nitric acid, and allow to stand for a few hours or overnight to ensure that the impurities dissolve completely. Dilute the acid solution with pure water to 5 ml. The sample is now ready for spectrographic examination. The samples are compared with standards by a solution-spark technique [1]. The spectrographic conditions are shown in Table I.

## STANDARDS

Typical compositions for standards for impurity elements found in chromium are shown in Table II. The standards are

## TABLE I
### Spectrographic Conditions*

| Electrical Parameters | |
| --- | --- |
| Discharge voltage | 15,000 |
| Capacitance | 0.007 $\mu f$ |
| Inductance | 50 $\mu h$ |
| Resistance | Residual |
| Discharges per second | 240 |
| Radio-frequency current | 8 amp |

| Exposure Conditions | |
| --- | --- |
| Spectral region | 2200-4500 A |
| Slit width | 0.02 mm |
| Spark preburn period | 10 sec |
| Spark exposure period | 50 sec |
| Neutral filter at slit | 50% |
| Film type | SA-1 |
| Electrode gap | 5 mm |
| Lower electrode | Lucite cup, porous cup or rotating disk |
| Upper electrode | $\frac{1}{4}$-in. diameter pointed electrode with $\frac{1}{16}$- in. radius at point |

*Photographic processing and photometry are carried out according to procedures outlined in "Methods for Emission Spectrochemical Analysis," ASTM, October, 1957.

taken through the same procedure, using the same quantities of reagents as for the samples.


## PHOTOMETRY

Working curves are made by plotting the intensity ratios of a selected element line and an internal standard line vs. the concentration of element in parts per million.

Analytical information can also be obtained by visual ob-

## TABLE II

Typical Standard Compositions Used with Solution-Spark Spectrographic Analysis (assuming impurities from 5 g of Cr are in 5-ml total volume)

| Standard | ppm Each Impurity, Based on Cr | mg/5 ml of Final Volume | mg Co/5 ml | Concn. $H_2SO_4$, % |
|----------|----------|----------|----------|----------|
| 1 | 30.0 | 0.15 | 0.15 | 20 |
| 2 | 10.0 | 0.05 | 0.15 | 20 |
| 3 | 3.0 | 0.015 | 0.15 | 20 |
| 4 | 1.0 | 0.005 | 0.15 | 20 |
| 5 | 0.3 | 0.0015 | 0.15 | 20 |
| 6 | 0.1 | 0.0005 | 0.15 | 20 |

## TABLE III

Spectrographic Detection Limits (PPM) of Various Elements in Chromium after Volatilization as Chromyl Chloride

| Element | Detection Limit, ppm | Notes |
|---------|---------|-------|
| Mg | 0.01 | Blank usually 0.5 ppm |
| Al | 0.5 | Blank usually 0.5 ppm |
| Ca | 0.01 | Blank usually 0.5 ppm |
| Mn | 0.05 | |
| Cu | 0.1 | Blank usually 0.5 ppm |
| Sn | Volatile | |
| Pt | 1.0 | |
| Y | 0.2 | |
| Ti | 0.3 | |
| V | 0.5 | |
| Fe | 0.1 | Blank usually 0.1 ppm |
| Ni | 0.5 | |
| Zn | 2 | |
| Mo | 0.3 | |
| Pb | 1 | |
| Nd | 30 | |
| Zr | 0.5 | |

servation of the film. This is effective if standards are used on each film—a recommended procedure in any case, unless a number of samples are being run at one time or on subsequent days.

The detection limits of elements of interest are given in Table III. Blank corrections are made where necessary, according to the method of Pierce and Nachtrieb [3].

## ACCURACY

An indication of the accuracy of this method is given in Table IV, which shows amounts of various elements recovered after addition to a chromium specimen of known purity.

A few elements, such as As, B, and Sn, are partially or completely volatile at temperatures of 200° to 220°C, the temperature at which chromium is evolved; therefore, they cannot be determined by this method. Other elements are not volatile or, as in the case of Mo, are lost by volatilization to the extent of only about 3% [4]. Precision data on four representative impurity elements are presented in Table V. The data were collected by analyzing six portions of a solution of 50 g of chromium. This method of sampling minimized any possible segregation of impurities in the chromium metal.

The wavelengths, in angstroms, of the spectral lines which were used for the precision study are 2598.37, Fe; 3383.76, Ti; 3414.76, Ni; and 2576.10, Mn. The cobalt line, 3405.12 A, was used as the internal standard.

## DISCUSSION

In performing this analysis, great care must be observed to prevent errors introduced by contamination. The chemical ware must be kept clean, the work must be done in a relatively dust-free area, and the reagents must be of high purity.

Perchloric acid should be handled with caution even though the reaction with chromium is not particularly hazardous. It is important to liberate perchloric fumes only in a hood free from organic material. In performing this analysis over the past several years, this laboratory has had no accidents.

More or less chromium could be used. Five grams was chosen because this amount is readily volatilized and results in sufficiently low detection limits for the impurities.

Cobalt should be added to the solution to be analyzed be-

## TABLE IV
### Recovery Data (values given in ppm)

| Description | Mn | Fe | Ni | Mo | Y | Cu | Ti | Zr | Pb |
|---|---|---|---|---|---|---|---|---|---|
| Analysis of Cr specimen | 0.2 | 19 | 2.7 | < 0.3 | < 0.2 | 0.9 | < 0.3 | < 0.5 | < 1 |
| Added, ppm | 1 | 3 | 3 | 3 | 3 | 3 | 3 | 3 | 3 |
| Total present | 1.2 | 22 | 5.7 | 3 | 3 | 3.9 | 3 | 3 | 3 |
| Detected ppm | 1.1 | 21 | 5.8 | 2.2 | 3 | 4 | 3 | 2.8 | 3 |
| Added ppm | 3 | 10 | 10 | 10 | 1 | 1 | 10 | 10 | 10 |
| Total present | 3.2 | 29 | 12.7 | 10 | 1 | 1.9 | 10 | 10 | 10 |
| Detected ppm | 3.2 | 29 | 11.5 | 4.5 | 0.9 | 1.8 | 9.4 | 9.6 | 7.2 |

## TABLE V
### Precision Data

| Aliquot | Spectro-graphic Run No. | Element, ppm | | | |
|---|---|---|---|---|---|
| | | Fe | Ti | Ni | Mn |
| I | 1 | 23 | 12 | 13.5 | 1.9 |
| | 2 | 23.5 | 12 | 12.5 | 1.95 |
| | 3 | 23 | 11.5 | 11 | 2.1 |
| | 4 | 22 | 10.5 | 10.8 | 2.0 |
| II | 1 | 24 | 12.2 | 13.5 | 1.85 |
| | 2 | 23.5 | 10 | 10.8 | 2.0 |
| | 3 | 23.5 | 11.2 | 12 | 2.1 |
| | 4 | 24 | 12.8 | 12 | 1.95 |
| III | 1 | 25 | 12 | 12.5 | 1.95 |
| | 2 | 24 | 12 | 14 | 1.9 |
| IV | 1 | 22.5 | 12 | 11.8 | 2.1 |
| | 2 | 24 | 11.4 | 12.5 | 2.15 |
| | 3 | 23 | 11.6 | 12.2 | 2.0 |
| | 4 | 23 | 10.8 | 11.8 | 1.9 |
| V | 1 | 24.5 | 11.8 | 11.6 | 1.85 |
| | 2 | 24.5 | 11.2 | 12.2 | 1.95 |
| | 3 | 25 | 11.9 | 12 | 2.05 |
| | 4 | 23 | 11.9 | 12 | 1.95 |
| VI | 1 | 25.5 | 11.9 | 12 | 1.9 |
| | 2 | 24.5 | 11.6 | 12 | 1.85 |
| | 3 | 25.5 | 11.6 | 12 | 2.1 |
| | 4 | 25.5 | 11.2 | 12.5 | 2.0 |

fore the fuming process to compensate for possible mechanical loss of the solution. Such loss of impurities would be accompanied by a proportional loss of cobalt and the line-intensity ratios would be unaffected.

## REFERENCES

1. W. K. Baer and E. S. Hodge, Appl. Spectroscopy 14, 141 (1960).
2. C. G. Baird, Appl. Spectroscopy 13, 29 (1959).
3. W. C. Pierce and N. H. Nachtrieb, Ind. Eng. Chem., Anal. Ed. 13, 774 (1941).
4. G. F. Smith, Mixed Perchloric Sulphuric, and Phosphoric Acids and Their Applications in Analysis, 2nd ed., G. Fredric Smith Chemical Co., Columbus, Ohio (1942).

# A FEATURE DIRECT-READING SPECTROGRAPH

## Walter G. Driscoll

Baird-Atomic, Inc.
Cambridge, Massachusetts

Major efforts in spectrochemical systems designs have for years been directed toward total automation. As an illustration of this observation, Fig. 1 shows the first Dow-Baird Direct-Reading Spectrograph which was introduced 17 years ago, and Figs. 2 and 3 show more recent installations of Direct Readers at the Bureau of Standards and at Pensacola Navy Base.

Although considerable progress has been made in automation programs, there are still some economic and technological items which indicate the advisability of investigating some of the problem areas in spectrographic analysis before it will be realistic to envision completely "closed loop" control.

Fig. 1. Original Dow-Baird Direct Reader.

Fig. 2. Baird-Atomic Direct Reader at the Bureau of Standards.

With this philosophy in mind, Baird-Atomic has built and installed in the Armco Research Laboratory in Middletown, Ohio, what is probably one of the most unique and versatile direct-reading spectrographs that has yet been assembled. Figures 4-6 are photographs of the Armco Laboratory installation and the Armco Direct Reader which will be discussed.

Figure 7 shows the Direct Reader set up at our plant just prior to delivery. I will discuss the components of this system proceeding from left to right as they are shown in the figure.

The optical arrangement of the instrument is basically similar to the several hundred Baird-Atomic Direct-Reading Grating Instruments which are presently operational throughout the world. As you know, Baird-Atomic Direct-Reading Instruments are modified Eagle mount spectrographs. The wavelength range is, to our knowledge, the widest ever covered with this type of direct-reading instrument. It is from 1930 A to 4200 A in the first order using a 30,000 line grating. The dispersion is 2.78 A/mm in this order. A modified Eagle mount is used; hence the only reflecting surfaces are a 45° entrance mirror and the concave grating. Since there are only two reflecting surfaces, this optical geometry allows only minimum light loss and permits analyses to be performed on elements at very low trace concentrations.

The Armco Research Direct Reader has a source unit (Fig. 8) which is external to the instrument. External sources

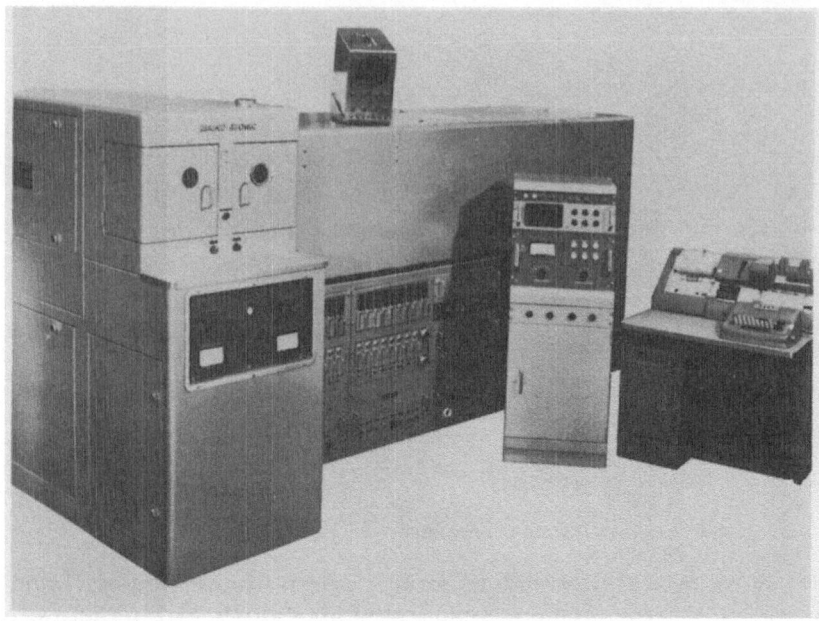

Fig. 3. Baird-Atomic Direct Reader at the Pensacola Naval Base.

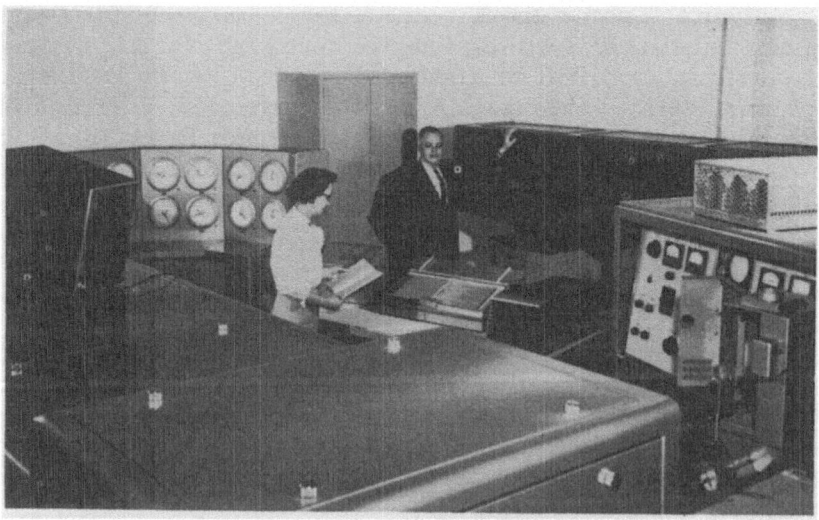

Fig. 4. The third Research Direct Reader manufactured by Baird-Atomic (located at Armco Research Labs.).

Fig. 5. The new Baird–Atomic Direct Reader at Armco Research.

Fig. 6. The new Baird–Atomic Direct Reader at Armco Research.

Fig. 7. The Baird–Atomic Direct Reader prior to shipment to Armco Research.

Fig. 8. Source unit.

are uncommon in most direct-reading instruments which are being used daily. Its high-voltage spark can supply RF currents from 0-40 amp in continually adjustable steps. The AC arc can be varied over ranges from 4800 v at 2.5 amp to 600 v at 10 amp. DC arc ranges are available from 0-50 amp in continuously overlapping steps. A low-voltage arc is also available which is spark-ignited from 0-10 amp with a wide range of spark parameters which can be introduced at the discretion of the operator. In this large source unit, it is anticipated that most types of excitation of primary interest at present can be established and applied in a direct-reading problem. This versatility contrasts with the limited number of excitations usually available in a direct-reading spectrograph for one or several specific industrial applications. This added flexibility will provide industry with an opportunity to optimize the solutions of their day-to-day problems to a greater extent than heretofore possible.

Turning again to some of the newer items, Fig. 9 shows an exterior optical bench and an enclosed stand. This variation

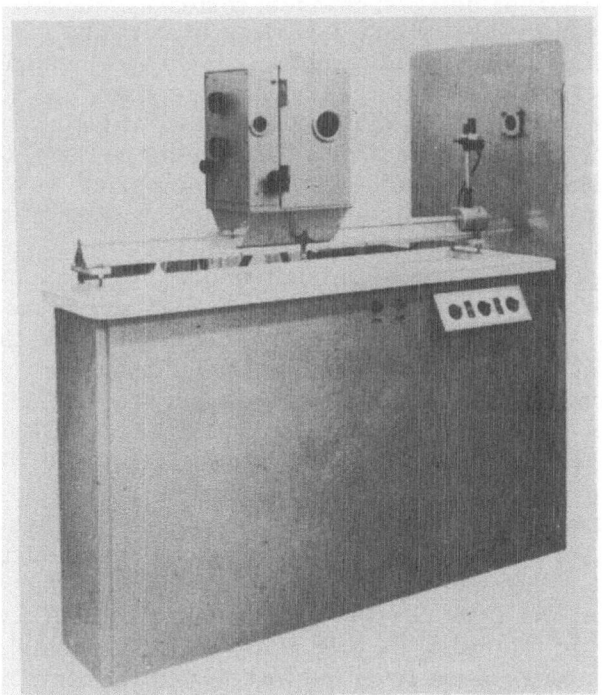

Fig. 9. External bench and source stand.

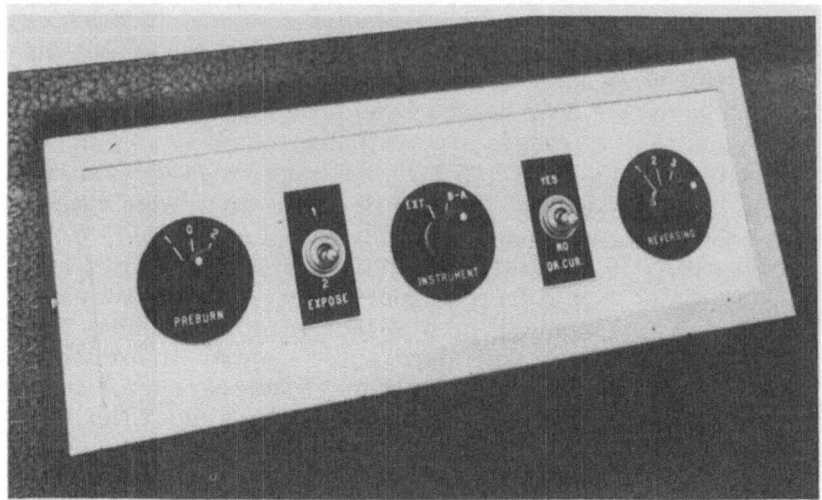

Fig. 10. Control panel for program timing.

was required by Armco and produced so that they would have
the flexibility of using such accessories as stepped filters,
stepped sector wheels, and a variety of combinations of entrance optics. The stand, which is of a new design, is extremely versatile in that it can handle pins, flats, and solutions
as well as pellets and powders. As a matter of fact, provisions
have been made to operate with the pins in a vertical or horizontal position or at a 45° angle. It is appropriate at this time
to say that the possibility and advantages of using other types
of sources, such as plasma jets, have not been overlooked. This
exterior type of configuration lends itself to this type of adaptability and has the added advantage that the source which
generates and radiates considerable heat, is outside the basic
spectrographic instrument housing, thus eliminating a possible
cause of undesirable temperature changes.

On the front of the optical bench is the control panel for
program timing. This panel is shown in Fig. 10. The programming subsystem of the instrument is unusually flexible
to the extent that many variations of programming are readily
available. For example, preburn timing can be varied from 0-5
min, as can the exposure time. Furthermore, if the operator
so desires, the dark current can be subtracted at either the
beginning or the end of a cycle, thus making it possible either
to perform integration from the instant that the start button is
pushed, or to delay integration until the volatilization has

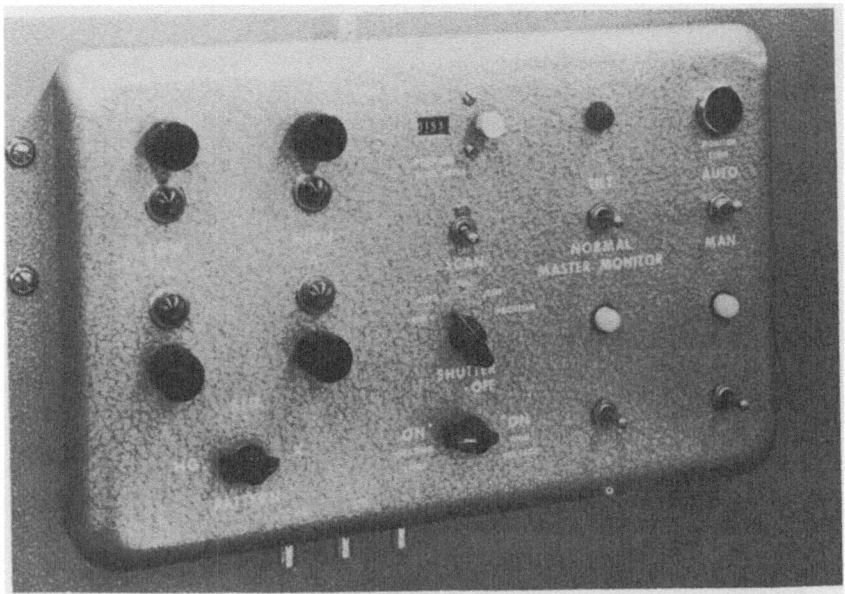

Fig. 11. Master monitor controls.

Fig. 12. Amplifier drawers.

Fig. 13. Clock console.

Fig. 14. "A" frame (showing grating mount).

reached a stable state. If desired, integration can also be made on alternate cycles of 1 sec.

The controls shown in Fig. 11 are conveniently located on the side of the spectrograph. They permit controlling and/or observing several situations: a) The position of each exit slit; b) the scope patterns of photomultiplier signals; c) scanning modes and tilt of the spinner plate; d) the shutter on the optical bench; e) the positive scan drive and setting of center reference line on scope; and f) automatic and manual program timing.

The electronics (see Fig. 12) for this direct reader are for the most part conventional with Baird-Atomic. However, two new options have been added to each amplifier channel. These options permit the selection, by the flicking of a switch, of one or two possible inputs for the amplifiers, and/or permit flexibility in the choice of integrating capacitors that are used with each amplifier channel.

The readout clock console (Fig. 13) is presently connected to provide for the simultaneous analysis of 18 elements, two clock positions reserved for spares. An additional feature of this Direct Reader is that the strip chart recorder, shown on the right-hand side of Fig. 7, may be connected in parallel

Fig. 15. "A" frame (facing rear of focal curve).

Fig. 16. "A" frame (side view of photomultipliers).

Fig. 17. End view of photomultipliers.

with the analytical clocks so that recorded information can also be available.

Turning now to the inner mechanisms of the Spectrograph, Fig. 14 shows the massive "A" frame structure which assures stability and rigidity. The 3-m diffraction grating is also shown positioned as the rear of the Spectrograph with the Baird-Atomic Optical Servo attached thereto. The two mirrors in the foreground of this figure are positioned to collect sodium illumination, if desired, and the small mirror and photomultiplier positioned in the center are for observing the central image that is received from the grating.

Figure 15 shows the opposite end of the "A" frame, facing the rear of the focal curve. In the center is a filter mechanism that can be used to control the intensity of specific spectral lines. Also, the master monitor control is positioned in the center of the focal plane and the spinner plate mechanism is mounted as usual on the left of the frame. The series of resistors shown temporarily mounted to the electrical mounting board are the resistors which will maintain the last dynode of each photomultiplier above electrical ground. This board will eventually be mounted in a permanent and more appropriate location.

The side view of this last collection of components (Fig. 16) shows how deflection mirrors are mounted behind the optical focal curve to reflect the light received from each element onto as many as 40 different photomultipliers. Again, in the center of this figure the equipment for driving the · spinner plate that is utilized in setting the exit slits may be observed.

There is one additional feature of this research instrument that can be pointed out in Fig. 16. In its center, the entrance slit to the Spectrograph and the 45° reflecting mirror can be seen. Mounted against the entrance slit is a removable entrance lens. This lens has been made removable so that other lens systems can be utilized with the instrumentation and mounted ahead of this position on the external optical bench.

The last photograph (Fig. 17) was taken with the rear panel removed from the Direct Reader and with all of the photomultipliers and ancillary optics properly positioned. The four mechanical units which facilitate positioning of the exit slits can be seen in this figure. These units control the individual slits separately from the top and bottom, and thus facilitate straightening a slit or positioning it to within microns. In the lower area are covered compartments which contain electronic circuitry for electrically gating the eighth dynode of the 2 photomultipliers which are used to servo-monitor this particular Direct Reader.

# CARBON INTERNAL STANDARD IN THE SPECTROCHEMICAL ANALYSIS OF LUBRICATING OIL

## J. H. LaSell

Cummins Engine Company
Columbus, Indiana

## ABSTRACT

The spectrographic analysis of lubricating oil for both wear metals and oil additive metals by the rotating disc procedure without the use of a buffer-internal standard is presented in this paper. In addition to the procedures used, the accuracy of the method is measured in terms of precision, comparison with other methods using round robin tests, and a determination of a confidence level in measuring engine wear based on actual engine tear-down inspection levels.

Some experimental work indicates that the carbon in the oil sludge formed on the graphite disc is the source or part of the source of the 2296.89 C III line. Limited tests indicate that this 2296.89 C III line could be used as the internal standard in the emission spectrographic analysis of other liquid organic materials for trace materials, and other corrosion inhibitor metals such as boron, sodium, and chromium.

## INTRODUCTION

The use of spectrochemical methods for measuring wear metals and oil additive metals in lubricating oil is not new. There are many references in the literature and several suggested standard methods listed by ASTM. This type of analysis as a tool for preventive engine maintenance is becoming very popular on diesel, gasoline, and jet engines today.

The general practice for this type of analysis is to use an added internal standard such as cobalt or nickel in the oil sample. One paper presented at the Chicago Symposium in 1958 showed good results on a direct reader using the hydrogen 4861 line as the internal standard.

239

The method to be presented here is a variation of the rotating disc technique using a high-voltage spark discharge for excitation. We have found it to be equivalent in accuracy to other photographic methods because it does not require the addition of an internal standard or buffer. The procedure is simple and does not require any more special care or technique than other spectrographic procedures.

## APPARATUS

1. Our power source is an Industrial Model Varisource. The excitation conditions used are shown in the upper section of Table I.

### TABLE I
Procedure for Spectrochemical Analysis of Lubricating Oil with a Carbon Internal Standard

| Excitation Conditions | |
|---|---|
| Capacitance | 0.005 $\mu$f |
| Inductance | 40 $\mu$h |
| R. F. amps | 5 |
| Breaks per $\frac{1}{2}$ cycle | 5 |
| Powerstat setting | 90-95 |
| Draft | 0.15 in., water |
| Spark gap | 3 mm |

| Exposure Conditions | |
|---|---|
| Spectral region | 1850 to 4350 A |
| Dispersion | 5.0 A/mm |
| Slit | 30 $\mu$ |
| Preburn | 40 sec |
| Exposure | 22 to 30 sec |
| Emulsion | SAE I 30-mm film |
| Filters | Fogged film for 3273 Cu and 3302 Zn |

| Photographic Processing | |
|---|---|
| Developer | D-19 68°F |
| Stop bath | SB - 1A |
| Water rinse | |
| Fixer | F 5 |
| Water rinse | |
| Drying | Sponge—warm air |

2. An enclosed arc-spark stand equipped with a rotating electrode assembly is connected to an exhaust system, which is equipped with a damper and draft gauge and is used to remove the carbon soot and other combustion products.

3. We use the No. 106 preformed graphite disc electrode made by United Carbon Co. and L-3809 graphite rods made by National Carbon Co. The rods are cut to a 120° included angle tip and are recut after each burn. The mandrel used for the disc electrode is stainless steel and is silver-soldered to the collet of the rotating electrode assembly. The rotation speed is 5 rpm.

4. The spectrograph is a JACO 3.4-m Ebert with a 15,000-lines/in. grating.

The long preburn will be explained in the discussion. With these conditions there is a minimum of continuum but we do make a correction.

5. Densitometry is performed on a Model 2400 JACO Micro Photometer which is a 35-mm film instrument.

6. Calibration of the emulsion is by means of a seven-step sectored DC arc spectrum of iron. An average curve is constructed for iron lines in each of the 2300 A, 2400 A, 2700 A, 2800 A, and 3100 A spectral regions.

7. Analytical curves are prepared by using the log of the ratio of the intensities as abscissa and the concentration as the ordinate. Shifts of the curve are adjusted according to the control standards used on each film.

## STANDARDS

We prepare all of our own standards, using certified metalloorganic compounds blended with USP mineral oil with a viscosity of 335 to 350 SUS. The first attempts to prepare different standards by dilution or aliquoting methods were unsatisfactory. Our working curves covered only a short concentration range and we experienced poor reproducibility on some elements. We attributed this to the wide variation in the total element concentration of the standards.

Our method of preparation of standards involves the proportionate blending of base stock solutions. One base stock contains all the wear metals in the highest concentration covered by the working curves. We then blend two oil additive metal base stocks. One is high in calcium and phosphorus with low zinc and barium. The other is low in calcium and phosphorus, but high in zinc and barium. The total element contents of these oil additive base stocks are approximately equal.

By blending the wear metal base stock with either or both of the oil additive base stocks, we have a final standard which has a total metal content similar to that of the new or used oil to be analyzed.

## ANALYTICAL LINES

In Table II are shown the analytical lines used, their concentration ranges, the concentration index, and the slope of the working curve. The 2296.89 C III line is used as the internal standard line for all analytical curves. By increasing the ex-

## TABLE II
### Analytical Lines

| Elements | Line | Conc. Range (ppm) | Conc. Index (ppm) | Curve Slope |
|----------|------|-------------------|-------------------|-------------|
| Fe | 2404 | 50-800 | 200 | 48° |
|    | 2598 | 1-100 | 30 | 53° |
| Mg | 2801 | 1-100 | 21 | 58° |
| Pb | 2833 | 10-100 | 95 | 57° |
| Cr | 2843 | 1-50 | 22 | 58° |
|    | 2860 | 20-110 | 110 | 58° |
| Si | 2881 | 2-50 | 28 | 53° |
|    | 2516 | 25-150 | 68 | 52° |
| Sn | 3034 | 10-150 | 130 | 60° |
|    | 2839 | 10-110 | 110 | 60° |
| Al | 3082 | 1-100 | 28 | 60° |
| Cu | 3273 (filtered) | 1-50 | 15 | 51° |
| B | 2496 | 10-100 | 38 | 53° |
| Zn | 3302 (filtered) | 150-1500 | 470 | 56° |
| Ca | 3006 | 200-7000 | 2800 | 58° |
| P | 2536 | 200-3000 | 1500 | 50° |
| Ba | 2335 | 50-700 | 380 | 50° |
|    | 2347 | 600-7000 | 2200 | 58° |
|    | 2304 | 50-500 | 200 | 58° |

2296.89 C III internal standard used with all analytical lines.

TABLE III

Comparative Analyses of Three Used Oil Samples

| Element | Fe | Si | Mn | Pb | Cr | Si | Sn | Al | Cu | Ca | Zn | P | Ba |
|---|---|---|---|---|---|---|---|---|---|---|---|---|---|
| Line | 2404 | 2881 | 2801 | 2833 | 2843 | 2516 | 3034 | 3082 | 3273 | 3006 | 3302 | 2536 | 2335 |
| | | | | | | | *Sample No. 1* | | | | | | |
| Conc. | 68 | 9 | 22 | 16 | 3 | | 41 | 5 | 13 | 1800 | 580 | 1600 | 180 |
| | 66 | 9 | 22 | 16 | 3 | | 39 | 5 | 13 | 1900 | 510 | 1700 | 180 |
| | 67 | 10 | 24 | 21 | 3 | | 42 | 5 | 17 | 1600 | 530 | 1700 | 190 |
| | 67 | 9 | 22 | 16 | 3 | | 41 | 5 | 14 | 1900 | 520 | 1600 | 180 |
| | 66 | 10 | 23 | 16 | 3 | | 42 | 5 | 14 | 1800 | 500 | 1700 | 170 |
| | 72 | 12 | 26 | 21 | 3 | | 52 | 6 | 18 | 1700 | 490 | 1700 | 170 |
| | 63 | 10 | 23 | 16 | 3 | | 42 | 5 | 14 | 1900 | 510 | 1600 | 190 |
| Average | 67 | 9.9 | 23 | 17 | 3 | | 43 | 5 | 15 | 1800 | 520 | 1643 | 180 |
| | | | | | | | *Sample No. 2* | | | | | | |
| Conc. | 450 | 38 | 31 | 21 | 11 | 41 | 72 | 34 | 18 | 2000 | 530 | 1500 | 150 |
| | 470 | 40 | 31 | 21 | 12 | 42 | 70 | 37 | 18 | 1900 | 540 | 1350 | 160 |
| | 465 | 38 | 33 | 21 | 12 | 41 | 74 | 37 | 17 | 2200 | 490 | 1400 | 160 |
| | 475 | 45 | 36 | 22 | 13 | 46 | 83 | 43 | 22 | 2200 | 510 | 1500 | 150 |
| Average | 465 | 40 | 33 | 21 | 12 | 42.5 | 75 | 38 | 19 | 2100 | 520 | 1475 | 155 |
| | | | | | | | *Sample No. 3* | | | | | | |
| Conc. | 105 | 10 | 9 | 5 | 3 | | 19 | 10 | 3 | 2400 | 570 | 1100 | <50 |
| | 96 | 9 | 8 | 5 | 3 | | 12 | 9 | 3 | 2700 | 630 | 1250 | <50 |
| | 100 | 9 | 8 | 5 | 3 | | 13 | 9 | 3 | 2400 | 660 | 1250 | <50 |
| Average | 100 | 9 | 8 | 5 | 3 | | 14.7 | 9 | 3 | 2500 | 620 | 1200 | <50 |

TABLE IV
Day-to-Day Variations in Analysis of a Used Oil Sample

| Date | Element | | | | | | | | | | | |
|---|---|---|---|---|---|---|---|---|---|---|---|---|
|  | Fe | Mn | Pb | Cr | Si | Sn | Al | Cu | Zn | Ca | P | Ba |
| 1-22-60 | 58 | 20 | 26 | BD | 10.5 | 33 | 12.0 | 34 | 720 | 1600 | 1800 | < 50 |
| 1-29-60 | 59 | 22 | 23 | 2 | 10.0 | 35 | 8.0 | 27 | 660 | 1500 | 1500 | < 50 |
| 2- 1-60 | 70 | 25 | 26 | BD | 9.0 | 35 | 7.5 | 30 | 720 | 1900 | 1600 | < 50 |
| 2- 3-60 | 62 | 24 | 19 | BD | 8.0 | 19 | 7.5 | 25 | 740 | 2000 | 1350 | < 50 |
| 2- 4-60 | 66 | 28 | 21 | 2.2 | 9.5 | 32 | 8.5 | 27 | 800 | 2100 | 1600 | < 50 |
| 2- 5-60 | 66 | 20 | 20 | BD | 9.0 | 28 | 8.5 | 24 | 620 | 1650 | 1400 | < 50 |
| 2- 8-60 | 70 | 22 | 17 | BD | 8.0 | 27 | 7.5 | 27 | 720 | 1700 | 1300 | < 50 |
| 2-10-60 | 58 | 20 | 20 | BD | 8.5 | 24 | 7.0 | 23 | 640 | 1800 | 1750 | < 50 |
| 2-11-60 | 62 | 26 | 18 | BD | 11.0 | 30 | 7.5 | 25 | 800 | 2000 | 1500 | < 50 |
| 2-12-60 | 70 | 28 | 25 | BD | 11.0 | 32 | 12.0 | 34 | 750 | 2100 | 1750 | 50 |
| 2-13-60 | 70 | 20 | 25 | 2.1 | 12.0 | 33 | 9.0 | 29 | 800 | 2100 | 1800 | 50 |
| Average | 64.5 | 23 | 22 |  | 9.6 | 30 | 8.6 | 28 | 727 | 1860 | 1577 | < 50 |
| Av. Dev. | 4.0 | 3.0 | 3.0 |  | 1.0 | 4.0 | 1.3 | 3.0 | 49 | 172 | 155 |  |

posure time to 60 sec, we can increase the sensitivity of the tin and lead to the 1–5 ppm range. Of course, this does increase the continuum.

We have not found any serious interfering lines with the exception of the iron at or above 100 ppm on the 2536 phosphorus line. The filter used on the 3273 copper and 3302 zinc lines is a fogged film strip located on the film holder rack just in front of the film.

PREPARATION OF SPECTROGRAM

Approximately 1.5 to 2.0 ml of the thoroughly mixed oil sample is transferred to a Coors No. 2 porcelain combustion boat with a soda straw used as a pipette. The quantity of oil is not critical, but should cover the OD of the graphite disc to a depth of about 1/16 in. when the sample is in sparking position.

With the sample and electrodes in position, the spark stand is closed, the air draft gauge is checked, the graphite disc electrode is rotated at least one revolution, and the burning cycle is started. During the burning, the upper rod electrode is adjusted to maintain the 3-mm gap, and the excitation conditions are controlled throughout the burning.

After the sample burn is complete, the sample and elec-

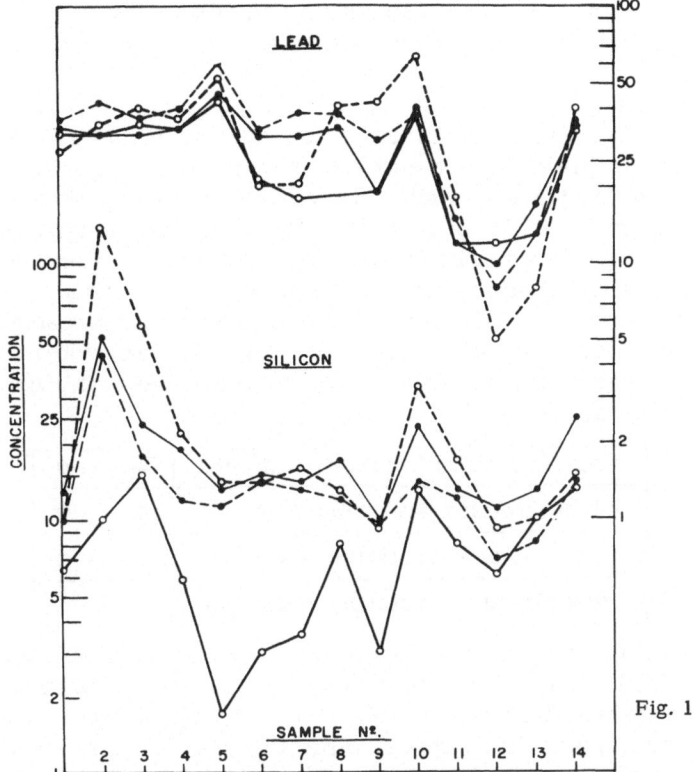

Fig. 1

trodes are removed, the sample table and mandrel are wiped clean with a cleaning tissue, and the setup is ready for installing the sample and electrode for the next burning.

For each group of spectrograms on a single film strip, at least two standards are run to cover two points of the working curve on each metal to be determined. Some parallel curve shift is experienced; this is believed to be due to the age of the developer, the condition of the auxiliary electrodes of the power source, or variations in atmospheric conditions.

## PRECISION AND ACCURACY

Since it is difficult to analyze lube oils chemically for the actual concentration of the different elements, we have resorted to comparative tests and analyses of synthetic samples. The reproducibility of our results for the same sample of oil is shown in Table III. These sets of results were from different exposures on the same film strip. The differences are probably due to changes in parameters or segregation in the sample.

The day-to-day variations in analytical results are shown in Table IV. There is an emulsion change in this group. The curve shift previously mentioned and its effect are included in this table. On the whole, the variations are not excessive.

In Figs. 1-3 is presented a graphic comparison of our method with other methods. Analyses of fourteen different used lube oils were obtained from three laboratories for an evaluation of our method. The data in these figures are representative of those for the other elements which were determined.

The legend for identifying the methods used in Figs. 1-3 is as follows:

| Carbon Internal Standard Method | ●——● |
|---|---|
| Direct reader—rotating disc | ●----● |
| Photographic—rotating disc | ○——○ |
| Photographic—rotating platform | ○----○ |

All of the procedures except our own use an added internal standard. In general, there is considerable variation in results from different laboratories for such metals as iron, silicon, aluminum, lead, and copper, while the zinc and other additive metals show better agreement. The difference in agreement is probably a function of the physical state of the element, i.e., suspended solid, emulsified, or dissolved state.

The values obtained for a new oil sample are presented in Table V. Something is known of the element concentration in such samples. All wear metals should be in the "trace" or "not detected" range. Silicone defoamer is added in the range of 5 to 10 ppm as silicon. Zinc, calcium, and phosphorus have nominal values of 600 ppm, 2500 ppm, and 1000 ppm, respectively.

Analyses of six oil samples prepared by blending known weights of certified metalloorganic compounds in mineral oil are presented in Table VI. Results of the synthesized analysis, our own analysis, and the analysis obtained by a commercial laboratory using a spectrographic procedure are shown. Note that there is fairly good agreement between our results and the calculated results except for low concentrations of lead and tin. Our copper curve does not cover the range above 50

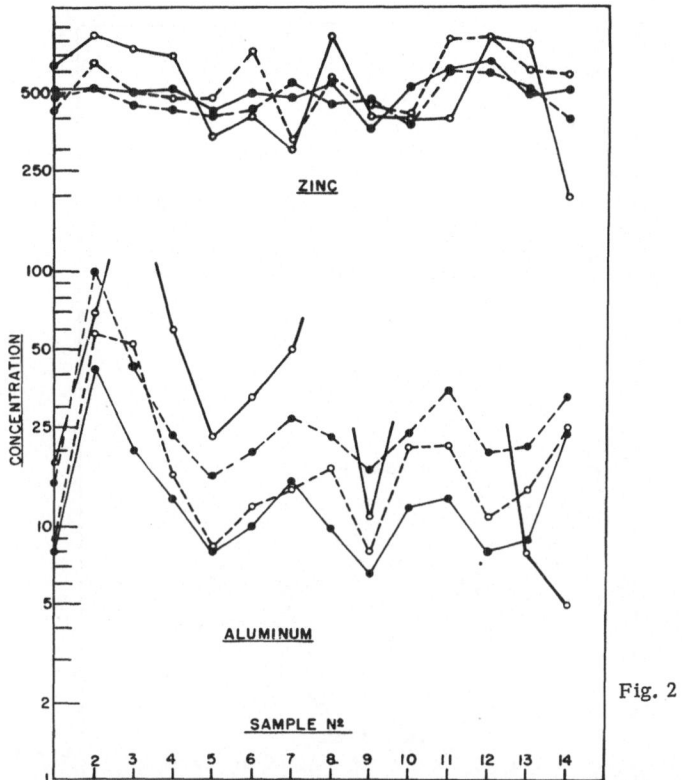

Fig. 2

ppm, as we have no need for the actual concentration in this range.

Since our original purpose in this study was to detect wear, I would like to cite the conclusions reported by our Quality Control Group. When actual engine tear-down inspection reports were compared to our engine evaluation on the basis of the lube oil analysis, the laboratory findings were more than 80% correct. Five percent of the engines were called for excess wear, but were passed by the engine inspector. With better control over engine operating conditions, oil sampling, and wear evaluation standards, our degree of accuracy would have been improved.

## INVESTIGATION OF THE 2296.89 C III LINE

The question immediately arises in this work as to what the source of the carbon line is. We have proved the accuracy

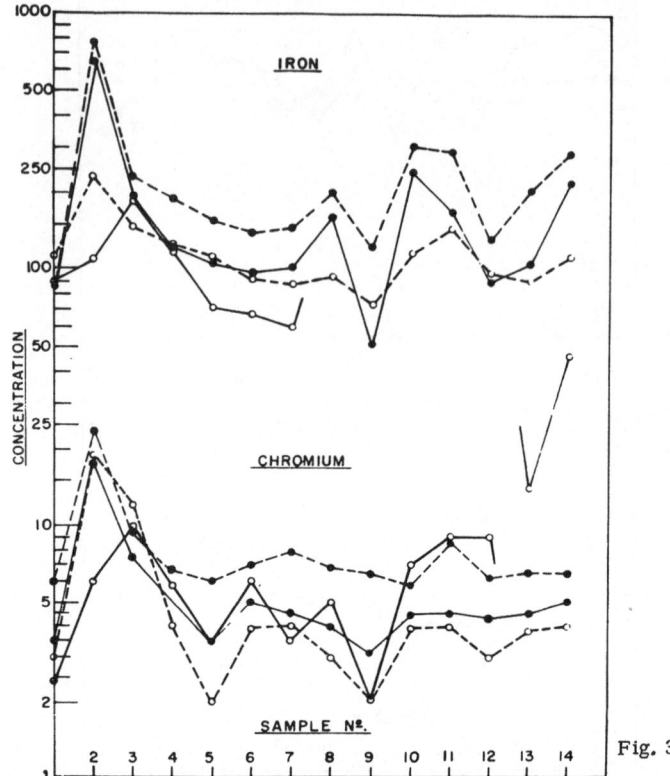

Fig. 3

of the method, both by comparison and by application. Nevertheless, since we are burning an organic material containing approximately 85% carbon between two graphite electrodes, it is desirable to determine which material is the source of the carbon line.

For our first test we machined both disc and rod electrodes out of a pure copper bar and ran a sample of used oil with these electrodes. The spectrogram had a very dark continuum over most of the wavelength range, but the 2296.89 C III line was visible. The 2404 Fe line and the 2516 Si line were then measured and the intensity ratios checked on our working curves. The concentration values were in close agreement with those obtained on the same sample with graphite electrodes.

This test indicated that the excitation potential, at the parameters used, was sufficient to excite the carbon in the oil

## TABLE V
### Comparative Analyses of a Sample of New Lubricating Oil by Four Different Methods

| Element | Rotating Wheel Photographic | Direct Reader | Cummins | Rotating Platform Photographic |
|---------|-----------------------------|---------------|---------|--------------------------------|
| Fe | 4 | < 5 | ND | 1 |
| Mn |  |  | ND |  |
| Pb | ND | 2.5 | ND | 1 |
| Cr | ND | 2.5 | ND | 1 |
| Si | 17 | 2.5 | 7 | 2 |
| Sn |  |  | ND | 1 |
| Al | 7 | 2.0 | BD | 1 |
| Cu | 0.5 | < 1 | ND | 1 |
| Zn | 300 | 420 | 500 | 450 |
| Ba | < 100 |  | 70 | 93 |
| Ca | > 2500 |  | 2200 | 2900 |
| P | Present |  | 1000 | 940 |

## TABLE VI
### Comparative Synthesized, Cummins, and Commercial Laboratory Analyses of Six Oil Samples

| Element | Syn.* | Cum.* | Com.* | Syn. | Cum. | Com. | Syn. | Cum. | Com. |
|---------|-------|-------|-------|------|------|------|------|------|------|
|  | Sample No. 1 | | | Sample No. 2 | | | Sample No. 3 | | |
| Fe | 130 | 130 | 127 | 60 | 58 | 50 | 10 | 9 | 8 |
| Al | 31 | 34 | 17 | 8 | 8 | 2 | 100 | 110 | 47 |
| Cr | 46 | 47 | 34 | 27.5 | 28 | 19 | 9 | 10 | 5 |
| Cu | 55 | gt 50 | 48 | 40 | 46 | 33 | 20 | 19 | 16 |
| Si | 27 | 29 | 11 | 9 | 9 | 2 | 137 | 145 | 66 |
| Pb | 67 | 66 | 80 | 49 | 56 | 48 | 30 | 37 | 27 |
| Sn | 73 | 68 | 39 | 55 | 41 | 29 | 37 | 34 | 13 |
|  | Sample No. 4 | | | Sample No. 5 | | | Sample No. 6 | | |
| Fe | 320 | 300 | 230 | 260 | 249 | 205 | 190 | 197 | 162 |
| Al | 77 | 70 | 55 | 62 | 54 | 55 | 46 | 46 | 36 |
| Cr | 101 | 101 | 83 | 83 | 80 | 75 | 64 | 68 | 52 |
| Cu | 5 | 5.5 | 6 | 100 | gt 50 | 85 | 75 | gt 50 | 57 |
| Si | 133 | 131 | 97 | 82 | 76 | 59 | 55 | 55 | 30 |
| Pb | 10 | 18 | 12 | 108 | 100 | 130 | 88 | 89 | 84 |
| Sn | 18 | 20 | 7 | 9 | 14 | 3 | 92 | 76 | 45 |

*Syn. = Calculated analysis for synthesized sample; Cum. = analysis by the Cummins method; Com. = analysis by a commercial laboratory.

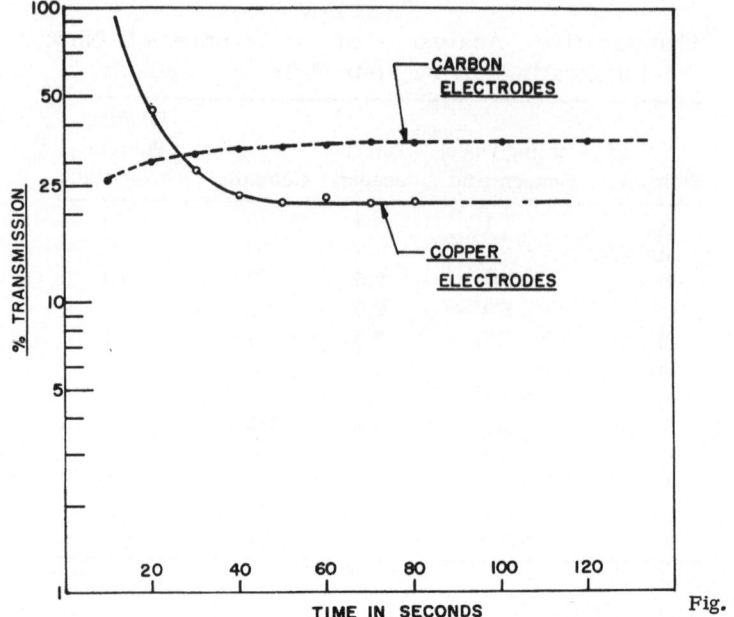

Fig. 4

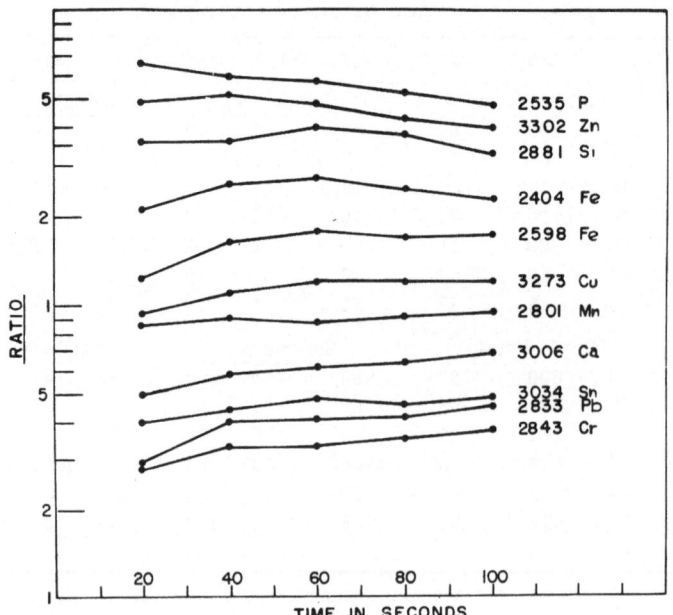

Fig. 5

sample. Also, the electrode material had little effect on the excitation of the sample.

To study this further, we ran moving plate or time-weight studies, using the copper electrodes with an oil sample. A comparison of this study with a similar study on graphite electrodes is shown in Fig. 4. Note the difference in the intensity of the carbon line during the early part of the burn and the leveling out at about 40 sec. After completion of the burn, the copper disc had a hard lacquer or varnish type residue on its OD.

Additional tests with mixed copper and graphite electrodes were also run. The intensity of the 2296.89 C III line showed very little variation in all cases. This further substantiated our belief that the carbon line was independent of the electrode material, and was therefore due to the excitation of the carbon in the oil.

In Fig. 5 is presented the results of a time-weight study on the intensity ratios of the different elements in an oil sample. The slope of many of these lines indicates a build-up in concentration on the rotating disc during the early part of the burn and then a constant emission over a long period. The more volatile elements such as zinc and phosphorus tend to drop off in time.

In another test we tried to determine whether this carbon line was only what might be called a monitoring line and whether the intensity of the element line was a direct function of its concentration. By plotting the percent transmission vs. concentration for the elements in our different standards for several spectrograms, we found considerable scattering. When these percent transmissions were changed to intensity ratios using the 2296 carbon line, we obtained a narrow band on both sides of our working curve. This was proof that the carbon line functioned as a true internal standard.

Since most of our tests were on oil samples of fairly uniform carbon content, we tried to evaluate the effects of varying the carbon content. Tests were made on various SAE grades of new and used oils in which the benzene insolubles varied from zero to as much as 15% by volume. Some increase in intensity of the carbon line was detected, but the over-all variation for all oil samples was less than 15% transmission.

Some samples of silicone oils with carbon contents of 32.4% and 57.8% were run to study the effects of this lower carbon content on the intensity of the 2296 C line. The intensities were somewhat less, but unfortunately there was either a de-hydration or polymerization reaction which took place and loaded the wheel with a gel. This made it difficult to maintain the gap spacing. Furthermore, the amount of build-up on the wheel made the results doubtful.

Some time ago we were asked to determine the sodium metaborate content in a proprietary brand of permanent anti-freeze. Standards were prepared by adding aliquots of a con-centrated water solution of sodium perborate to ethylene glycol. These standards and a sample of the antifreeze were excited in the same way as our oil samples and the sodium and boron contents determined by the same method used for oil. Calculating the sodium and boron to sodium metaborate in the antifreeze sample gave a value within 20% of the nominal concentration specified by the antifreeze manufacturer.

The results of this antifreeze analysis may open up a new approach to the problem of analyzing for metal content in liquid organic materials. We feel, moreover, that using carbon as an internal standard will eliminate variations in analytical results which result from the introduction of other internal standard metals in some types of analysis.

We believe our laboratory investigations substantiate our theory that the 2296 C line is due to the excitation of the car-bon in the oil. However, we are not convinced that the graphite electrodes are not at least part of the source of the 2296 C line. We welcome any comments or suggestions relative to this method of analysis or our investigation of the 2296 C line.

# X-RAY SPECTROSCOPY

# SOME UNSOLVED PROBLEMS IN X-RAY ABSORPTION SPECTROMETRY

## H. P. Hanson
University of Texas
Austin, Texas

This X-ray session of the Thirteenth Annual Mid-American Spectroscopy Symposium has brought together a relatively large group of people who are interested in the spectroscopy of X-ray absorbers and emitters *in situ*. It is probably the largest such group assembled since the conference organized by Professor Beeman and held at the University of Wisconsin about a dozen years ago. I hope we can profit from this assemblage.

While certain advances in theory and experiment have been made since the Wisconsin conference, and while there has been clarification and reformulation of the problems, there have been no dramatic breakthroughs. The reasons for this desultory pace of progress on what are intrinsically challenging and interesting problems are perhaps manifold, but I shall comment briefly on only two.

First, the problems have been around too long. In the early days of crude theory and cruder experiment, some of the greatest names in physics concerned themselves with these problems. Meager data adequately checked the available theory; and the luminaries of physics went on to other fields, perhaps little realizing they were leaving pockets of unsolved problems.

Second, there seems to be a sad lack of acrimonious controversy. There have been a number of basically divergent points of view propounded in the literature but we all seem merely to say, "you may possibly be right," and then return to our own attitudes and our own particular experiments. I know of no other field of physics in which there has been a greater tendency to ignore what has been done by other workers in the field. In many cases this may be a wholly justified procedure (perhaps, especially when it is my work that is involved), but I believe that most of you will agree that the discussions in the literature are frequently too specialized and too *ad hoc*. I think we should engage in more solid argument so that we will all be forced to re-examine some of the things we have said—re-examine, defend, and possibly discard.

Therefore, since there are so many people gathered here who are interested in this problem, we could hope for some concrete advances. Rather than presenting some completely noncontroversial measurements, with a minimum of interpretation representing the modest amount we think we understand about this field, I want to challenge the group with some problems that I feel need solving before we can really make progress and with some statements that should elicit some rebuttals.

To me it is very interesting and grafifying that we have arrived at a stage where X-ray absorption spectrometry (and I shall primarily concern myself with absorption rather than emission) can command a whole day's program at an Applied Spectroscopy Symposium. But surely it must be true that in order to realize the ultimate potential of the technique, we must have a better understanding of what physical processes are involved in the production of the absorption edges as we measure them.

There are numerous problems associated with the interpretation of edge structures, not the least of which is the fundamental difficulty of obtaining good data. For years the data put into the literature were not plotted as the absorption coefficients but rather as photometric traces of transmitted intensities of a photographic plate which had been exposed to the dispersed beam with an appropriately placed absorber. Part of this difficulty was eliminated when it became rather standard to plot log $I_0/I$. Then it was shown by Parratt [1] and co-workers that data obtained with thick samples were washed out because of the shape of the spectral window. This accounted for an effect which many of us had noted but had not regarded with the proper curiosity.

There is still another simple effect that must be considered, sample uniformity. In many of the salts it is difficult to get a really uniform sample, and what is worse, we have no quantitative way of expressing the lack of uniformity or its effect on the edge profile. It is readily shown that the general result of nonuniformity is to bow the edge structure upward, eliminating contrast in the high absorption region. Some of my students, in the process of taking and analyzing their data, have investigated this point on the basis of various distributions of particle size,

but I am not particularly impressed by the generality or sophis-
tication of our approach.

All in all, there is probably a vast amount of really poor
data in the literature—some of which is mine. Then there is
also a considerable amount of data which is admittedly not
really good by the most stringent criteria, but which is probably not bad either—and this I rather suspect includes most of
mine. There is much of this type of data that has an almost
quantitative correctness—and these data should not be and
cannot be justifiably ignored. I wish to emphasize here that no
minor correction for resolving power or sample thickness or
sample nonuniformity can alter the fact that these data have
strong immutable features which need explanation.

Let us now consider what we can believe about the extant
theoretical interpretations. I believe that first of all we can
accept the fact that the high energy structure, the so-called
Krönig structure, is really due to an interference-diffraction
type phenomenon involving the passage of the ejected electron
through the crystalline lattice. We have seen good evidence
that the edge structure is related to the crystalline structure
per se, but it would probably be a pointless way of investigating
crystal structure, about which we get precise information from
other types of experiments. There may be something to learn
about the passage of electrons through crystalline material, but
by and large this Krönig structure, is not well defined—and it
is difficult to conceive that any theory would be critical enough
to demand precise and detailed agreement with experiment.

It is in the low energy region of the edge, then, where the
major puzzle lies for the physicist, and I believe for the chemist
as well. It is here that the structure is well defined and shows
an empirical dependence on the number, the nature, and the
binding of the nearest neighbors. This region of the edge is
generally referred to as the Kossel structure.

I believe that in this group there are only two or three basic
experiments about which we could possibly arrive at any agreement. I do not include for discussion the long-wavelength
studies which might appropriately be considered to be ultraviolet, for they involve different energies and different life-
times. At present there seems to be solid state information
to be gleaned from such studies.

Of the X-ray studies we might agree on, the first is the
well-known work of Parratt [2] on argon gas and related
studies. In these we examine the excitation from the $K$ shell into

empty levels which are truly $4p$ in character and are to be associated with the excited but otherwise unperturbed atom. Here the original analysis of Parratt cannot be otherwise than essentially true even though more recent study makes the agreement between theory and experiment less satisfying. The structure of the solidified gases is, of course, still unaccounted for.

The other basic work which is probably correct in its interpretation is that of the Hopkins school having to do with metals, particularly the work of Beeman and Bearden [3]. Slater [4] has pointed out that the relaxation time in metals is such that the edge structure probably does reflect characteristics of the empty bands.

It should be noted, however, that there are modern theories about the transition metals, notably those of Lomer and Marshall [5] and Mott and Stevens [6] which present quite a different picture of the population of the $d$ bands than was previously held. It would seem that this experiment should have something to say about whether the $d$ bands fill uniformly or undergo the abrupt population shift indicated by the experiments of Weiss [7] and perhaps those of Shull and Wilkinson [8]. It may be that these absorption edge measurements ought to be re-examined in terms of these more modern theories. It may, for example, account for the fact that there has never been any indication of the splitting of the $3d$ levels as is ordinarily considered to be the case in the conventional theory. Thus, even this experiment and particularly its analysis is by no means a completed chapter.

To me it would seem that the next problem most amenable to analysis should be that of the ion in solution. Here it is that the ion is most nearly like a free ion, unaffected by the repeating character of the lattice, although it is surrounded by a hydration layer which is quite strongly bound to the parent ion. As a first approximation, one can assume that the ion absorbs X rays as if it were truly free, so that the theory is no different than the picture developed by Parratt for argon. Beeman and Bearden [9] exploited this idea for the edge of the ions of certain divalent metals and found an impressive correlation of theory and experiment. However, there are other theories of edge structures which are essentially unrelated to the Beeman-Bearden approach. There is, for example, the interference type theory such as is reasonably successful for polyatomic gases. There might also be an application of the electron plasma oscillation concept advocated by Marton [10]. One thing is certain—there are no inferences to be drawn here about solid state phenomena.

Recently Van Nordstrand [11] has contended that the interference concept is the one that should be invoked. He points out that the Kossel structure in argon occupies an energy range of only a few volts. Now if the ion in solution truly absorbed as if it were free, Van Nordstrand's argument would be specious because it really should take some 20 or so volts merely to remove the outer two electrons of the parent atom, and this certainly would give the type of edge structure found. But Van Nordstrand questions whether it can readily be defended that an unperturbed potential well of about 40 volts for the $K$ excited ion could really exist in solution. There has already been reference to the hydration layer in which the water molecules with their strong dipole moments attach themselves to the central ion—the bonding being presumably of a simple electrostatic character. A calculation of the radial distribution for the $4p$ state to be associated with the $K$ excited state of the ion shows the position of the maximum radial probability to be as large as the distance of the water molecules from the central ion. In other words, it is necessary either to consider that the energy levels are greatly modified by the presence of the ligands, or to introduce an effective dielectric constant. In either event, the depth of the well for the electron in the field of the $K$ excited ion is probably considerably less than could account for the structure in the way we have done, thus supporting Van Nordstrand's point of view.

Thus, even though there may be apparent qualitative agreement between theory and experiment for the cations, one must investigate the situation for anions. Certainly here there will be no potential well of the same nature as might exist for the cations. The negative charge on the anion would indicate that an ejected $K$ electron could be only very loosely bound. However, when one makes the measurement on bromine, for example, there is a strong absorption peak some ten or more volts wide. This despite the fact that the neighboring element gaseous krypton shows no structure at all for quite understandable reasons. This may well indicate that the structure is due to interference—certainly it casts doubt on the interpretation of the cation edge structure. Whatever phenomenon is involved in producing the edge in the negative ion must surely contribute to the structure found in the metal ions as well.

If it thus turns out that the free ion explanation advanced for the edge structures in solutions is invalid, then this certainly will invalidate the discussions by Cotton and me [12] of the application of crystal field theory to the edge. The extension of the Beeman-Bearden concept to salts of high symmetry is basic to our treatment. Presumably in cases where the nearest neighbors of the metal atom are bound to it largely by ion-ion or ion-dipole forces, we should have a tractable problem, since a fairly detailed picture of the organization of the optical levels of the metal atom can be obtained by the application of crystal field theory, in which the Stark perturbations of the metal orbitals are calculated to the first order. If, however, the $K$ electron is not sent into bound states, as Van Nordstrand suggests, or even if the levels are greatly modified, then our treatment is nonsense.

Let us grant that there is serious reason for doubting our model of ionic absorption, and consider the presumed alternative. I find it hard to accept the interference model because in its simplest aspects it does not involve the atomic selection rules. Yet it is possible to cite numerous examples of the fact that the selection rules do play a prominent role. Comparing the $L_1$, $L_{11}$, and $L_{111}$ edges of tungsten [13], for example, one finds the $L_{11}$ and $L_{111}$ edges are virtually identical, with a sharp peak brought about by the $p \rightarrow d$ transition, while the $L_1$ edge is essentially structureless. Furthermore, energy cycles show that the $L_{11}$ and $L_{111}$ edges occur comparatively at lower energies.

Another thing which I consider to be significant in this regard is the shape of the edges in zinc. Even in the pure metal one finds quite a strong sharply peaked initial absorption compared to the transition metals which precede it in the periodic table, and in the ions the effect is even more pronounced. Ultimately this must be because in the zinc atom we have a closed $3d$ shell in an $S$ state which does not interact extensively with the $4p$ levels. These effects are shown in Fig. 1, particularly the sharpness of the peaks.

Further, we note the similarity of the three edges. In the instance of $ZnF_2$, the $Zn^{++}$ ion in aqueous solution, and $ZnSO_4 \cdot 7H_2O$, the zinc is in a position of octahedral symmetry. The edges are quite typical of ionic compounds.

In Fig. 2, a similar set of data is seen for nickel. The only essential difference lies in the fact that the initial peak is not so prominent, which concurs with my previous statement. The essential point is this: The ion in solution and the hydrated salt

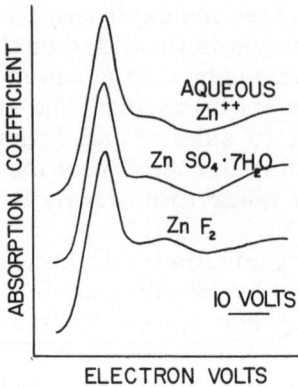

Fig. 1. Zn $K$ absorption edges (octahedral symmetry).

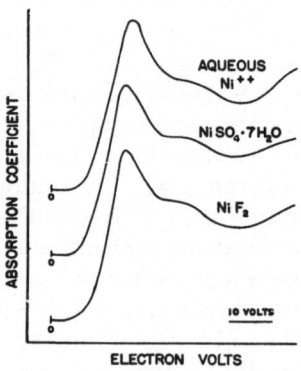

Fig. 2. Ni $K$ absorption edges (octahedral symmetry).

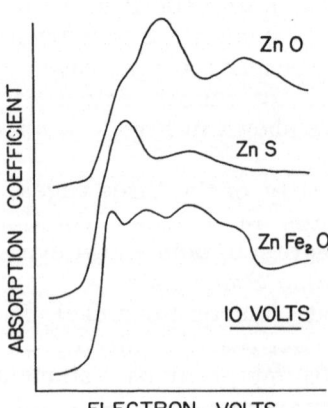

Fig. 3. Zn $K$ absorption edges (tetrahedral symmetry).

have essentially identical structures (which is not so unexpected), but the well-defined fluoride salt also shows the structure.

It is for this reason that I have never been quite able to accept without considerable reservation the interpretations which have been placed on the edge structure in the painstaking KCl measurements. On the basis of our measurements on Ni and Zn, for example, I would not be surprised if the edge structure of $K^+$ in a water solution would be similar to that of solid KCl. I truly hope that my impressions are wrong on this score. But if my conjecture proves to be the case, then there is little solid state information to glean from even the best of such measurements. It would be a difficult experimental problem and as far as I know the only data on KCl in solution were obtained by Japanese workers [14], and their photometric traces are not at all conclusive.

We have considered cases of high symmetry of the ligands. Let us next consider some cases of slightly lower symmetry, namely, the tetrahedral structure. Within the framework of crystal field theory, the edges should not be greatly different from the octahedral case.

In point of fact we find the interesting group shown in Fig. 3. In each of these structures the Zn is in a site of tetrahedral symmetry. The edge for ZnO may not be much different from the $Zn^{++}$ ions shown in Fig. 1, but the ZnS is rather different and the $ZnFe_2O_4$ is certainly passing strange.

I must make a parenthetical note here. I mentioned the existence of bad data in the literature, and one of my contributions to it has been a curve of ZnS which I published [15]. Probably from a combination of poor crystals, sample thickness, sample nonuniformity and perhaps again even sample impurity [11], the published curve of ZnS shows these two first peaks as being about the same height, which is quite wrong.

The $Zn_2FeO_4$ data are particularly interesting. This ferrite is a normal spinel in which the zinc has the same immediate surroundings as it has in ZnO. The peak looks almost as if it were split by the action of a magnetic field, but calculations show that the local fields must be too small by at least an order of magnitude.

This is not the only remarkable edge structure which a tetrahedron of oxygen molecules can produce. A number of years ago, in Professor Beeman's laboratory I measured the edge structure of manganese in $KMnO_4$. Here one gets a narrow peak which is only a bit wider than the crystals.

We find this same symmetry in $K_2CrO_4$. Van Nordstrand has grouped the edges for these two complexes together, which certainly must be proper. Nevertheless, they are different in at least one respect, which may or may not be significant. In the chromate, there is a second small hump, first observed I believe by Sanner [16], between the first peak and the main edge. This brings me to the final point I would like to make.

I cannot subscribe to an opinion expressed in the published discussion which followed one of Van Nordstrand's [11] presentations in which it was stated that there was an advantage to the low resolution of the single crystal work. These edge structures are already so lacking in detail that one can scarcely afford to lose any—and it is possible to mechanize a two-crystal spectrometer so as to painlessly get information out to several hundred volts beyond the initial edge. If we do not take all the resolution we can, I feel we are merely proving the old saw that in the dark all cats are grey.

## REFERENCES

1. L. G. Parratt, C. F. Hempstead, and E. L. Jossem, Phys. Rev. 105, 1228 (1957).
2. L. G. Parratt, Phys. Rev. 56, 295 (1939).
3. W. W. Beeman and J. A. Bearden, Phys. Rev. 56, 455 (1942).
4. J. C. Slater, Phys. Rev. 98, 1039 (1955).
5. W. M. Lomer and W. Marshall, Phil. Mag. Ser. 8, 3, 185 (1958).
6. N. F. Mott and K. W. H. Stevens, Phil. Mag. Ser. 8, 2, 1364 (1957).
7. R. J. Weiss, Rev. Mod. Phys. 30, 59 (1958).
8. C. G. Shull and M. K. Wilkinson, Phys. Rev. 97, 304 (1955).
9. W. W. Beeman and J. A. Bearden, Phys. Rev. 61, 455 (1942).
10. L. B. Leder, H. Mendlowitz, and L. Marton, Phys. Rev. 101, 1460 (1956).
11. R. A. Van Nordstrand, Conference on Non-Crystalline Solids, John Wiley & Sons, Inc., New York (1960), p. 168.
12. F. A. Cotton and H. P. Hanson, J. Chem. Phys. 25, 619 (1956).
13. J. A. Bearden and T. M. Snyder, Phys. Rev. 59, 162 (1941).
14. S. Kiyono, Sci. Rep. Tohoku Univ. 37, 250 (1953).
15. F. A. Cotton and H. P. Hanson, J. Chem. Phys. 28, 83 (1958).
16. V. H. Sanner, Thesis, Uppsala University, Sweden (1941).

# X-RAY ABSORPTION EDGE STUDIES
# OF SUPPORTED COBALT CATALYSTS

## Rolland O. Keeling, Jr.*

Gulf Research and Development Company
Pittsburgh, Pennsylvania

## ABSTRACT

Evidence is presented which suggests that the adsorption sites necessary to initiate the formation of the $\delta$-phase ($Co^{+2}$ ions occupying positions in the structure of the support) in supported cobalt catalysts are acid sites. In addition the $\delta$-phase is found to be more difficult to reduce than the $\beta$-phase ($Co_3O_4$), and the presence of a competing ion ($Ni^{+2}$) in the impregnating solution is found to increase the proportion of cobalt in the $\beta$-phase. The assignment of octahedral coordination to the $Co^{+2}$ ions in the $\delta$-phase follows from measurements of the Co $K$-absorption edge in spinel-type compounds.

## INTRODUCTION

In a previous study [1] of cobalt $K$-absorption edges in supported cobalt-on-alumina ($Co/Al_2O_3$) catalysts, the shape and position of the edge was found to vary with concentration over the range from 0.8 to 7.5 wt % cobalt. The edge corresponds to that of $Co_3O_4$ at 7.5% and shifts to lower energy with decreasing concentration (Fig. 1). This observation was attributed to an increased proportion of divalent cobalt at the low concentration [2]. Similar measurements [1] on supported cobalt-on-silica ($Co/SiO_2$) over the same concentration range gave only the $Co_3O_4$ edge (Fig. 2). Subsequent magnetic susceptibility and chemical kinetic studies [3] confirmed the absorption edge results and extended our knowledge of the structure of these catalysts. The resulting model [3] of the $Co/Al_2O_3$ system postulates the existence of two cobalt-containing "phases." One of these is a dispersed ($\delta$) phase predominant at low con-

*Now with Department of Physics: The Michigan College of Mining and Technology; Houghton, Michigan.

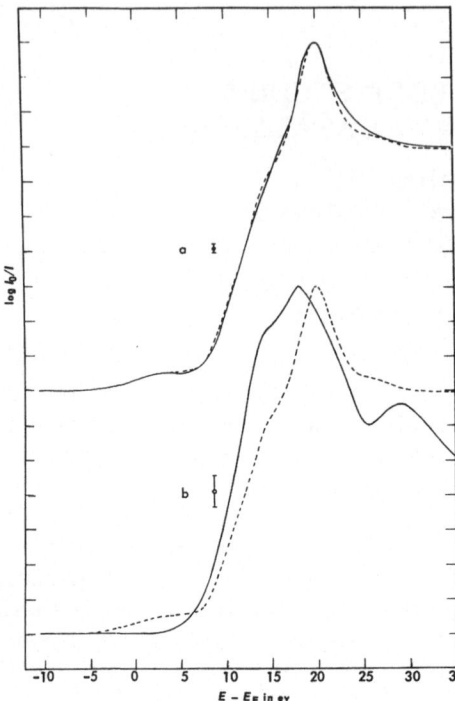

Fig. 1. Cobalt $K$-absorption edges in (a) 7.5% $Co/Al_2O_3$, and (b) 0.8% $Co/Al_2O_3$. The dotted-line curve is $Co_3O_4$. The mean standard deviation of experimental points from the curves is shown here and in succeeding curves (where significant) by the customary vertical bar.

centrations (<1.5%) and consisting of divalent cobalt associated with the alumina support. The other is a nondispersed ($\beta$) phase predominant at higher concentrations and consisting of $Co_3O_4$. When impregnation is carried out in relatively weak solutions [below about 0.2 M $Co(NO_3)_2$], most of the cobalt is adsorbed by the alumina surface. Subsequent calcination of the dried catalyst causes this adsorbed cobalt to diffuse into the alumina structure, giving the $\delta$-phase. In stronger impregnating solutions the adsorption sites are saturated, leaving a relatively large portion of the cobalt in solution in the pores of the support. This solution dries to give a hydrated cobalt nitrate which in turn converts to $Co_3O_4$ during calcination. One interesting result of the magnetic measurements was that the absolute amount of cobalt in the $\delta$-phase did not increase to a maximum with increasing total cobalt concentration and then remain constant as might be expected but, after attaining a maximum at about 1.2% total cobalt, began to decrease until it was virtually undetectable at 7.0%. This result is explainable if, during calcination, the occluded cobalt nitrate incorporates all adsorbed $Co^{+2}$ ions with which it is in contact.

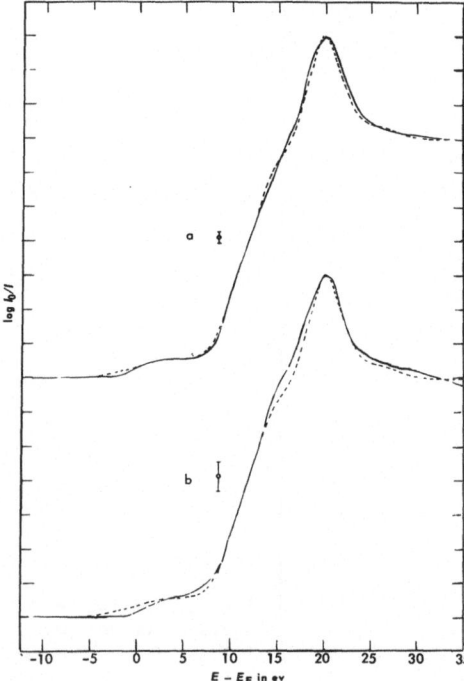

Fig. 2. Cobalt $K$-absorption edges in (a) 6.2% $Co/SiO_2$, and (b) 0.8% $Co/SiO_2$. The dotted-line curve is $Co_3O_4$.

In $Co/SiO_2$ there is no detectable $\delta$-phase above 0.5 wt % cobalt. Below this concentration the evidence is inconclusive. In any event the tendency for the formation of a $\delta$-phase on silica is much less than on alumina. The calcined silica-supported catalyst is viewed as a simple distribution of particles of $Co_3O_4$ within the pores of the support.

The present work is concerned with pursuing certain implications of the model of these catalysts by the X-ray absorption edge technique. Points covered in this study are:

1.  The relation between the acidity of a support and its ability to form a $\delta$-phase.

2.  The relative reducibilities of the $\delta$- and $\beta$-phases.

3.  The effect of a competing ion ($Ni^{+2}$) in the impregnating solution.

4.  The coordination of the $Co^{+2}$ ion in the $\delta$-phase.

## EXPERIMENTAL

The absorption edges reported here were recorded with a two-crystal X-ray spectrometer employing the (111) planes of

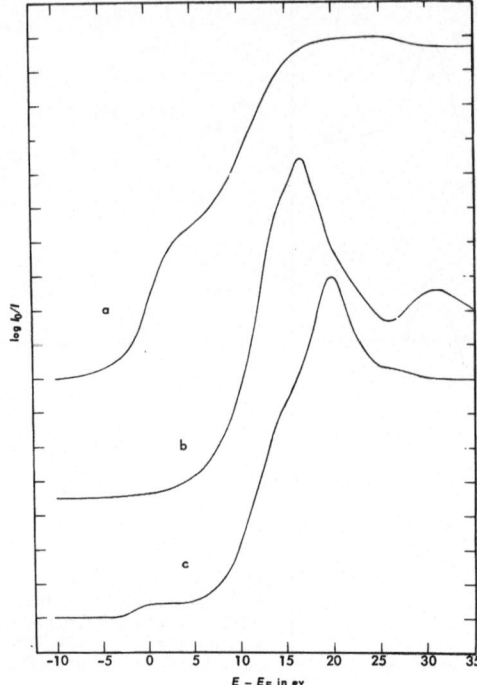

Fig. 3. Cobalt $K$-absorption edges in (a) metallic cobalt; (b) $CoO$; and (c) $Co_3O_4$.

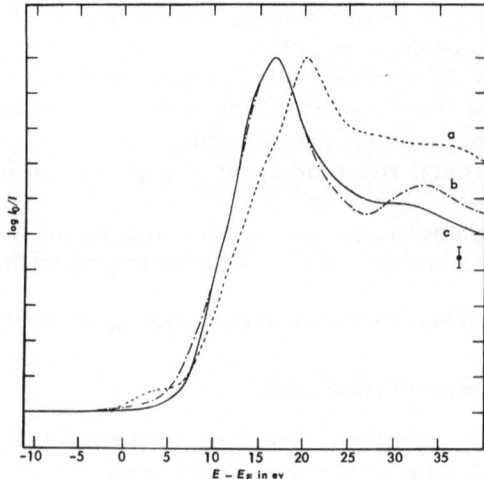

Fig. 4. Cobalt $K$-absorption edges in (a) $Co_3O_4$; (b) $CoO$; and (c) $1.0\% Co/SiO_2-Al_2O_3$.

silicon crystals in the (1,-1) position. The X-ray source was a Machlett OEG-50, tungsten-target tube powered by a Norelco X-ray generator with current and voltage stabilization. The data were collected automatically in stepwise intervals of approximately 0.5 or 1.0 ev from about 15 ev below the edge to 35 ev above it. An automatic sample changer made it possible to measure the transmitted intensity for each of three samples, one of which was a reference sample of CoO or $Co_3O_4$ (Fig. 3), and to monitor the incident beam at each spectrometer setting before proceeding to the next. Transmitted intensities were measured with a Geiger counter and the associated Norelco scaler. Elapsed times for a fixed count (usually 6400) were recorded by a Streeter-Amet printer. The final curves are averages of five independent runs. These curves are normalized plots of log $(I_0/I)$, where $I_0$ is the incident intensity and $I$ the transmitted intensity in counts/sec, vs. $E$-$E_F$, where $E_F$ is the energy of the first inflection point on the edge for cobalt metal (Fig. 3). The normalization is such as to make the height of all edges equal.

Samples of catalysts were in the form of thin pressed disks 1 in. in diameter and of such thickness as to give from 0.5 to 1.0 mg cobalt/$cm^2$ of the disk. Samples of compounds were prepared by spreading the powdered substance on cellophane tape to give a uniform thickness of from 1.0 to 3.0 mg cobalt/$cm^2$, depending on the compound.

## RESULTS

### Cobalt-on-Silica-Alumina

The presence of a $\delta$-phase on alumina and its absence (in detectable amounts) on silica suggests a direct connection between $\delta$-phase formation and the acidity of the support. If so, then a coprecipitated silica-alumina support, being even more acidic than alumina, should yield a more prominent $\delta$-phase. With this in mind, two samples of cobalt on a coprecipitated support of 50 wt % silica and 50 wt % alumina ($Co/SiO_2$-$Al_2O_3$) were examined by the absorption edge technique. One sample contained 1.0% cobalt and the other 3.2%. The results for the 1.0% sample are shown in Fig. 4. The initial rise of the absorption coefficient and the position of maximum absorption in this sample coincide with that of CoO, indicating divalent cobalt. There is no suggestion of the presence of $Co_3O_4$.

In the more concentrated sample (Fig. 5), the absorption

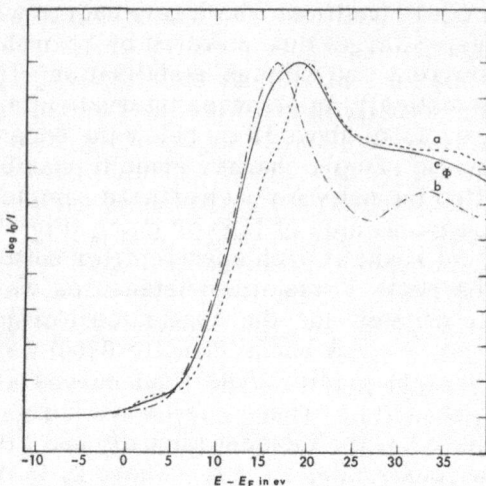

Fig. 5. Cobalt $K$-absorption edges in (a) $Co_3O_4$; (b) $CoO$; and (c) 3.2% $Co/SiO_2-Al_2O_3$.

maximum has shifted to higher energy and the decline of the absorption coefficient above the maximum is less rapid. Both of these features suggest the presence of some $Co_3O_4$. Thus, it follows that this system may be described in a manner analogous to that for $Co/Al_2O_3$. That is, a $\delta$-phase (consisting of divalent cobalt associated with the support) predominates at low concentration while a $\beta$-phase (consisting of $Co_3O_4$) predominates at higher concentrations. Assuming that within the limits of measurement the 1.0% sample consists of pure

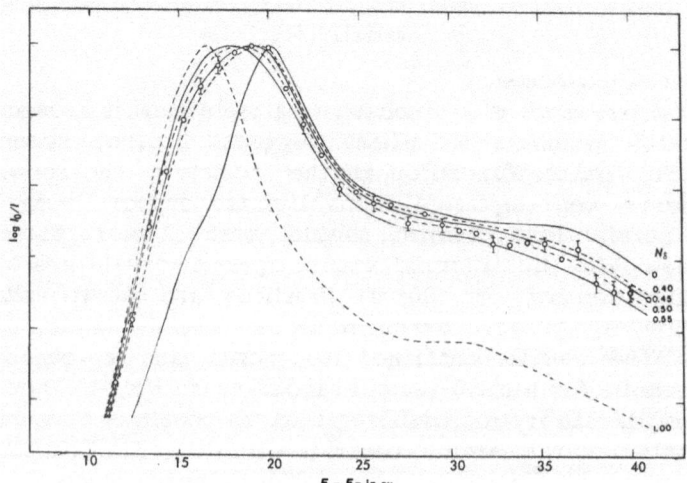

Fig. 6. Experimental points for 3.2% $Co/SiO_2-Al_2O_3$ compared to calculated curves for various values of $N_\delta$.

$\delta$-phase and that the 3.2% sample is a mixture of the two phases, a quantitative analysis of the 3.2% sample was carried out by comparing the experimental points with curves computed from the weighted averages of curves for the pure phases. A further assumption underlying this process is that the absolute absorption coefficients at maximum absorption in the two pure phases are equal. The comparison is shown in Fig. 6, in which it can be seen that the calculated curves for $N_\delta$ (the fraction of cobalt in the $\delta$-phase) = 0.45 and $N_\delta$ = 0.50 agree about equally well with the experimental points; the former gives the better fit in the low-energy region, and the latter the better fit in the high-energy region. From these curves we conclude that the fraction of cobalt in the $\delta$-phase in the 3.2% sample is 0.48 ± 0.05.

It is interesting to compare the results of this work with previous quantitative results on the $Co/Al_2O_3$ and $Co/SiO_2$ systems obtained from the magnetic susceptibility measurements [3]. This is done in Fig. 7, in which $N_\delta$ is plotted as a function of total cobalt concentration for each of the three systems. From this plot it is evident that the $\delta$-phase persists to much higher concentrations in $Co/SiO_2-Al_2O_3$ than in $Co/Al_2O_3$. A

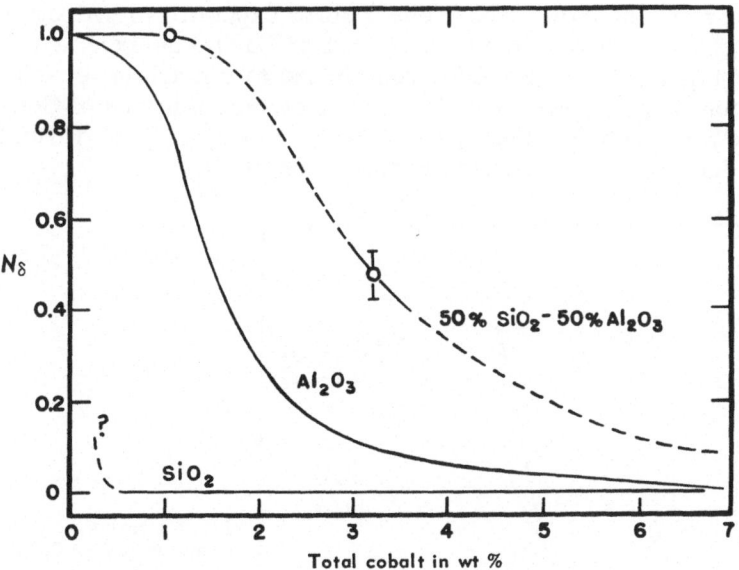

Fig. 7. Fraction ($N_\delta$) of cobalt present in the $\delta$-phase as a function of total cobalt concentration for each of the three supports: $Al_2O_3$, $SiO_2$, and $SiO_2-Al_2O_3$.

measure of the tendency of a support to form a $\delta$-phase may
be taken as the concentration at which $N_\delta = 0.5$. In $Co/SiO_2$,
the exact value of this concentration is unknown but it certainly lies below 0.5 wt % Co. In $Co/Al_2O_3$ it is 1.5% and in
$Co/SiO_2$-$Al_2O_3$ it is approximately 3.0%. These concentrations
are in the same order as the relative acidities of the supports
and therefore bear out the hypothesis of a connection between
the acidity of the support and its ability to form a $\delta$-phase.

### Relative Stability of the $\delta$- and $\beta$-Phases in $Co/Al_2O_3$ to Attempted Reduction in Hydrogen

Results of measurements to determine the relative stability
of the $\delta$- and $\beta$-phases to attempted reduction in hydrogen are
presented in Figs. 8 and 9. In Fig. 8, the edge for a sample consisting primarily of the $\delta$-phase is shown before (a) and after
(b) a 5-hr exposure to hydrogen at 500°C. This sample, with
1.73 wt % Co, was prepared by a triple impregnation in a solution of 0.05 M $Co(NO_3)_2$ with calcination at 500°C between successive impregnations. The similarity of these curves, particularly in the region of the initial rise, indicates that very little
reduction has occurred. In Fig. 9 the edge for a singly im-
pregnated sample having 5.0 wt % Co is shown before (a) and
after (b) the same treatment. Before exposure to hydrogen the
edge is virtually identical to that of $Co_3O_4$, as expected. Following exposure the edge has shifted to a much lower energy.
This curve is interpreted as arising from a mixture of CoO and
metallic cobalt. Hence, the $\delta$-phase in $Co/Al_2O_3$ is difficult to
reduce while the $\beta$-phase reduces readily.

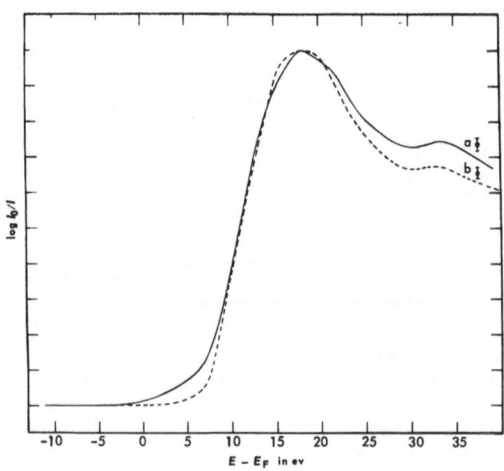

Fig. 8. Cobalt $K$-absorption
edges in 1.74% $Co/Al_2O_3$
(triple impregnation) be-
fore (a) and after (b) a
5-hr exposure to hydrogen
at 500°C.

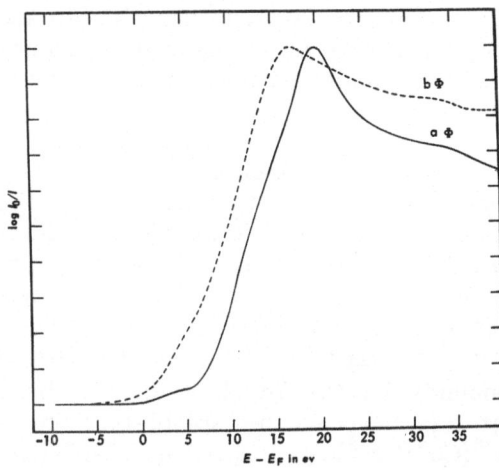

Fig. 9. Cobalt $K$-absorption edges in 5.0% Co/Al$_2$O$_3$ (single impregnation) before (a) and after (b) a 5-hr exposure to hydrogen at 500°C.

## Competitive Effect of Ni$^{+2}$ in the Impregnating Solution

The competitive effect of another cation in the impregnating solution is illustrated in Fig. 10. Curves (a) and (b) are for Co/Al$_2$O$_3$ catalysts containing 1.0 and 1.5 wt % Co, respectively. These differ in the expected way, with (b) showing a larger proportion of Co$_3$O$_4$ than (a). Curve (c) is from a sample containing 1.0 wt % Co + 0.5 wt % Ni. This curve, instead of being similar to (a), is more nearly the same as (b), sug-

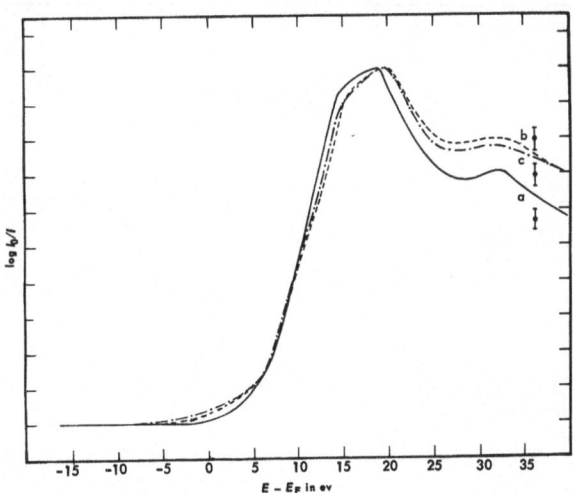

Fig. 10. Cobalt $K$-absorption edges in (a) 1.0% Co/Al$_2$O$_3$; (b) 1.5% Co/Al$_2$O$_3$; and (c) 1.0% Co + 0.5% Ni/Al$_2$O$_3$.

gesting nearly equal proportions of $Co_3O_4$ in samples (b) and (c). In other words, the presence of nickel has forced an abnormally large amount of the cobalt into the $\beta$-phase by tying up some of the adsorption sites.

### Effect of Coordination on the Cobalt K-Absorption Edge

The observed magnetic moment of the $Co^{+2}$ ion in the $\delta$-phase in $Co/Al_2O_3$ (4.94 $\mu_B$) was found to correspond more closely to that for octahedral coordination ($\sim 5.0 \mu_B$) than to that for tetrahedral coordination ($\sim 4.3 \mu_B$) [3]. In order to see whether or not the effect of coordination could be observed in absorption edges it was necessary to carry out a series of measurements on compounds having divalent cobalt in a known coordination. For this purpose the spinel-type compounds $Ge[Co_2]O_4$, $Fe^{+3}[Co^{+2}Fe^{+3}]O_4$, $Co[Al_2]O_4$, $Co[Cr_2]O_4$, and $Co[Mn_2]O_4$ were chosen.* The first two, together with the oxide $CoO$, provide three samples in which the $Co^{+2}$ ions occupy octahedral sites. The absorption edges for these three compounds are shown in Fig. 11. Note that the midpoint of the edge ($\sim 11.3$ ev) is virtually the same for all three compounds but that the peak position is shifted to higher energy for $Ge[Co_2]O_4$ and $Fe^{+3}[Co^{+2}Fe^{+3}]O_4$, than for $CoO$. Figure 12 com-

*In the spinel formulas the ions occupying octahedral sites are enclosed in brackets while those occupying tetrahedral sites are in front of the brackets.

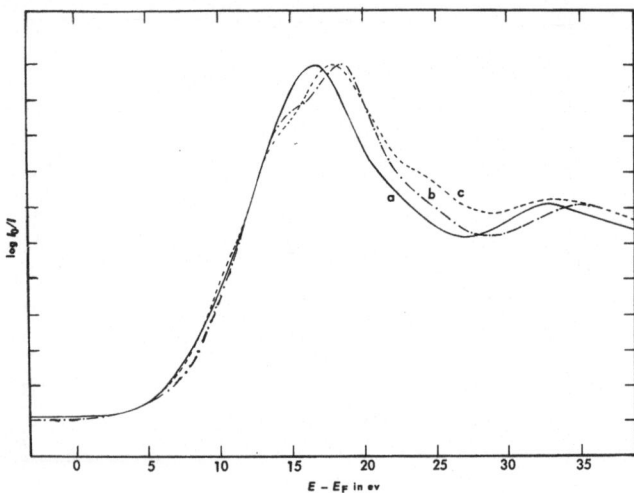

Fig. 11. Cobalt K-absorption edges for $Co^{+2}$ ions in octahedral coordination: (a) $CoO$; (b) $Ge[Co_2]O_4$; and (c) $Fe^{+3}[Co^{+2}Fe^{+3}]O_4$.

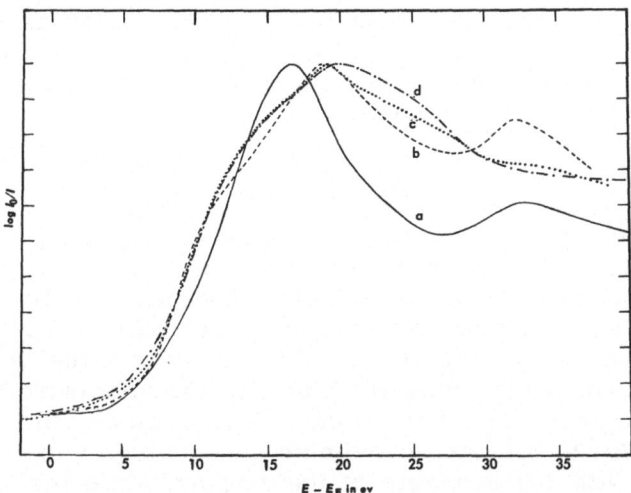

Fig. 12. Cobalt $K$-absorption edges for $Co^{+2}$ ions in (a) CoO and in tetrahedral sites of the spinel structure (b): $Co[Al_2]O_4$; (c) $Co[Cr_2]O_4$; and (d) $Co[Mn_2]O_4$.

pares the edges in $Co[Al_2]O_4$, $Co[Cr_2]O_4$, and $Co[Mn_2]O_4$ in which the $Co^{+2}$ ions occupy tetrahedral sites. Note that the midpoints of the edges for these compounds also occur at the same energy but that this energy ($\sim$10.1 ev) is significantly below that for octahedral coordination. Furthermore, the positions of maximum absorption for tetrahedral coordination occur at higher energies than for octahedral coordination. The curve for CoO is reproduced in Fig. 12 to illustrate this difference. If the difference in energy between the midpoint of the edge and the absorption maximum is taken as a measure of the "width" of the edge, then this width is significantly greater for the compounds containing cobalt in tetrahedral sites than for those containing cobalt in octahedral sites. Measured values of the width are given in the following table:

|  |  |  |
|---|---|---|
|  | CoO | 5.3 ev |
| Octahedral coordination | $Fe^{+3}[Co^{+2}Fe^{+3}]O_4$ | 6.8 ev |
|  | $Ge[Co_2]O_4$ | 7.1 ev |
|  | $Co[Al_2]O_4$ | 8.9 ev |
| Tetrahedral coordination | $Co[Cr_2]O_4$ | 9.0 ev |
|  | $Co[Mn_2]O_4$ | 9.7 ev |

The widths for the $\delta$-phase in $Co/Al_2O_3$ (taken as the 0.8% sample in Fig. 1) and in $Co/SiO_2-Al_2O_3$ (Fig. 4) are 7.0 ev and 5.9 ev, respectively. Both values are in the range corresponding to octahedral coordination.

## CONCLUSIONS

Results of this work are consistent with the previously proposed two-phase model of supported cobalt catalysts [3]. The fraction of cobalt present in the $\delta$-phase in the calcined catalysts is a function of the support in the order: $Co/SiO_2-Al_2O_3$ > $Co/Al_2O_3$ > $Co/SiO_2$. This observation leads to the suggestion that the adsorption sites which initiate formation of the $\delta$-phase are acid sites. The relative reducibilities of the $\delta$- and $\beta$-phases in $Co/Al_2O_3$ indicate an intimate association of the $\delta$-phase cobalt with the structure of the support, while the $\beta$-phase, consisting of particles of $Co_3O_4$, does not interact with the support. The competing ion effect shows that the adsorption sites are the same for $Ni^{+2}$ as for $Co^{+2}$ and that the presence of the nickel ions in the impregnating solution forces an abnormally large proportion of cobalt into the $\beta$-phase by tying up a portion of the sites. The measurements on spinel-type compounds lead to the assignment of octahedral coordination to the $Co^{+2}$ ions in the $\delta$-phase, in agreement with the magnetic susceptibility measurements.

## ACKNOWLEDGMENT

The author wishes to express appreciation to his former colleagues in the Physical Science Division of the Gulf Research and Development Company for many stimulating discussions and helpful suggestions. He also wishes to thank Dr. B. B. Wescott, Executive Vice President of the Gulf Research and Development Company, for permission to publish this work.

REFERENCES

1. R. O. Keeling, Jr., J. Chem. Phys. 31, 279 (1959).
2. W. W. Beeman and J. A. Bearden, Phys. Rev. 61, 455 (1942).
3. J. R. Tomlinson et al., Proceedings of the Second International Congress on Catalysis, Editions Technip, Paris (1961).

# SINGLE-CRYSTAL SPECTROMETERS
# FOR X-RAY ABSORPTION SPECTROSCOPY

## George J. Klems, Badri N. Das,
## and Leonid V. Azároff

Metallurgical Engineering Department
Illinois Institute of Technology
Chicago, Illinois

## ABSTRACT

A comparison study of the adaptability of several commercially available diffractometers for X-ray absorption fine-structure studies was carried out. The following experimental arrangements were tested:

1. Continuous scanning with a proportional counter recording total counts accumulated at predetermined intervals.

2. Point-by-point scanning (manual and automatic) using a scintillation counter, a proportional counter, and a geiger counter.

Several kinds of slit geometries, X-ray tubes, and sample thicknesses have been tested. All measurements were carried out using copper foils having minimum preferred orientation and silicon single-crystal analyzers. The results were compared to those obtained using a two-crystal instrument employing two parallel silicon crystals and a scintillation counter followed by a pulse-height analyzer. It is concluded that comparative measurement can be carried out with commercial instruments but that quantitatively accurate absolute determinations of the edge structure require more highly stabilized X-ray generators than those presently manufactured.

## INTRODUCTION

The present investigation was undertaken to determine whether commercially available X-ray diffraction equipment can be utilized successfully in X-ray absorption spectroscopy and what the best arrangement for such studies should be. Al-

This research was supported in part by the National Science Foundation.

though a number of commercial and custom-built instruments have been described in the literature [1-3], virtually no comparisons of their relative efficiency have been reported. The present comparisons were made in two ways: (1) Single-crystal spectrometers using different kinds of slit systems, detectors, and scanning procedures were examined; and (2) the results of single-crystal measurements were compared to double-crystal spectrometer measurements. In addition, the effect of using different crystals in a double-crystal spectrometer and the role of foil thickness also were examined. Because the absorption-edge structure of copper has been previously studied quite carefully [4, 5], all reported measurements were made using copper foils. (Similar results were obtained in a parallel investigation of nickel foils.)

## SINGLE-CRYSTAL SPECTROMETERS

The X-ray diffraction instruments of three manufacturers were examined. Their selection was determined primarily by their availability in our laboratory.

The General Electric instrument (Fig. 1) was operated in two ways: (1) The analyzing crystal was driven continuously with a slow-speed motor and the counts accumulated in a proportional counter were printed out at predetermined time intervals; and (2) the angular interval thus determined was subsequently used in a manual point-by-point counting over the same angular range. No significant difference was observed in the results obtained by the two counting procedures. It was not possible, however, to slow down the continuous scanning sufficiently to accumulate enough counts for good counting statistics.

The Norelco instrument (Fig. 2) was equipped with geneva gears which advanced the angular setting of the analyzing crystal in equal steps of 0.01°. At each position, the time required to accumulate a predetermined total count in a geiger counter was printed out. The instrument was actuated by a special programmer built for this purpose, based on the design of Van Nordstrand [6].

The Hilger instrument (Fig. 3) was manually advanced in regular angular increments and the total counts accumulated in a scintillation counter during a fixed time were recorded. Because this instrument became available only recently, its full evaluation was not completed in time for inclusion in this report. Our interest in this instrument was primarily one of

Fig. 1

testing the advantage of using a very narrow focal spot $(10\mu)$ to define the angular width of the beam.

The actual arrangements used and their respective resolving powers expressed as $\Delta\lambda/\lambda$ are summarized in the table. The pulses from the proportional and scintillation counters were recorded after passage through a pulse-height discriminator; those coming from the argon-filled geiger counter were counted directly. No significant difference attributable to instrument differences was detected in the results obtained with the three instruments used. In Fig. 4, the results of the three basically different arrangements are compared to those obtained by Krogstad [5], using a double-crystal spectrometer and by Hayasi [4] with a bent-crystal spectrometer (photographic). The latter two are representative of the best published curves for copper and are included for that reason.

As can be seen in Fig. 4, the single-crystal spectrometers show poorer resolution immediately at the edge than the two "reference" curves. The extended structure beginning about 30 ev from the edge, however, is very nearly the same in all five curves; any visible differences are of the same order as

Fig. 2

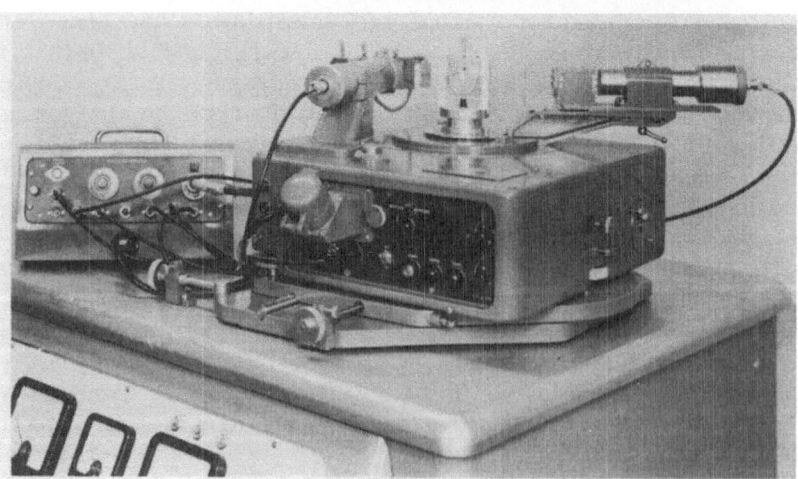

Fig. 3

## Single-Crystal Spectrometer Arrangements

| Crystal | Rock-ing-curve width (sec)* | Slit width (degrees) | Counter used | Counting interval (degrees 2θ) | Resolution (Δλ/λ)§ |
|---|---|---|---|---|---|
| Silicon (111) | 18 | 0.05 | Proportional | Continuous† | 1000 |
| Silicon (111) | 18 | 0.05 | Proportional | 0.02 | 1000 |
| Silicon (111) | 5.8 | none‡ | Scintillation | 0.01 | — |
| Silicon (111) | 5.8 | 1/30 | Geiger | 0.01 | 1700 |
| Topaz (101) | 28 | 1/30 | Geiger | 0.01 | 800 |

*For Ag $K_\alpha$ radiation.
†Read at 0.002 or 0.02° intervals.
‡Focal spot width of 10 μ used as defining slit.
§Resolution of double-crystal instrument used in this investigation was 7000.

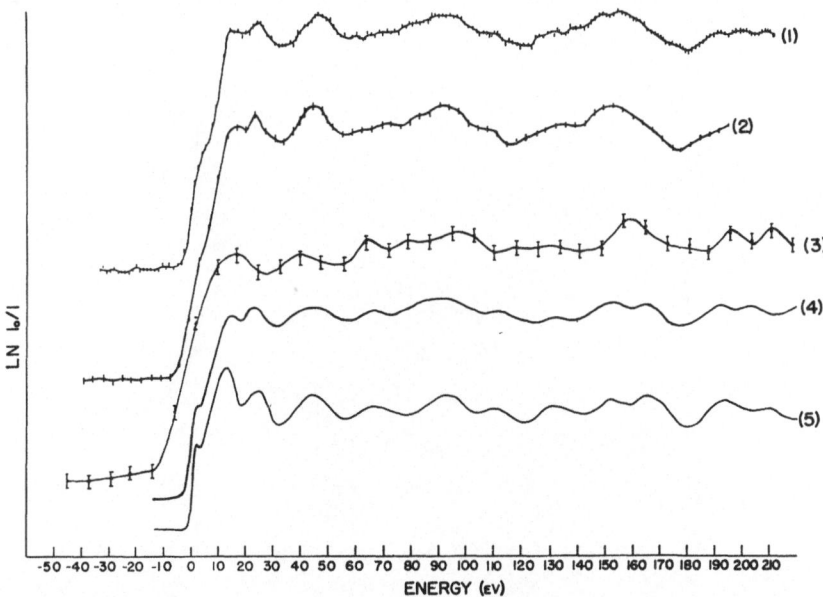

Fig. 4. X-ray absorption-edge fine structure of copper foils approximately 7.6 μ thick. The single-crystal spectrometer curves were obtained with: (1) Topaz (101), step-scanning at 0.01° intervals; (2) silicon (111), step-scanning at 0.01° intervals; (3) silicon (111), continuous scanning, counts recorded at 0.02° intervals. Curves (4) and (5) are the results of measurements by Krogstad [5] and Hayasi [4], respectively.

those between the two reference curves. Curve (3) shows the poorest resolution, but this is because of the much poorer counting statistics obtained in the time interval used.

## DOUBLE-CRYSTAL SPECTROMETER

The double-crystal spectrometer used in this study (Fig. 5) was built by Van der Hyde to fit a standard Norelco diffractometer. Since the extended structure can be recorded quite accurately with a single-crystal instrument, the present investigation was limited to studies of the structure in the immediate vicinity of the edge. Three pairs of crystals were used: Si-Si, Si-topaz, and topaz-topaz. The rocking-curve widths determined with Ag $K_a$ radiation were 8 sec for the silicon crystals and 28 sec for the topaz crystals. Because the interplanar spacing of silicon ($d_{111}$ = 3.138 A) is larger than that of topaz ($d_{101}$ = 1.356 A), the much sharper rocking curves of silicon are relatively diminished in effectiveness by the greater dispersion of topaz. Nevertheless, the resolution obtained with the Si-Si

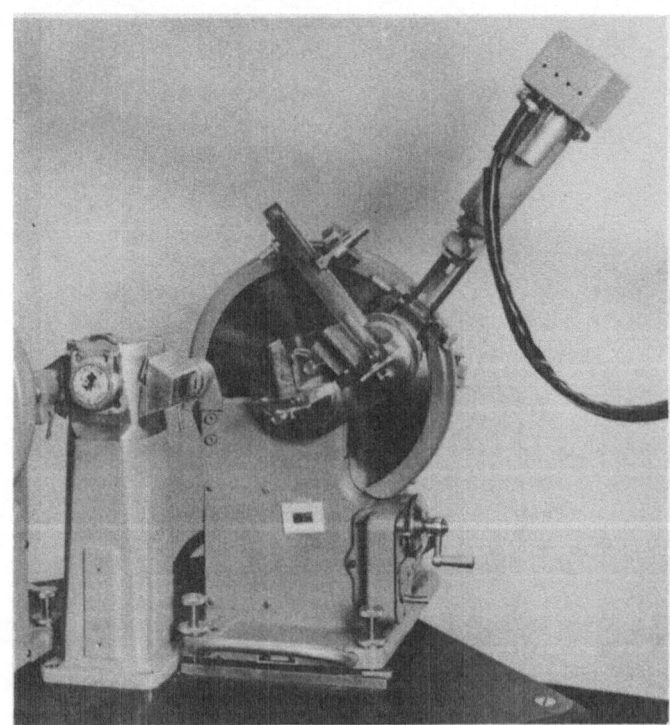

Fig. 5

combination is superior, as can be seen in Fig. 6, in which our curves are compared to those of Krogstad and Hayasi. The curves in Fig. 6 show clearly that the resolution at the edge is notably improved in our double-crystal instrument over that obtained with the single-crystal instruments (Fig. 4), but that it is not as good as that shown in the two reference curves. There are several possible reasons for this, including poorer instrumental stability, lower X-ray tube intensities from standard diffraction tubes, and fluctuations in tube intensities during the counting intervals. It is hoped to remove some of these difficulties in the immediate future by the utilization of a more accurately stabilized constant-potential X-ray generator.

## EFFECT OF FOIL THICKNESS

It is generally believed that the thinner the foils used in absorption spectroscopy, the better the results. This belief is

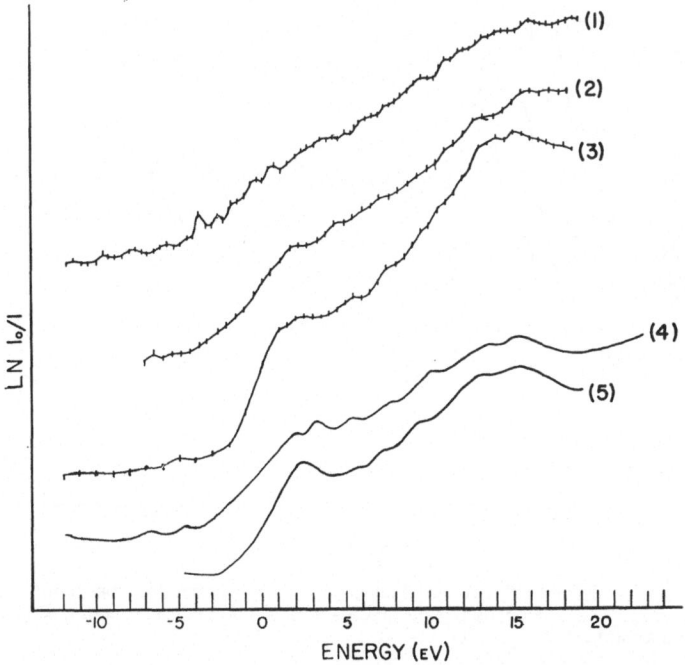

Fig. 6. Fine structure near the absorption edge of copper foils approximately 7.6 μ thick. Curve (1) was obtained with two topaz crystals in the (1, +1) position; (2) with silicon-topaz; and (3) with silicon-silicon. Curves (4) and (5) are from measurements by Krogstad [5] and Hayasi [4], respectively.

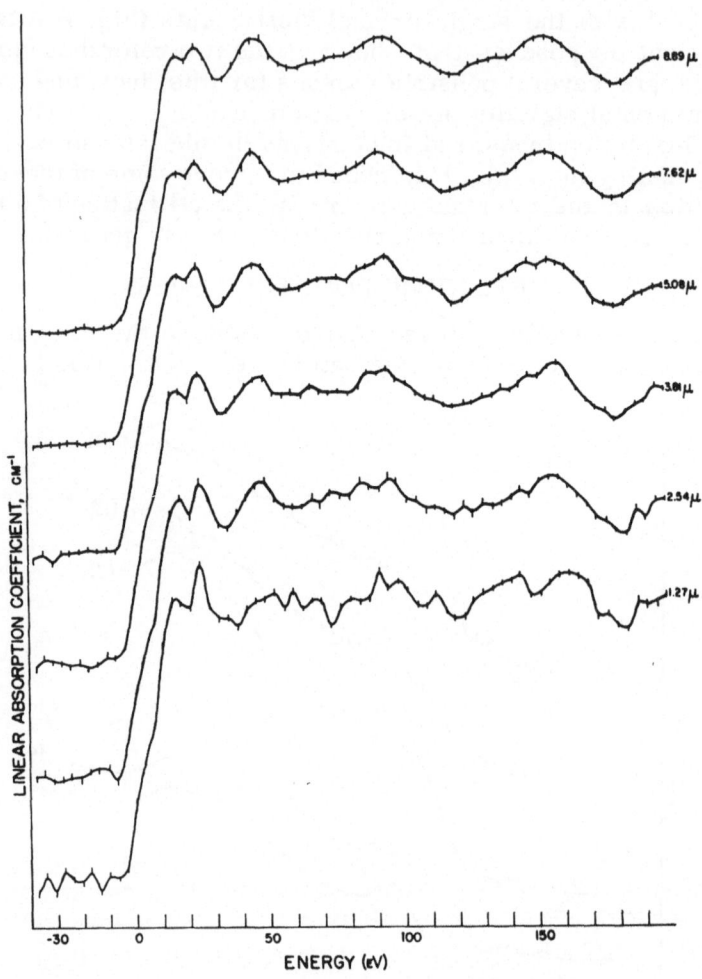

Fig. 7. Effect of foil thickness on X-ray absorption spectrum. The curves were recorded on a single-crystal spectrometer using Si (111) as an analyzer.

based partly on the increased transmissivity of such foils and the resultant improvement in the counting statistics [7]. To this end, six copper foils ranging in thickness from 8.89 to 1.27 $\mu$ were examined in the single-crystal spectrometer using the best silicon crystal as an analyzer. The results of these measurements are shown in Fig. 7, in which $\mu$ rather than $\mu x = \ln I/I_0$ is plotted so that the comparison is more meaningful. As can be seen in Fig. 7, the gross structure of the absorption spectrum does not seem to depend significantly on foil thickness. With a foil thickness of less than 3.81 $\mu$, however, additional "fine structure" appears superimposed on this gross structure. Whether this structure is physically meaningful or whether it simply reflects experimentally erratic scatter due to the decreased volume of the absorber is not clear at present. A similar increase in the structure visible at the edge has been observed in double-crystal measurements in going from 7.5 $\mu$ to 2.5 $\mu$. A complete evaluation of these results, however, requires a consideration of the "window-effect" pointed out by Parratt [7], and will be described in a separate publication.

## DISCUSSION

The foregoing analysis essentially confirms previously reached conclusions [1-3] regarding the relative suitability of single-crystal and double-crystal spectrometers. Until proven differently, it appears that the extended fine structure beginning about 30 ev from the main edge varies so gradually that single-crystal instruments are adequate for its study. Because such instruments are easily devised by suitably modifying commercial X-ray diffractometers, it is concluded that this is one area in which such instruments can be utilized. It also has been demonstrated that the quality of the analyzing crystal is an important factor in such analyses and that both narrow rocking curves and high dispersion (small $d$) are necessary for optimum performance. Another important factor not directly evident in this analysis is the need for high-intensity and well-regulated X-ray generators.

For purposes of studying the fine structure in the immediate vicinity of the absorption edge, it is necessary to utilize the greater resolving power of a double-crystal spectrometer. Here, the rocking-curve widths of the crystals appear to be more significant than the dispersion power of the crystals. As first pointed out by Parratt [7], the experimental measurements must also be corrected for asymmetries in the shapes of the rocking curves before they can be properly interpreted. This can only be done using a double-crystal spectrometer. It follows from this that a double-crystal instrument should be used whenever quantitatively accurate absorption-edge structures are sought. Two important criteria govern the choice of instrument, namely, mechanical stability and X-ray generator stability, so that it is questionable whether commercially available instruments can be used for high quality investigations.

REFERENCES

1. C. H. Shaw, Theory of Alloy Phases, American Society for Metals, Cleveland (1956), p. 13.
2. A. E. Sandstrom, Handbuch der Physik, Vol. 30, Springer Verlag, Berlin (1957), p. 78.
3. M. Sawada et al., Ann. Rep. Scient. Works, Fac. Sci. Osaka Univ. 7, 1 (1959).
4. T. Hayasi, Sci. Rep. Tohoku Univ. (Section I), 25, 661 (1936); 36, 225 (1952).
5. Reuben S. Krogstad, Ph. D. thesis, State College of Washington (1955).
6. R. Van Nordstrand, Private communication.
7. L. G. Parratt, C. F. Hempstead, and E. L. Jossem, ASTIA Document No. AD-95207, July 15, 1956.

# X-RAY ABSORPTION FINE-STRUCTURE INVESTIGATIONS AT CRYOGENIC TEMPERATURES

Farrel W. Lytle

Boeing Scientific Research Laboratories
Seattle, Washington

## ABSTRACT

The temperature dependence of the extended X-ray absorption fine structure was investigated in the range from 9-573°K. It was found that the temperature dependence of the intensity of the fine structure features was quantitatively accounted for by applying the Debye-Waller factor. A change in position of the features with the expansion and contraction of the lattice was also observed. These experimental data suggest a simple diffraction (Krönig theory) mechanism; however, the observed absorption minima do not agree with those predicted by the Krönig equation. The single-crystal, X-ray spectrometer, X-ray cryostat, and experimental conditions are also discussed in some detail.

## INTRODUCTION

A generally acceptable theory of X-ray absorption fine structure was first proposed by Krönig [1]. In his derivation the photoelectron ejected in the X-ray absorption process moves through the crystal as a free electron. By assuming that the only influence on the electron is a small perturbation due to the periodic field of the lattice, it is found that certain energy regions occur which are forbidden to the electron; a forbidden energy region is that combination of momentum and direction for which the Bragg diffraction condition is fulfilled (Brillouin zone boundary). An allowed energy region will result if the electron can travel freely without diffraction. Since energy must be conserved in the X-ray absorption process, an X-ray will not be absorbed if the energy of the resulting

photoelectron corresponds to a forbidden energy region. Therefore, the forbidden zones are areas of minimum X-ray absorption.

Comparison of Krönig's theory with experiment results in
partial agreement, but there are evidently other competing
and/or simultaneous processes involved. Efforts to modify
Krönig's theory have been made and other energy loss mechanisms such as collision ionization or excitation of plasma
oscillations proposed. (For a discussion and original references
see Parratt [2].)

Hanawalt [3] first determined the general effect of temperature on the X-ray absorption fine structure. He showed
that as the temperature increases the extended fine-structure
features move toward the absorption edge and decrease in
amplitude, disappearing at the melting point. More recently
Borovskii et al. [4-6] have investigated the effect of temperature on the structure in and very close to the edge. Although
Borovskii's experiments were not directly applicable, in that
the present investigations were concerned with the extended
fine structure, the arguments of the authors concerning the
temperature sensitivity of electron transitions to unfilled bands
and the possibility of a collective plasma oscillation "image"
of the absorption edge appearing in the fine structure were
quite interesting. Parratt [2] has discussed the effect of temperature broadening on the widths of X-ray states and given
some experimental data on the temperature effect at the
chlorine absorption edge in KCl. In none of the above experiments was the effect at very low temperatures on the extended
fine structure investigated.

The purpose of the present investigation was to determine
the effect of temperature in the range from liquid helium to
room temperature on extended X-ray absorption fine structure.
The Krönig theory is developed in order to introduce a quantitative temperature dependence term, although complete adherence to the Krönig theory was not found.

## THEORY

In the following discussion the Krönig assumption of a free
photoelectron which is not influenced by the X-ray excited
state is accepted. It is generally agreed that this must become
very nearly true as the electron energy becomes large, i.e.,
greater than 100 ev, and it is this region in which we are

interested. After a simple diffraction argument which derives the Krönig relationship, the temperature correction (Debye-Waller) factor commonly used to predict the variation in diffracted intensity with temperature is introduced.

Consider a photoelectron of wavelength

$$\lambda = \frac{h}{(2mVe)^{\frac{1}{2}}} \tag{1}$$

where $h$ is Planck's constant; $m$ is the electron mass; $V$ is the accelerating potential; and $e$ is the electron charge.

Furthermore, if $E$ equals the electron energy in electron volts, then

$$E = \frac{h^2}{2m\lambda^2} \tag{2}$$

This electron travels as a plane wave through a cubic crystal in which the relationship between the unit cell distance $a$ and the interplanar spacing $d$ is

$$a^2 = d^2 \, (h^2 + k^2 + l^2) \tag{3}$$

where $h$, $k$, and $l$ are the Miller crystal indices. Now we impose the Bragg diffraction condition

$$n\lambda = 2d \sin \Theta \tag{4}$$

where $n$ is the order of diffraction and $\Theta$ is the angle between the electron direction and the crystal plane. Solving (2), (3), and (4) for $E$, we obtain the energy of a forbidden zone (absorption minima):

$$E = \frac{n^2 h^2 (h^2 + k^2 + l^2)}{8ma^2 \sin^2 \Theta} \tag{5}$$

In equation (5) $\Theta$ applies to only one electron direction. This condition cannot be realized in an X-ray absorption experiment. With the nonpolarized X rays and polycrystalline samples usually used, the distribution of electrons will be approximately spherical. When integrated over all directions, as Krönig [1] did, the following equation results (the factor $n^2$ is included in the Miller crystal indices):

$$E = \frac{h^2 (h^2 + k^2 + l^2)}{8ma^2} \tag{6}$$

In our simplified derivation this amounts to neglecting all planes except those perpendicular to the electron direction where $\sin^2 \Theta = 1$. A more simple form follows if $K = 1/d$:

$$E = \frac{h^2 K^2}{8m} \tag{7}$$

The effect of atomic thermal vibrations on diffracted intensities [7] is to introduce the factor $\exp(-2M)$ into the intensity expression where

$$2M = \frac{12h^2}{m_a k \Theta_D} \left\{ \frac{\Phi(X)}{X} + \frac{1}{4} \right\} \left( \frac{\sin \Theta}{\lambda} \right)^2 \tag{8}$$

in which $m_a$ is the atomic mass; $k$ is Boltzmann's constant; $\Theta_D$ is the characteristic (Debye) temperature; $T$ is the absolute temperature; $X$ is equal to $\Theta_D/T$; $\Phi(X)$ is the Debye function; and other quantities are as previously defined. In the application to Krönig structure it is not possible to determine the $\sin \Theta/\lambda$ part of equation (8) directly from the experimental arrangement as in X-ray diffraction. However, from (4) we see that

$$\frac{\sin \Theta}{\lambda} = \frac{K}{2} \tag{9}$$

Thus, the Debye-Waller factor becomes

$$2M = \frac{3h^2 K^2}{m_a k \Theta_D} \left\{ \frac{\Phi(X)}{X} + \frac{1}{4} \right\} \tag{10}$$

To apply this equation, $E$, the energy of an absorption minimum, is determined experimentally and $K$ evaluated from equation (7). Then $I_1$ and $I_2$, the intensities of an absorption minimum at temperatures $T_1$ and $T_2$, can be examined for adherence to the theory. If

$$I_1 = I_0 \, e^{-2M_1} \quad \text{and} \quad I_2 = I_0 \, e^{-2M_2} \tag{11}$$

then

$$\ln \frac{I_1}{I_2} = 2M_2 - 2M_1 \tag{12}$$

## EXPERIMENTAL

The absorption data were obtained with a conventional horizontal diffractometer used as a single-crystal spectrometer. The single crystal was placed in the sample holder and small slits used to obtain a satisfactory resolving power. The absorber was held perpendicular to the X-ray beam within a cryostat placed between the entrance slit and the diffracting crystal. The operating conditions were: Molybdenum target tube at 25 kv and 30 ma, 0.05-mm entrance and exit slits, 2° vertical divergence Soller slits, 37-cm diameter focusing circle, NaI scintillation counter, single-channel pulse-height analyzer, and automatic strip chart recording at $1/16°$ $2\Theta$/min.

With this arrangement the resolving power $\lambda/\Delta\lambda$ was approximately 1500 and sufficient for the experiment under consideration. The X-ray intensity varied from 10,000 cpm

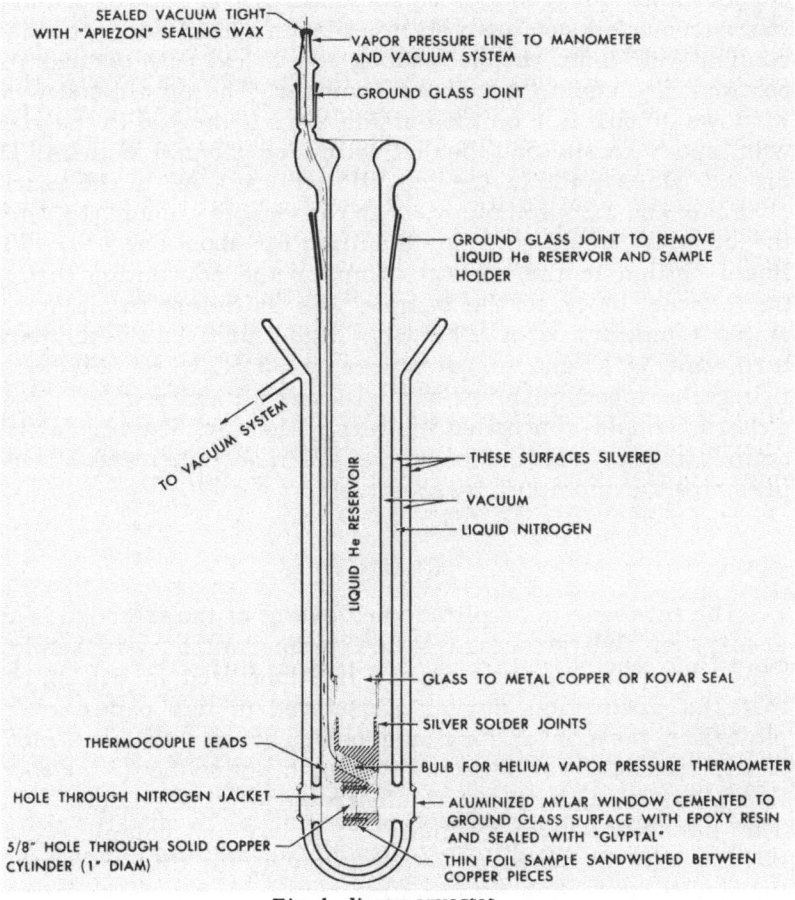

Fig. 1. X-ray cryostat.

(counts per minute) in the region of highest absorption to 300,000 cpm on the long-wavelength side of the absorption edge. The copper and nickel absorbers were prepared from at least 99.9% pure material by rolling and electropolishing to the desired thickness (about 12 $\mu$) and then annealing to remove any work hardening.

The absorbers were mounted in the cryostat shown in Fig. 1. The cryostat was constructed of 2-mm pyrex tubing with over-all dimensions of about $2\frac{3}{4}$ in. diameter and 14 in. height. The liquid coolant was contained in the central chamber, which terminated in a glass-to-kovar metal seal silver soldered to a cylindrical copper block with a hole through it for the X-ray beam. The absorber was clamped tightly to the copper block for good thermal contact. The central chamber was surrounded successively by a vacuum chamber, a secondary cooling well (note the hole along the X-ray path through this portion), and again the vacuum chamber. The aluminized mylar windows (0.001 in.) on the outside were cemented to the glass with epoxy resin and sealed with red Glyptal cement. The ground glass joint in the top allowed removal of the central chamber and sample holder. With reasonable vacuum ($10^{-4}$ mm) the cryostat would hold liquid helium for about one hour. With liquid helium in the central chamber the conduction through the copper block to the sample was sufficient to maintain a stable temperature of 9°K. With liquid nitrogen the temperature was 78°K and temperatures to 300°C were obtained by filling the inner chamber with silicone oil and regulating with a thermocouple-controlled immersion heater. The high and low temperatures were measured by thermocouples and a vapor pressure thermometer, respectively.

## RESULTS

The increase in amplitude and extent of the extended X-ray absorption fine structure at low temperatures was remarkable. A typical recorder trace is illustrated in Fig. 2. Note that the absorption curves illustrated in this paper are not plotted in the conventional manner. The curves are a plot of X-ray intensity vs. X-ray wavelength where the X-ray wavelength has been converted to electron energy in electron volts. The intensity scale (ordinate) is linear and completely arbitrary with the zero of X-ray intensity far off the bottom of the graph. It is evident from the asymmetry and detail within

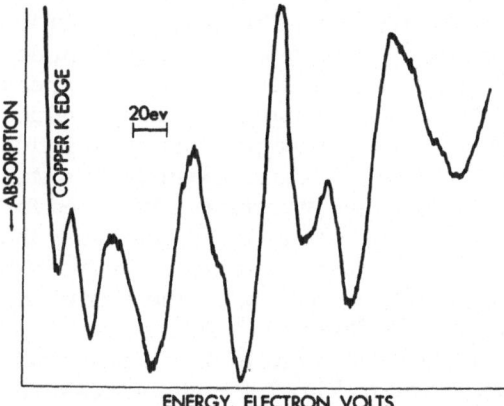

Fig. 2. X-ray absorption fine structure in copper at 78°K.

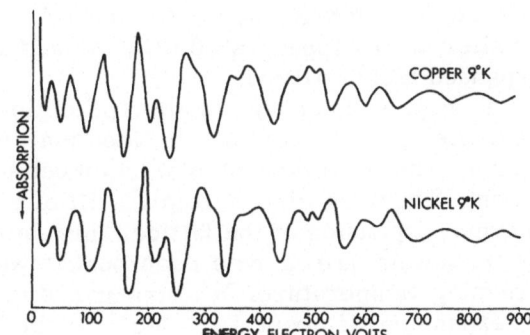

Fig. 3. X-ray absorption fine structure in copper and nickel at 9°K.

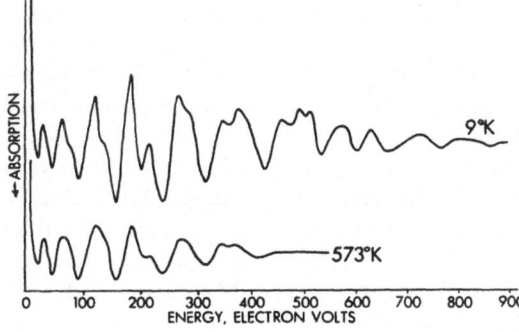

Fig. 4. Temperature dependence of the X-ray absorption fine structure in copper.

each peak that the structure is quite complicated, a convolution of many peaks. Because of this there is uncertainty in determining the energy position of each feature since the peak position and centroid do not always coincide. The energy values which are tabulated apply to the peak position (absorption maxima or minima) of the fine structure. Several tungsten lines recorded along with the fine structure were used for calibration. The background was obtained by removing the sample and adjusting to comparable X-ray intensity with aluminum absorbers. Any features present in the continuum could then be subtracted from the fine structure.

The fine structure for copper and nickel at 9°K after subtracting the background is shown in Fig. 3. Previous measurements at room temperature demonstrated the structure to about 400 ev. At 9°K the structure is much sharper and extends to at least 900 ev. The very similar features for copper and nickel are expected from the Krönig theory. Both metals are face centered cubic with nickel, $a_0 = 3.5238$ A, slightly smaller then copper, $a_0 = 3.6150$ A, and this causes the more expanded nickel pattern.

In Fig. 4 the change in absorption fine structure with temperature is illustrated. An identical effect occurred with nickel. The structure at 9°K has much greater amplitude and extent. There is also a small shift of each feature due to the thermal expansion of the lattice. This shift (which was observed by Hanawalt [3]) moves each peak toward the edge with increasing temperature. For copper and nickel with coefficients of expansion of about $10^{-5}$ deg$^{-1}$ in the temperature range of interest, this energy shift is quite small; $\Delta E/E = 10^{-5}$ ev-ev$^{-1}$-deg$^{-1}$. For example, a peak at 100 ev should shift 0.5 ev toward the absorption edge as the temperature changed from 9 to 573°K. This small change would be hidden by the inaccuracy of measurement (±1 ev). A peak at 500 ev which should move about 2.8 ev tends to become so broad and indistinct as to make accurate determination of peak position difficult. However, a change of the right order of magnitude was observed for both copper and nickel.

The integrated intensity of each absorption minimum was determined by planimetering the recorder traces. The relative intensities (normalized to make the maximum amplitude unity) at various temperatures were then plotted against peak energy. In Fig. 5 the experimental points were obtained at 9°K and 298°K. The Miller crystal indices scale indicates that high

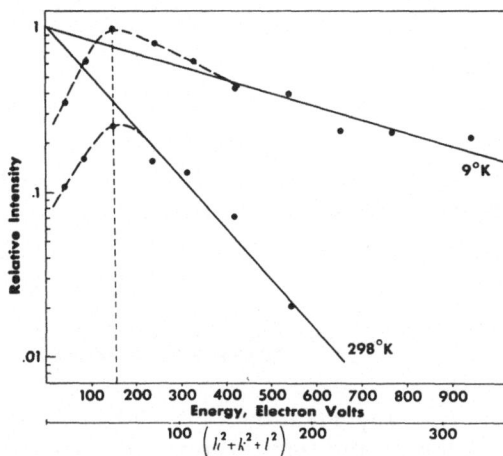

Fig. 5. Temperature dependence of the intensity of X-ray absorption fine structure.

numbered planes participate in the diffraction. The solid lines connect the calculated Debye-Waller factors, which include the appropriate temperature and energy position from equations (7) and (10). The agreement is very good at high electron energies with a break in the experimental curves at about 150 ev. Evidently this is the region at which the diffraction mechanism becomes less important and indeed is approximately what would be expected considering the basic assumption of a free electron.

Agreement with the Debye-Waller theory was also checked by another method which has been used by James and Brindley [8]. If equation (12) is written in the form of (10) and rearranged, then

$$\frac{4}{K^2}\ln\frac{I_1}{I_2} = \frac{12h^2}{m_a k\Theta_D}\left\{\frac{\Phi(X_2)}{X_2} - \frac{\Phi(X_1)}{X_1}\right\} \tag{13}$$

A plot of either side of equation (13) should be linear in temperature, where the left-hand side is determined experimentally and the right-hand side calculated theoretically. The effect of the $4/K^2$ factor makes the intensity of the peak independent of its energy position. In Fig. 6, $T_2$ is 298°K and $T_1$ varies from 9 to 573°K. Each point on the graph is an average for a number of peaks in the 200–400 ev range. The solid line is a plot of the right-hand side of equation (13). The deviation about the line is rather large and probably represents the difficulty of making quantitative intensity measurements on com-

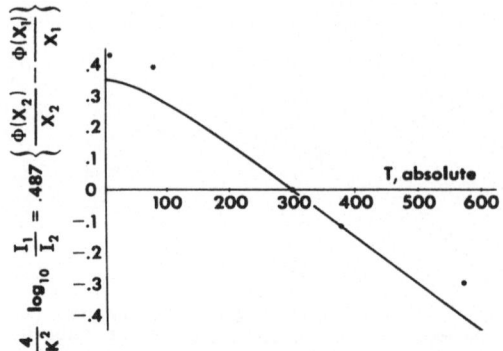

Fig. 6. Variation of intensity with temperature.

plicated structure rather than any meaningful deviation from theory. Figures 5 and 6 apply to the temperature dependence of the fine structure as observed in both copper and nickel.

The energies of absorption maxima and minima for copper and nickel as measured at 78°K are presented in the table. The first point of inflection of the $K$ edge was used as the zero energy (Cu $K$ edge = 8980.7 ev; Ni $K$ edge = 8331.4 ev) [9]. The accuracy of the energy position for the features within 200 ev of the edge was ±1.5 ev. However, the increasingly broad and poorly defined peaks at higher energies caused an uncertainty of ±5 ev at 900 ev.

## DISCUSSION

The purpose of this investigation was to determine the temperature dependence of the extended X-ray absorption fine structure. Experimental results indicate that the usual Debye-Waller factor may be applied. However, comparison of the experimentally determined absorption minima (table) with the energies predicted by the Krönig equation result in no satisfactory agreement. The Krönig theory predicts much more structure than is actually observed, although most of the other features of the absorption fine structure are accounted for. Either the theory has been oversimplified or some other mechanism should be postulated.

The change in absorption energy due to the thermal expansion and contraction of the lattice was observed and, although difficult to measure exactly, was of the correct order of magnitude. The dependence of the intensity of the fine structure on temperature was particularly striking. For elec-

Energies of Absorption Maxima and Minima in Copper
and Nickel

| Copper | | Nickel | |
|---|---|---|---|
| Maxima, ev | Minima, ev | Maxima, ev | Minima, ev |
| 21 | 9 | 27 | 10 |
| 45 | 30 | 49 | 37 |
| 89 | 60 | 98 | 72 |
| 153 | 117 | 167 | 129 |
| 199 | 180 | 215 | 193 |
| 233 | 213 | 251 | 231 |
| 312 | 263 | 332 | 289 |
| 349 | 343 | 375 | 360 |
| 418 | 369 | 435 | 404 |
| 467 | 450 | 491 | 468 |
| 493 | 488 | 507 | 501 |
| 526 | 503 | 556 | 532 |
| 591 | 562 | 628 | 593 |
| 642 | 636 | 688 | 646 |
| 751 | 711 | 827 | 787 |
| | 863 | | 901 |

The temperature of measurement was 78°K; for comparison to measurements at other temperatures, $\Delta E/E = 10^{-5}$ ev-ev$^{-1}$-deg$^{-1}$

tron energies large enough so that the free electron approximation was appropriate, the agreement with the Debye-Waller factor was good. This evidence demonstrates a scattering mechanism for the extended X-ray absorption fine structure. The experimental data (Fig. 5) also indicate the region (electron energy less than 150 ev) where this mechanism may become complicated by other phenomena. The structure in this region has been compared to characteristic electron energy loss spectra with some degree of success [10, 11].

It was originally proposed that the intensities of the absorption fine structure in different elements should vary approximately as the melting points [12]. This association was fortuitous and has been shown to be generally invalid [13, 14]. In light of the experimental data presented herein, it would appear that for measurements made at room temperature, all other factors being equal, the intensity of the fine structure of different materials should be proportional to the characteristic temperature ($\Theta_D$) of the materials involved. This suggests the possibility of obtaining more clearly defined patterns by al-

loying or otherwise combining the element of interest with a
material of high characteristic temperature.

An important consideration to the experimenter is the comparative ease with which the absorption fine structure may be
recorded using commonly available equipment. By incorporating a cryostat and making measurements at low temperatures,
a considerable range of the fine structure may be studied.
For most purposes a cryostat for use with liquid nitrogen is
adequate.

Although X-ray absorption spectroscopy will always be
complicated by the preparation of uniformly thin absorbers,
the possibility of making rapid, extensive fine-structure
measurements suggests development as an analytical tool. The
sensitivity of the structure to parameters such as particle
size, degree of crystallinity, chemical combination, and order-
disorder transformations has not been extensively investigated
and may prove interesting and fruitful.

## REFERENCES

1. R. deL. Krönig, Z. Physik 70, 317 (1931); 75, 91 (1932); 76, 468 (1932).
2. L. G. Parratt, Revs. Mod. Phys. 31, 616 (1959).
3. J. D. Hanawalt, Z. Physik 70, 293 (1931); J. Franklin Institute 214, 569 (1932).
4. I. B. Borovskii and G. N. Ronami, Izvest. Akad. Nauk SSSR, Ser. fiz. 21, No. 10 (1957).
5. I. B. Borovskii and B. A. Batyreu, Izvest. Akad. Nauk SSSR, Ser. Fiz. No. 4 (1960).
6. I. B. Borovskii and V. V. Schmidt, Doklady Akad. Nauk SSSR 129, No. 4 (1959).
7. R. W. James, The Optical Principles of the Diffraction of X-Rays, G. Bell and Sons, London (1954).
8. R. W. James, op. cit., p. 233.
9. A. E. Sandstrom, Handbuch der Physik, Vol. 30, Springer Verlag, Berlin (1957).
10. L. B. Leder, H. Mendlowitz, and L. Morton, Phys. Rev. 101, 1460 (1956).
11. P. E. Best and J. L. Robins, Proc. Phys. Soc. 77, 1046 (1961).
12. V. Coster and J. Veldkamp, Z. Physik 70, 306 (1931); 74, 191 (1932).
13. W. H. Zinn, Phys. Rev. 46, 659 (1934).
14. W. W. Beeman and H. Friedman, Phys. Rev. 56, 392 (1939).

# APPLICATIONS OF X-RAY ABSORPTION EDGE ANALYSIS

**E. A. Hakkila and G. R. Waterbury**

University of California
The Los Alamos Scientific Laboratory
Los Alamos, New Mexico

## ABSTRACT

The X-ray absorption edge technique using secondary targets as radiation sources was applied to the determination of some elements, including rhodium, uranium, yttrium, tantalum, cobalt, and cerium, that have absorption edges in the wavelength region from 0.53 to 2.2 A. Polystyrene absorption cells having path lengths between 0.12 and 3.0 cm were found to be satisfactory. For the longer wavelength radiation, the cells of shorter path length were used, and the sensitivity and precision were increased by using a low-density low-molecular weight diluent such as isopropyl alcohol. At optimum concentrations, relative standard deviations in the range of 0.2 to 0.7% were obtained. This secondary target technique requires only standard X-ray fluorescence equipment with simple and inexpensive modifications that can be attached or removed in a few minutes.

## INTRODUCTION

The need for rapid methods of determining various elements in unirradiated reactor fuels and cladding materials led to the investigation of X-ray spectrographic methods for this purpose. X-ray fluorescence methods for the determination of many elements are available, but their application to these materials of widely varying composition would require a series of calibration curves for each element determined. Therefore, the X-ray absorption edge technique, being less affected by matrix composition, was investigated for possible

Work performed under the auspices of the U. S. Atomic Energy Commission.

applications to this problem. In X-ray absorption edge analysis, the transmitted intensities for two X-ray energies, one located on each side of a suitable absorption edge of the element being determined, are related to the concentration of that element using accepted absorption principles. The continuous radiation from the X-ray tube has been used by various workers [1, 3, 4, 10-12] in conjunction with a crystal spectrometer and pulse-height analyzer to obtain the desired X-ray energies. In a few cases, such as the determination of lead with a thorium target X-ray tube, the primary X-ray lines from the tube target may be employed [9]. Engstrom [6] utilized secondary radiation from selected pure elements in the analysis of biological specimens. Because this technique requires a minimum of instrument modifications and provides high line intensities with a minimum of second order interference, its application was investigated further.

## THEORY

X rays are attenuated by matter according to the equation

$$I = I_0 \, e^{-(\mu/\rho) C L} \qquad (1)$$

where $I_0$ is the incident intensity; $I$ is the transmitted intensity; $(\mu/\rho)$ is the mass absorption coefficient of the material; $C$ is the concentration (in g/cc); and $L$ is the path length of the cell (in cm). The concentrations and mass absorption coefficients for all elements in the material must be considered. If $I_0$ is measured for the empty cell, the maximum cell path length that will give an intensity through the solvent sufficiently large to provide reasonable counting statistics can be calculated from the mass absorption coefficients for the solvent. Count rates greater than 1000 cps are preferred.

Mass absorption coefficients for any element can be related to wavelength by the equation

$$(\mu/\rho) = k \lambda^n \qquad (2)$$

In this equation $k$ is a constant for the element over a wavelength region where no absorption edge occurs, and is different for each element; $\lambda$ is the wavelength (in A); and $n$ is a constant for the element over a limited wavelength region. At the absorption edge the mass absorption coefficient changes ab-

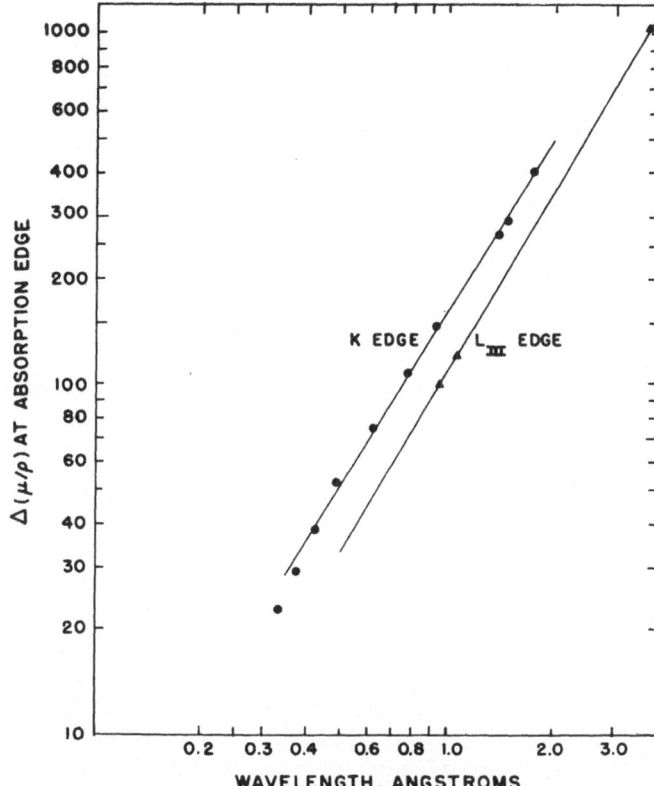

Fig. 1. Increase in mass absorption coefficient at $K$ and $L_{III}$ edges as a function of wavelength.

ruptly, the magnitude of this change being dependent on the absorption edge used and the wavelength, as shown in Fig. 1. The $L_{III}$ edge is more sensitive than the $L_{II}$, $L_{I}$, or $K$ edges for most elements but is not applicable in solution analysis to wavelengths longer than approximately 2 A. Therefore, for elements of lower atomic number than approximately 57 (lanthanum) the $K$ or $L_{II}$ edge must be used. The increase in mass absorption coefficient at the absorption edge increases with the wavelength of the edge. However, because the mass absorption coefficient for solvent and cell materials also increases with wavelength, the cell path length must be decreased correspondingly to prevent complete attenuation of the X-ray beam by the solvent.

When intensity measurements are made on each side of the absorption edge through a reference solution and through the

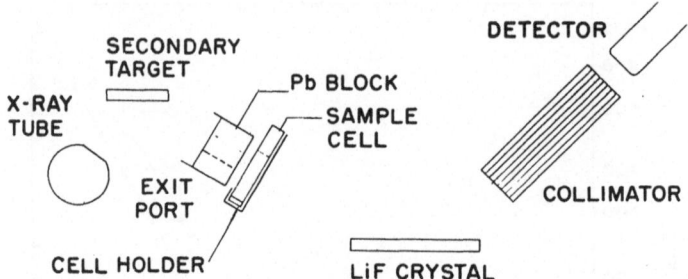

Fig. 2. Optical arrangement for X-ray absorption edge analysis
using secondary targets.

sample, the concentration can be determined using the equation

$$c = k_1 \log l_1^0/l_1 - k_2 \log l_2^0/l_2 \qquad (3)$$

This equation has been derived by Dodd [4], Engstrom [5], and
Peed and Dunn [10]. In this equation $c$ is the concentration
(in mg/ml); $l_1^0$, $l_1$, $l_2^0$, and $l_2$ are intensities transmitted through
the solvent and sample at the respective wavelengths; and $k_1$ and
$k_2$ are values determined empirically using solutions of known
concentration. Two solutions are prepared: the first contains
the solvent and a known concentration of impurity element,
and the second contains the solvent, impurity element, and a
known concentration of the element being determined. Because $c$ for the first solution is zero, a value for $k_1/k_2$ can be
calculated. This value is then used to determine values for
$k_1$ and $k_2$ for the second solution.

## APPARATUS

In order to apply this technique, only slight modifications
to a Philips Electronics Instruments X-ray spectrograph with
lithium fluoride analyzing crystal were necessary (Fig. 2). The
exit collimator was replaced by a lead block having a hole
$1/2$ in. in diameter drilled to coincide with the normal collimator window. A plastic cell holder to accommodate cells
$7/8$ in. wide and not greater than 1 cm in path length was mounted
in front of the exit port, using the screw holes for the exit
collimator. This holder can be replaced readily by a similar
one that accommodates cells having path lengths no greater
than 3 cm. About 10 min is required for conversion of the

spectrograph from X-ray fluorescence to absorption edge analysis.

Cells are constructed of polystyrene (Fig. 3) with windows $\frac{1}{32}$ in. thick; thinner windows are required for long wavelength radiation. The path length of the cell for any particular analysis is selected from a consideration of the increase in mass absorption coefficient at the absorption edge of the element being determined, and the mass absorption coefficient of the solvent.

The choice of counter to be used depends on the wavelength region being studied. For wavelengths shorter than approximately 1 A, high intensities are transmitted through most samples, and a scintillation detector is preferred because a counting efficiency of almost 100% is obtained at low operating voltages. However, for wavelengths greater than approximately 1.2 A, the voltage applied to the scintillation detector must be increased to such a degree that the noise level of the counter reduces analytical precision. Therefore, a proportional counter, which has a much lower noise level at longer wavelengths, is preferred.

## APPLICATIONS

The absorption edge technique has been applied in the Los Alamos Scientific Laboratory to the determination of several elements, including rhodium, yttrium, uranium, tantalum, cobalt, and cerium, that have absorption edges in the wavelength region from 0.534 to 2.16 A. Of these elements, rhodium and yttrium are two of the metals added to uranium or plutonium in the preparation of alloys known as "fissium" which have compositions approximating the compositions of actual fuels following a definite percentage burnup. These "fissium" alloys are used in the investigation of various fuel purification schemes. Uranium is a prime reactor fuel material which must be determined in the presence of fission product elements or various elements added in fuel preparation. Alloys of tantalum with various metals, including yttrium, tungsten, zirconium, molybdenum, and hafnium, are being considered as container or cladding materials in the Los Alamos Molten Plutonium Reactor Experiment (LAMPRE) program. Cobalt and cerium are considered as possible diluents for plutonium reactor fuels in the LAMPRE program. In many cases starting materials and final products must be analyzed. From these

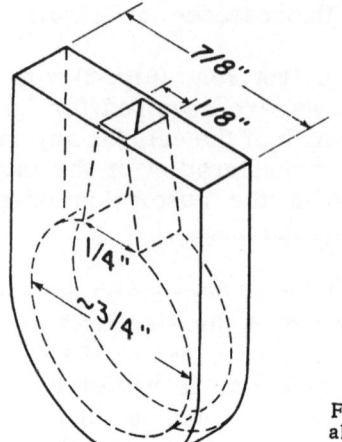

Fig. 3. Cell used in X-ray
absorption edge analysis.

## TABLE I
### Applications of X-Ray Absorption Edge Analysis

| Element Determined | Absorption Edge Used | Secondary Target | X-ray Lines (A) | | Cell Path Length (cm) | Solvent Used |
|---|---|---|---|---|---|---|
| Rh | K 0.533 | Cd- In | $K_\alpha$ 0.535 | $K_\alpha$ 0.512 | 1.0 or 3.0 | HCl |
| U | $L_{III}$ 0.722 | Nb- Mo | $K_\alpha$ 0.746 | $K_\alpha$ 0.709 | 1.0 or 3.0 | $HNO_2$ |
| Y | K 0.728 | Nb- Mo | $K_\alpha$ 0.746 | $K_\alpha$ 0.709 | 1.0 or 3.0 | $HNO_3$ |
| Ta | $L_{III}$ 1.255 | Hg- Au | $L_{\alpha1}$ 1.241 | $L_{\alpha1}$ 1.276 | 0.25 | 25%HF |
| Co | K 1.608 | Ni- Cu | $L_\alpha$ 1.658 | $K_\alpha$ 1.541 | 0.16 0.34 | Dilute $HNO_3$ – 70% i-$C_3H_7OH$ |
| Ce | $L_{III}$ 2.164 | Cr- Mn | $K_\alpha$ 2.290 | $K_\alpha$ 2.102 | 0.12 | Dilute $HNO_3$ – 70% i-$C_3H_7OH$ |

considerations, it can be seen that the analytical procedures must be applicable to materials having wide concentration ranges of numerous impurity elements. Detailed procedures for determining uranium [7] and cobalt [8] in these reactor materials by the absorption edge technique have been described previously.

Table I summarizes absorption edges, secondary targets, cell path lengths, and solvent systems used in these analyses. Note that as the wavelength of the absorption edge increases, not only must the path length of the cell be decreased but solvents of lower density and lower mass absorption coefficients must be found. In the X-ray absorption edge determination of cerium, fewer than 500 cps are transmitted through low density solutions of 70% isopropyl alcohol in a cell having a path length of only 0.12 cm. In this case, care must be taken in applying corrections for second order intensities and electronic noise of the counting circuit.

The precision of the absorption edge method compares favorably with precision obtainable by X-ray fluorescence methods. Relative standard deviations obtained by the absorption edge procedure for several elements are listed in Table II.

## INTERFERENCES IN ABSORPTION EDGE ANALYSIS

The most serious interference in absorption edge analysis is that caused by elements having absorption edges between the X-ray lines selected for analysis. Because each element has only a single $K$ absorption edge and three $L$ edges, the number of elements interfering with a particular analysis will usually be small. The $M$ series edges need be considered as a possible source of interference only for elements having atomic numbers greater than 92 (uranium). Elements interfering with the analyses previously described are listed in Table III.

A second source of error in absorption edge analysis is the matrix effect. Values for $k_1$ and $k_2$ in equation (3) are theoretically related to wavelength by equations (4) and (5).

$$k_1 = \frac{2.303\lambda_2^n}{L[\lambda_2^n (\mu/\rho)_{\lambda_1} - \lambda_1^n (\mu/\rho)_{\lambda_2}]} \tag{4}$$

$$k_2 = \frac{2.303\lambda_1^n}{L[\lambda_2^n (\mu/\rho)_{\lambda_1} - \lambda_1^n (\mu/\rho)_{\lambda_2}]} \tag{5}$$

## TABLE II
Precision Obtained for Several Elements Using the Absorption Edge Technique

| Element Determined | Cell Path Length (cm) | Conc. (mg/ml) | Counts Taken | Std. Dev. (mg/ml) | Rel. Std. Dev. (%) |
|---|---|---|---|---|---|
| Rh | 1.0 | 1.00 | $10^6$ | $0.04_4$ | 4.4 |
|    | "   | 4.10 | "      | $0.07_4$ | 1.8 |
| U  | 3.0 | 1.00 | $2 \times 10^5$ | $0.01_9$ | 1.9 |
|    | 1.0 | 10.00 | $10^6$ | $0.06_5$ | 0.65 |
|    | "   | 40.00 | "     | $0.13_6$ | 0.34 |
| Y  | 1.0 | 1.00 | $10^6$ | 0.028 | 2.8 |
|    | "   | 10.00 | "    | 0.033 | 0.33 |
|    | "   | 30.00 | "    | 0.051 | 0.17 |
| Ta | 0.25 | 0 | 128,000 | $0.10_8$ | — |
|    | "    | 5.00 | "    | $0.09_7$ | 1.9 |
|    | "    | 10.00 | "   | $0.09_9$ | 0.99 |
|    | "    | 20.00 | "   | $0.09_6$ | 0.48 |
| Co | 0.34 | 0 | 256,000 | $0.01_6$ | — |
|    | "    | 2.00 | "    | $0.03_0$ | 1.9 |
|    | " .  | 10.00 | "   | $0.04_6$ | 0.46 |
| Ce | 0.12 | 0 | 128,000 | $0.07_1$ | — |
|    | "    | 1.00 | "    | $0.04_1$ | 4.1 |
|    | "    | 5.00 | "    | $0.06_4$ | 1.3 |
|    | "    | 10.00 | "   | $0.06_8$ | 0.68 |

## TABLE III
Absorption Edge Interference for Determination of Rh, U, Y, Ta, Co, and Ce

| Elements Determined | Interfering Elements and Edges |
|---|---|
| Rh | Pu, $L_I$; Am, $L_I$; Cm, $L_{II}$ |
| U  | Y, K; Po, $L_I$; As, $L_{II}$; Rn, $L_{II}$; Pa, $L_{III}$ |
| Y  | Po, Li; As, $L_{II}$; Rn, $L_{II}$; Pa, $L_{III}$; U, $L_{III}$ |
| Ta | Er, $L_I$; Yb, $L_{II}$ |
| Co | Sm, $L_I$; Eu, $L_{II}$; Gd, $L_{II}$; Tb, $L_{III}$; Dy, $L_{III}$ |
| Ce | V, K; Cs, $L_I$; Xe, $L_I$; Ba, $L_{II}$; La, $L_{II}$, $L_{III}$; U, $M_I$; Pu, $M_{II}$; Am, $M_{II}$ |

Because values for $n$ in equations (4) and (5) vary from approximately 2.3 to 3.0 for different elements, values for $k_1$ and $k_2$ will also vary, and must be determined in the presence of the impurity elements expected in the samples. The $k_1$ and $k_2$ values are relatively insensitive to changes in concentration of the impurity element and are applicable over a wide concentration range.

Dividing equation (4) by equation (5), we obtain

$$k_1/k_2 = (\lambda_2/\lambda_1)^n \tag{6}$$

From experimentally determined values for $k_1/k_2$, values for $n$ can be estimated. It can be seen from equation (6) that as $\lambda_1$ approaches $\lambda_2$, $k_1/k_2$ approaches unity and varies less with changes in $n$. For minimum matrix error, and also to reduce the number of elements having interfering absorption edges, it is advantageous to select $\lambda_1$ and $\lambda_2$ as close together as possible. The limiting factors in selecting these wavelengths are the resolving power of the spectrometer and the availability of suitable target materials.

In the absorption edge determination of tantalum, the $L_{\alpha_1}$ lines for mercury and gold were used. Because of the proximity of these lines, a change in the value of $n$ from 2.3 to 3.0 should theoretically increase $k_1/k_2$ from 1.067 to 1.086, a difference of 1.8%. Experimentally, values ranging from 1.067 to 1.090 with an estimated error of 0.010 were obtained, indicating only a small matrix error and therefore little need for adding impurity elements to standard solutions. For example, between 0 and 20 mg/ml of tungsten had no effect on the determination of 0 or 10.00 mg/ml of tantalum. In contrast, the $K_\alpha$ lines for chromium and manganese, used in determining cerium at the $L_{III}$ edge, are located at 2.290 and 2.102 A, and varying $n$ from 2.3 to 3.0 will change $k_1/k_2$ by 6.2% (from 1.218 to 1.294). Values for $k_1/k_2$ between 1.238 and 1.300 were determined experimentally, indicating that a significant matrix error exists and that the major elements expected in the sample must be added to the standards.

Another source of interference in absorption edge analysis is the absorption edge maximum a few electron volts in width and having a much higher mass absorption coefficient than would be normally expected. This peak or maximum must be considered if it approaches or overlaps the lines used in an analysis. For example, in the determination of lanthanum and cerium using the $K_\alpha$ lines for chromium and manganese, approximately the same sensitivity should be obtained for each element. Actually, lanthanum absorbs the $K_\alpha$ line of manganese twice as strongly as does cerium because the $L_{II}$ absorption edge for lanthanum at 2.103 A overlaps the $K_\alpha$ line of manganese at 2.102 A. In the determination of tantalum using the $L_{\alpha_1}$ lines of gold and mercury, abnormally high values for $k_1/k_2$ (1.13 compared to a theoretical 1.08) were caused by overlap of the $L_{\alpha_1}$ line of gold at 1.276 A with the absorption edge maximum of zinc at 1.283 A.

## CONCLUSIONS

This investigation of X-ray absorption edge analysis indicates that the procedure is applicable to determining any element having a suitable absorption edge in the wavelength region from 0.5 to 2.25 A. Elements of atomic number 23 (vanadium) to 46 (palladium) and 58 (cerium) to 94 (plutonium) may be determined using the $K$ and $L_{III}$ edges, respectively. The absorption edge technique is generally less susceptible to interference effects than is X-ray fluorescence spectroscopy and is much less affected by wide variations in matrix composition. The magnitude of the matrix error is related to the wavelengths at which measurements are made and may be eliminated by adding major impurity elements to standards used in analysis. The absorption edge technique is especially suited to the determination of elements such as hafnium, tantalum, and tungsten, for which second order $K$ series radiation of zirconium, niobium, or molybdenum, normally present as contaminants, cause serious interference in X-ray fluorescence analysis.

## ACKNOWLEDGMENTS

The authors gratefully acknowledge the assistance and encouragement of Dr. C. F. Metz, under whose direction this work was done. Their appreciation is also expressed to Dr. J. M. Dickinson and Vernon Struebing for the preparation of the alloys used as secondary targets, and to William Whitworth for performing some of the X-ray absorption edge determinations.

REFERENCES

1. R.H. Barieau, Anal. Chem. 29, 348 (1957).
2. A.H. Compton and S.K. Allison, X-Rays in Theory and Experiment, 2nd ed., D. Van Nostrand, Inc., New York (1935), pp. 800-806.
3. W.C. Dietrich and R.E. Barringer, U.S. Atomic Energy Commission Report Y-1153 (1957).
4. C.G. Dodd, Advances in X-Ray Analysis, Vol. 3, Plenum Press, New York (1960), pp. 11-39.
5. A. Engstrom, Acta Radiol. Supp. Vol. 63 (1946).
6. A. Engstrom, Rev. Sci. Instr. 18, 681 (1947).
7. E.A. Hakkila, Anal. Chem. 33, 1012 (1961).
8. E.A. Hakkila and G.R. Waterbury, Advances in X-Ray Analysis, Vol. 5, Plenum Press, New York (1962), pp. 379-388.
9. H.R. Hughs and F.P. Hochgesang, Anal. Chem. 22, 1248 (1950).
10. W.F. Peed and H.W. Dunn, U.S. Atomic Energy Commission Report ORNL-1265 (1952).
11. J.H. Stewart, Jr., Anal. Chem. 32, 1090 (1960).
12. W.B. Wright and R.E. Barringer, U.S. Atomic Energy Commission Report Y-1095 (1955).

# X-RAY ABSORPTION FINE STRUCTURE IN GLASS*

## W. F. Nelson, I. Siegel, and R. Wagner

Fundamental Research Section
Owens-Illinois Technical Center
Toledo, Ohio

## INTRODUCTION

When considering such things as electronic behavior or structure-dependent phenomena in solids (including glass), one is faced with the problem of finding ways to make measurements which yield information about events and mechanisms on a microscopic or atomic scale. In other words, although a great deal can be learned from the many macroscopic measurements available, there comes a time when the results must be relatable to behavior and mechanisms involving the individual entities making up the solid.

As an illustration of the point, consider the many diffraction techniques which have been used to investigate the structure of solids. Thus, we have X-ray diffraction, electron diffraction, neutron diffraction, etc. However, all of these methods suffer from the same limitations: (1) They are essentially cooperative phenomena in the sense that they become more and more difficult to interpret as the number of participating particles is reduced, i.e., as the size of the region in which the phenomena take place becomes smaller. Thus, in the case of X-ray diffraction, as we reduce the particle size of a crystalline sample, its diffraction pattern becomes less and less "crystalline" and more ambiguously like an "amorphous" or liquid diffraction pattern. (2) It should also not be forgotten that the interpretation of diffraction measurements has been greatly dependent for its success on the rather restrictive stipulation of crystalline order and symmetry which has made much of the mathematics of the theories involved tractable. On an anthropomorphic basis, this might be one more example of the limitations of "togetherness" and the need for a certain amount of atomic independence.

*This paper was published in part in The Physical Review 127, 2025 (1962).

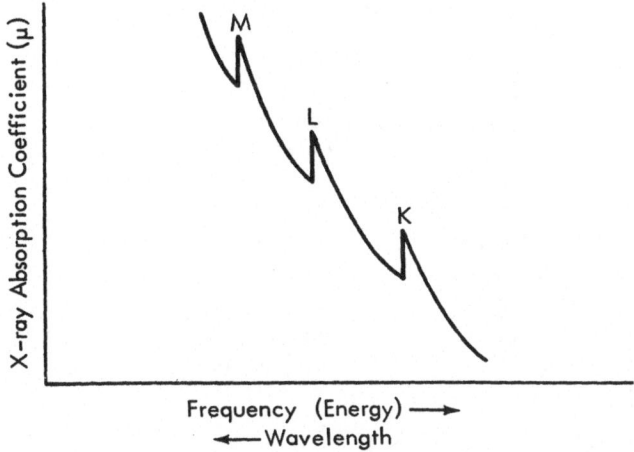

Fig. 1. X-ray absorption coefficient vs. frequency (energy).

With the advent of such tools as nuclear magnetic resonance and electron spin resonance (paramagnetic resonance), techniques have become available in which the phenomena involve only highly localized regions (few entities), such as specific nuclei, specific atoms, or at most specific atomic groups. In effect, one has available a sensitive probe, capable of investigating very limited regions in the solid.

The purpose of this paper is to discuss a similar probe which, although perhaps less well known, is probably more readily available to many laboratories since it utilizes an old phenomenon, viz., X-ray absorption.

If one looks at the behavior of the X-ray absorption coefficient of a material as a function of the frequency (energy) of the incident X rays, one finds a relation somewhat as indicated in Fig. 1. In this figure, the discontinuities are explained as being due to the X-ray ionization (or ejection) of an electron from the $K, L$, or $M$ shell of the atom into bands of unoccupied levels. With improved resolution, it is found that the absorption coefficient in the vicinity of one of these discontinuities is not a simple step-function followed by a monotonic decrease as the energy increases (as shown in Fig. 1), but rather that on the short-wavelength side of the discontinuity, the absorption coefficient fluctuates, giving rise to what has been described as a "fine structure."

Figure 2 shows examples of such fluctuations for the cases when the absorber is (a) a monatomic gas, (b) a polyatomic gas, and (c) a crystalline solid [1, 2].

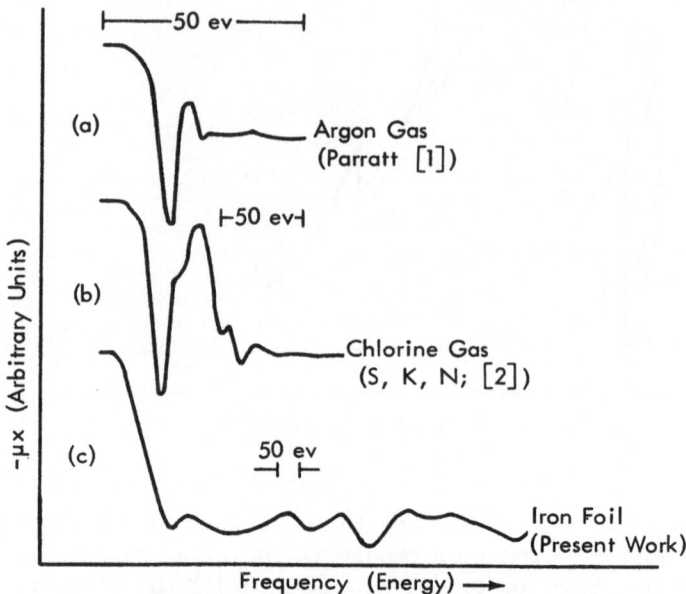

Fig. 2. X-ray absorption coefficient vs. frequency: (a) Monatomic
gas; (b) polyatomic gas; (c) crystalline solid.

The explanation for these fluctuations is relatively simple
for the monatomic gas (a). The fluctuations can be related to
the transitions of the X-ray ejected electron to unoccupied optical levels of the atom. The fluctuations in such a case have
never been observed for energies beyond the ionization energy
of the atom; in the case of argon this is ~ 16 ev.

The fluctuations for the slightly more complicated case of
the polyatomic molecule (b) extend over a somewhat larger energy region. The general approach used to explain the fluctuations in this case is to consider the problem as a scattering
problem in which the ejected electron is scattered by the atoms
adjacent to the atom from which the electron is ejected. The
success of this approach for polyatomic molecules has been
only mediocre.

For the third case (c), that of the solid, as far as we know
both experiments and theoretical considerations have been
confined to crystalline solids. After a brief discussion of the
experimental method used to obtain the data, we shall return
to a consideration of extant theories about the fluctuations observed for solid materials and, particularly, how they relate
to the results we have obtained for $GeO_2$ glass.

## EXPERIMENTAL PROCEDURE

It will be recalled that, for a given wavelength $\lambda$ of X rays, the absorption properties of a material can be represented by an expression of the form

$$I = I_0\, e^{-\mu x} \qquad (1)$$

where $I_0$ is the incident intensity; $I$ is the transmitted intensity; $x$ is the thickness of the absorber; and $\mu(\lambda)$, which is a function of wavelength, is the linear absorption coefficient. Since we are interested here in how $\mu$ varies with $\lambda$, it is essential for this experiment to have a uniform and fixed thickness for the absorbing sample. Furthermore, it turns out that it is desirable to have the sample as thin as possible while still maintaining a uniform, homogeneous character over its cross section.

The preparation of such a sample can be accomplished in several ways. If the material to be studied is a metal, for instance, the sample can be rolled or etched to a foil of the necessary thinness. Glasses and single crystals can be ground to the necessary thinness. For nonmetallic, polycrystalline materials, a fine powder can be suspended in some carrying medium such as plastic or castor oil.

With the sample properly prepared, the method of taking data is simply to measure the intensity of a monochromated beam in a diffraction unit, insert the sample into the beam, and measure the transmitted beam intensity. This process is repeated for each wavelength selected and either the ratio $I/I_0$ or, more commonly, $\ln I/I_0$ is plotted as a function of $\lambda$. The reason for plotting $\ln I/I_0$ is that, for a given sample thickness $x$, $\ln I/I_0$ is directly proportional to $\mu$, the proportionality constant being simply $-x$.

A schematic diagram of the experimental setup is shown in Fig. 3. The simplest way to select the energies of the X rays incident on the absorber is to use a single crystal and collimating slits as the monochromator. As a result of Bragg's law for X-ray reflection from a crystal, $n\lambda = 2d \sin \theta$, we can spread the X rays of different energies into a fairly large angular region and then, with a collimating slit, select a very narrow frequency range to illuminate the sample. It should be noted that, in addition to the desired wavelengths, it is also possible to have additional wavelengths present at the same angle $\theta$ by letting $n = 2, 3, \ldots$ in Bragg's law. Since these X rays are not

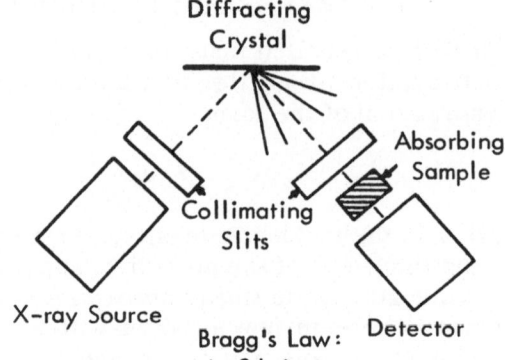

Fig. 3. Schematic diagram
of experimental arrange-
ment.

characteristically absorbed by the sample, they add to the background, thus reducing the contrast of the resulting absorption spectrum. The usual procedure followed to eliminate these extraneous wavelengths is to operate the X-ray tube at a voltage just below the value which would excite them according to the restriction $E = h\nu$. Another way is to use a pulse-height analyzer, if available.

When making absorption measurements of this type, consideration must be given to the method of detecting the X rays. In the early measurements, all of the absorption spectra were photographed. Although most workers today utilize some sort of electronic detector such as GM tubes or scintillation detectors, measurements are still made photographically. The main advantage of the photographic method is its convenience since the whole spectrum can be photographed at once (although requiring a very long exposure time). The disadvantages of the photographic method, however, are many. The three main sources of difficulty are: (1) absorption by elements in the emulsion; (2) uncertainties in knowing the amount of darkening of the film at a given wavelength and exposure time; and (3) the necessity of making a densitometer measurement of the actual intensities recorded. Another minor drawback of the photographic method is the difficulty involved in rechecking small regions of a given spectrum.

When using an electronic detector such as a GM tube, the usual procedure is to set the instrument for a given X-ray energy ( $\lambda$ or $\theta$ ) and record the intensity. The sample is then inserted in the beam and the intensity is again recorded. After correcting for background intensity, the ln of the ratio of these two readings is plotted for that wavelength and the process is repeated step by step over a range of wavelengths.

If one is fortunate to have an X-ray unit whose stability is excellent, it is frequently possible to record a single $I_0(\lambda)$ curve for a given region. This curve can then be used for all absorption measurements in this region and one can record $I(\lambda)$ throughout the region without the necessity of continually removing and replacing the absorber. In this case, the measurements are more quickly completed; moreover, it is quite easy to automate the whole measurement by utilizing a step-scanner as has been done by Van Nordstrand as reported in "Non-Crystalline Solids," John Wiley (1960).

The measurements being reported here were made using a GE XRD-3 diffraction unit set for a 2° take-off angle. The monochromator was a single crystal of NaCl. The exit slit from the X-ray tube was a 1° slit and the detector slit was 0.02°. The unit was operated at 13 kv and 29.5 ma when studying the iron edge and at 20° kv and 41.4 ma when studying the Ge edge. A total of 10,000 counts was accumulated for each $I_0$ and $I$ value for a given angular setting and each curve is the average of three such measurements. For the iron data, readings were taken for angular increments of $2\theta = 0.01°$ corresponding to ~1.93 ev. For the Ge data, readings were taken for angular increments of $2\theta = 0.005°$, corresponding to ~2.415 ev.

## RESULTS AND DISCUSSION

Figure 4 shows the results of measurements made of the absorption spectrum of iron in the pure metal and in several of its oxides. This set of curves is typical of what is found for crystalline material. One observes, for example, pronounced fluctuations in the absorption coefficient over a range of energies of the incident X rays of several hundred electron volts. Furthermore, the differences in the patterns obtained suggest a strong dependence on environment of the absorbing iron atom.

The first successful explanation of these observed fluctuations in the absorption coefficient was given by Krönig in 1923 [3]. He assumed that the electron was ejected from the absorbing atom into one of the conduction bands of the crystal. That is, instead of the electron being raised to one of the unoccupied optical levels where all of its energy would still be potential energy, he considered the electron as having received enough kinetic energy to be essentially free to move around in the lattice and no longer bound to its parent atom.

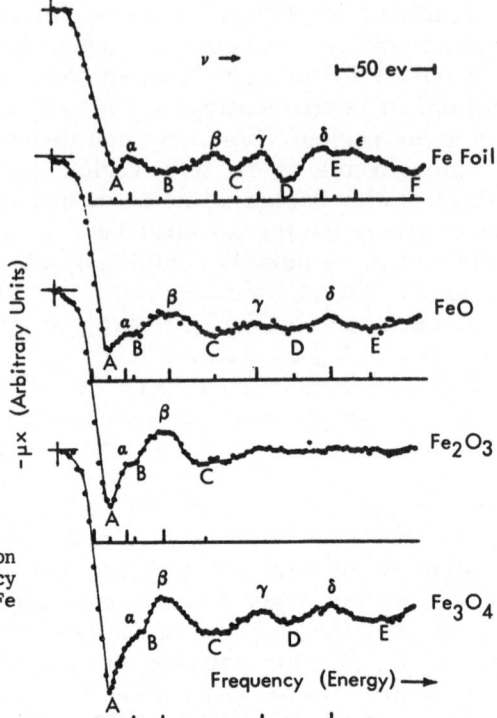

Fig. 4. Fe X-ray absorption coefficient vs. frequency for Fe and several Fe oxides.

This assumption leads to a description of the excited electron by a plane wave of the form $\exp(i\bar{k}\cdot\bar{r})$ modulated by the potential in the crystal. Krönig calculated the wave function of an electron in such a situation by solving the Schrödinger equation using a perturbation technique. Thus, if one writes down the Schrödinger equation for such an electron, it has the form

$$\nabla^2\psi = \frac{8\pi m}{h^2}\left(E - \sigma V(x,\ y,\ z)\right)\psi = 0 \qquad (2)$$

where $m$ is the mass of the electron; $h$ is Planck's constant; $V(x,\ y,\ z)$ is a periodic potential function with the same period as the lattice; and $\sigma$ is a parameter indicating the degree of binding. When $\sigma = 0$, the electron is free.

Equation (2) has a solution of the form

$$\psi = u(x,\ y,\ z)\ \exp\left(\frac{2\pi i(l x + m y + n z)}{Gd}\right) \qquad (3)$$

where $l$, $m$, $n$ are Miller indices; $u(x, y, z)$ is a periodic function with the same period as the lattice; and $G$ is an integer. After carrying out the calculation, one finds that certain energies are forbidden to the electron. These are the same energies which would correspond to the electron wave being reflected from crystallographic planes for which a Bragg angle condition is satisfied. For a cubic lattice, these energies are given by the expression

$$E = \frac{h^2}{8md^2} \cdot \frac{(\alpha^2 + \beta^2 + \gamma^2)}{\cos^2 \theta} \tag{4}$$

where $\alpha$, $\beta$, $\gamma$ are related to the plane whose Miller indices are $h$, $k$, $l$; $d$ is the interplanar spacing; $\theta$ is the angle between the momentum vector $\bar{k}$ of the electron and the vector normal to the planes $(h, k, l)$.

If one used polarized X rays (i.e., X rays with a known direction of the momentum vector) and a single crystal absorber, the electron energies given by equation (4) would be forbidden. In this case, the incident X rays could not be absorbed and there would be "spikes" of low values of the absorption coefficient in the absorption spectrum. Since measurements are usually made with unpolarized X rays and polycrystalline absorbers, however, it is necessary to integrate over all angles $\theta$. The result is that, instead of "spikes" one observes only fluctuations in $\mu$ rather than discontinuities.

Krönig's theory was capable of giving a fairly good qualitative explanation of the experimental data available at the time. The theory is particularly useful in explaining the observed dependence of the fine structure on temperature and on crystalline structure. However, its limitations were severe: (1) It is only applicable for fluctuations at a considerable distance (in energy) from the main absorption edge. (2) It did not take account of the transition probabilities between energy states but only considered the existence and multiplicity of the energy states. (3) It did not take account of the fact that electrons with these energies (several hundred electron volts) do not travel very far in a solid without losing their energy. That is, the lifetime of the energy state of the ejected electron may be short.

Several alternative explanations or modifications of Krönig's theory have been proposed [4-6]. The one which seems most promising at the present time is the one given by Shiraiwa et al. [6]. In this theory, the ejected electron is considered to be scattered by the adjacent atoms in the solid, i.e., the phenomenon is treated as a collision scattering problem. The electron is again represented by a plane wave function which is, however, multiplied by an attenuating factor to take account of the energy losses incurred in the inelastic collisions of the electron with neighboring atoms.

The consequence of this treatment is that the fine structure observed may be related only to the atoms in the neighborhood of the absorbing atom. This would be true if the interactions between the ejected electron and the adjacent scattering atoms were largely inelastic, i.e., if the electron is never really "free" in the lattice, as is required by the Krönig assumption. The theory of Shiraiwa et al. is also able to account for the observed dependence of the absorption fine structure on crystal structure and temperature since, for a crystal, the local environment (both the symmetry and spacing) is determined by the crystal structure and the temperature of the sample.

In order to test the applicability of these theories to the study of glasses, we measured the X-ray absorption coefficient in glass samples for two different elements—Fe and Ge. Meas-

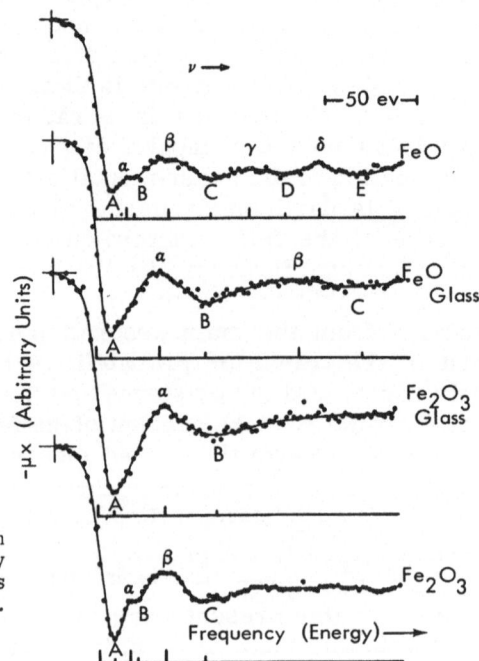

Fig. 5. Fe X-ray absorption coefficient vs. frequency for crystalline Fe oxides and glasses containing Fe.

urements were made in a sodium disilicate glass containing approximately 5% Fe by weight in both the $Fe^{2+}$ and $Fe^{3+}$ valence states. The results are shown in Fig. 5. As can be seen, the results are somewhat ambiguous, primarily as a result of the small amount of iron present and the instability of the X-ray unit at the time the measurements were made. This small amount of iron made it difficult to separate the part of the absorption due to the iron alone from the absorption due to the rest of the material in the sample. Despite the difficulty of attaching too much meaning to the fluctuations for the iron case, we were still able to measure a shift in the relative position of the main absorption discontinuity as shown in Fig. 6. The values of the shift measured for the crystalline oxides are in general agreement with other measurements in the literature [7]. Although we have not tried it, our measurements suggest that, with the use of a curved crystal or two-crystal spectrometer (i.e., a higher resolution capability), the extent of the shift could be used to measure the amount of $Fe^{2+}$ or $Fe^{3+}$ present in a sample. Such an arrangement could be used for any element for which there is a measurable shift.

Because of the difficulties encountered with the iron measurements, we measured the absorption coefficient in the vicinity of the Ge $K$-absorption edge in two crystalline forms and in an amorphous form of $GeO_2$ which is a good analogue of $SiO_2$. These measurements had the following advantages: (1) The germanium present in the sample accounted for the largest part of the total absorption. (2) There were no chemical changes to complicate the comparison of the measurements. (3) The germanium $K$-absorption coefficient in pure germanium metal has been extensively studied [9-15] and the absorption coefficient in at least one of the crystalline forms of the oxide has been pre-

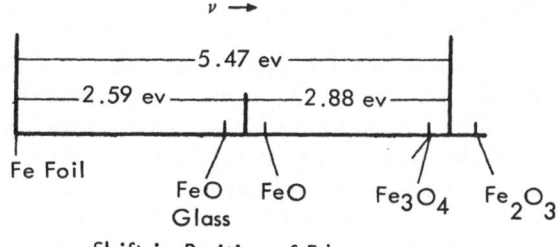

Fig. 6. Shift in portion of main Fe X-ray absorption discontinuity as a function of Fe valence.

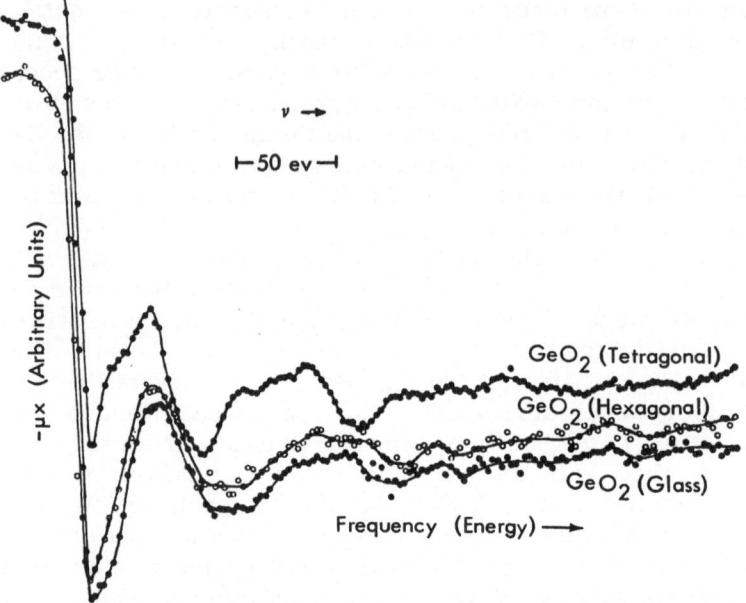

Fig. 7. Ge X-ray absorption coefficient vs. frequency for crystalline and
amorphous GeO₂.

viously measured [16]. The results of our $GeO_2$ measurements
are shown in Fig. 7.

From the figure, we see that the two crystalline forms give
fluctuations in the absorption coefficient which are markedly
different from each other. This result can be expected from
the viewpoint of either the original Krönig theory or from its
modifications due to Shiraiwa et al. The most interesting result
is the fluctuation pattern for the $GeO_2$ glass or amorphous form.
From the figure we see that, within the limits of the measurement, the pattern is identical to that for the hexagonal crystalline modification.

On the basis of the Krönig theory, this is a somewhat surprising result since the $GeO_2$ glass was distinctly "amorphous"
as measured by ordinary X-ray diffraction and optical microscopy, while the hexagonal crystalline modification was indeed
crystalline. Thus, although it may not be surprising that there
are fluctuations for the glassy $GeO_2$ out to 300 ev in energy, it
is surprising that the fluctuations are identical with those for
the crystalline form since electrons of this energy should begin
to be affected by the differences in "long-range order" between

the two materials if the "free electron" approximation of Krönig is valid.

On the other hand, on the basis of the Shiraiwa modification of Krönig's theory, which allows for all of the "fine structure" to be caused by nearest or next-nearest neighbors, the result is reasonable. But even here, the fact that the results are identical is still a little surprising because this would seem to indicate a requirement for not only the same coordination but, in fact, exactly the same symmetry and spacing for at least the first or second coordination shells in both the crystalline $GeO_2$ and the $GeO_2$ glass. In other words, a truly random network concept does not seem applicable to the $GeO_2$ glass, at least as far as can be determined from a measurement of this type. At the same time, the ordinary X-ray diffraction measurements indicate that there is no really extensive long-range order in the sample.

Although we do not make the unequivocal claim that the $GeO_2$ glass is "crystalline" when examined on a sufficiently localized scale, the present results do suggest such a possibility. Furthermore, infrared measurements made on the two crystalline forms and on the amorphous form of $GeO_2$ by Obukhov-Denisov et al. [17] give similar results; i.e., the infrared and Raman spectra of the glass and of the hexagonal form are quite similar while the tetragonal form is completely different.

Recent measurements by Lytle [18] of the X-ray absorption spectra of Cu and Ni at cryogenic temperatures led him to the conclusion that if the Krönig explanation is correct, it is necessary to consider the Bragg reflection of the ejected electron to occur at planes of extremely high Miller indices, i.e., very small interplanar spacings and therefore rather short distances from the absorbing atom. If the $GeO_2$ glass is, in fact, crystalline on a localized scale, the Krönig explanation might still be valid if one accepts Lytle's interpretation since reflection from higher-index planes would not require any long-range order in the sample. Furthermore, utilizing the higher-order planes makes the Krönig and Shiraiwa theories essentially equivalent since diffraction is only a special type of scattering.

In summary, we can make the following statements:

1. X-ray absorption phenomena can give useful information about the amorphous or glassy state as well as about the crystalline state of solids.

2. The information about such phenomena is readily obtain-

able in many laboratories since it requires only a conventional X-ray diffraction unit and some patience.

3. Measurements made on glasses would be of particular value to the physics community, since the data obtained would help to clarify the theoretical picture of X-ray absorption in solids.

4. The possibility exists of utilizing X-ray absorption phenomena as a nondestructive means of measuring valence states in a solid.

5. The results of measurements made on the $GeO_2$ system in the amorphous state (prepared from hexagonal crystal form) suggest that the glass has a degree of order and symmetry very similar to, or identical with, the hexagonal crystalline modification of $GeO_2$. The notion of a true random network does not seem applicable to the $GeO_2$ glass.

6. Our results on the $GeO_2$ system are more consistent with the modified, rather than with the original, Krönig theory.

## ACKNOWLEDGMENTS

The authors would like to acknowledge the kindness of T. Baak, E. J. Hornyak, and G. R. Whittaker of the Fundamental Research Section in preparing the iron glasses and the tetragonal crystalline modification of the $GeO_2$ used in this study and the kindness of Doris Stifel, Fundamental Research, in making the X-ray diffraction measurements. They would also like to thank Prof. J. J. Turin of the University of Toledo Physics Department for making the X-ray spectrometer available.

## REFERENCES

1. L. G. Parratt, Phys. Rev. 56, 295 (1939).
2. S. T. Stephenson, R. Krogstad, and W. Nelson, Phys. Rev. 84, 806 (1951).
3. R. de L. Krönig, Z. Physik 75, 191 (1932).
4. T. Hayashi, Sci. Repts. Tohoku Univ., Ser. 1, 33, 123, 183 (1949); 34, 185 (1950).
5. A. I. Kostarev, Zhur. eksptl. teoret. Fiz. SSSR. 11, 60 (1941); 19, 413 (1949); 21, 917 (1951).
6. T. Shiraiwa, T. Ishimura, and M. Sawada, J. Phys. Soc. Japan 13, 8, 847 (1958).
7. A. H. Barnes, Phys. Rev. 44, 141 (1933).
8. C. Kurylenko, Cahiers de phys. 70, 35 (1956).
9. W. W. Beeman and H. Friedman, Phys. Rev. 56, 392 (1939).
10. M. H. Hulubei and Y. Cauchois, Compt. rend. 211, 3161 (1940).
11. W. F. Nelson, Ph. D. Thesis, Washington State College (1956), University Microfilms, Ann. Arbor.
12. D. G. Doran and S. T. Stephenson, Phys. Rev. 105, 1156 (1957).
13. J. M. El-Hussaini and S. T. Stephenson, Phys. Rev. 109, 51 (1958).

14. K. Tsutumi and M. Obashi, J. Phys. Japan 13, 591 (1958).
15. J. N. Singh, Phys. Rev. 123, 1724 (1961).
16. M. H. Hulubei and Y. Cauchois, Compt. rend. 211, 316 (1940).
17. V. V. Obukhov-Denisov, N. N. Sobolev, and V. P. Cheremisinov, Opt. i spekt. 8, 4, 267 (1960); see also: E. R. Lippincott et al., J. Res. Nat. Bur. Std. 61, 61 (1958).
18. F. W. Lytle, this volume, p. 285.

# A SUGGESTED APPROACH TO THE
# TEACHING OF X-RAY SPECTROSCOPY

## William F. Loranger

Picker X-Ray Corporation
White Plains, New York

It has become increasingly apparent that one of the most important devices for instrumental chemical analysis is the X-ray emission spectrometer. It should now be considered as sharing in importance in the teaching laboratory with infrared and ultraviolet spectrophotometers, the optical emission spectrograph, the polarograph, the chromatograph, and nuclear paramagnetic resonance instrumentation. Indeed, present commercial usage (over and above research and university laboratories) of the X-ray emission spectrometer has grown to the point that the shortage of scientists and technicians previously trained and currently being trained in this specialty presents a serious problem to industry.

For fifty years or more, X-ray diffraction techniques have been taught (and considerable research undertaken) at many university laboratories in various departments: geology, metallurgy, ceramics, agronomy and soil sciences, chemistry, physics, mineralogy, etc. Only in the last decade or so, however, has any great importance been ascribed to X-ray emission techniques. This importance is primarily from a research aspect, however, rather than a formal classroom approach. Even though X-ray emission, like diffraction, transcends departmental boundary lines, it appears that the main approach to the teaching of this vital analytical technique should be the prime responsibility of the university's chemistry department. The responsibility seems clear-cut when X-ray emission is simply described as another form of instrumental chemical analysis, and the approach to teaching the techniques involved is similar in many respects to what has been used for many years for other instrumental methods. The stumbling block may have been, and may continue to be, the relative high cost of X-ray instrumentation. There are a number of commercial

322

instruments available on the market today, and the cost of these ranges from $18,000 to $25,000, depending, of course, on the number and variety of accessory items ordered. What does seem surprising, however, is that although many laboratories already have diffractometer systems, the possibility of converting from diffraction to emission with a modest expenditure for the proper accessories has been somewhat overlooked. When this has been done, it has been done primarily for research and not for teaching purposes.

It is certainly unnecessary to review for the participants in this Symposium the theory of X-ray emission or the principals of operation of a particular type of spectrometer. It seems sufficient merely to reflect that X-ray emission techniques for elemental analysis have been demonstrated for nearly every element in the periodic chart from magnesium to uranium on a "dry" basis, to replace the "wet" gravimetric and volumetric routine analysis. This does not mean that classical methods of analytical chemistry will ever be completely replaced. To the contrary, it gives impetus to classical research to improve wet methods to provide better instrumental calibration data.

In the past few years, hundreds of papers on X-ray emission analytical procedures and two worthwhile texts have appeared in the literature. Application has become so diversified as to know practically no limitations as to type of analysis. Trace analysis by X-ray emission is becoming increasingly important with some laboratories reporting less than one part per million concentration routinely detected and analyzed quantitatively.

X-ray emission instrumentation has found its way in the past several years into routine industrial chemical laboratories as a substitute for routine wet analytical methods for process control. With the speed of X-ray analysis, analytical data cease to be historical data on a product, but data with which a process can be controlled. (I must digress here for a moment to say that at present I personally am dubious of so-called "on-line control" only from a standpoint of cost and difficulty encountered with instrument down-time. At present I feel that laboratory analyzers, either singly or in groups, can handle any process analytical problems more than adequately, at considerably less cost and a higher degree of reliability.) More and more, X-ray analyses are becoming the responsibility of industry's chief chemist, who after training himself in X-ray spectrometer operation, eventually finds himself in the role of teacher, training technicians to do routine X-ray analysis. Fortunately, the chief chemist has had in most instances, previous experience with instrumental chemistry, either during his university training or in his own laboratory. He has generally developed a feeling for instrumental analysis, in that all instruments yield useful results only when successfully calibrated. A further appreciation for the fact that instrumentation yields a high degree of precision with speed unobtainable by wet methods is the chemist's prime motivation in selecting X-ray emission as an analytical procedure.

I purport not to be a professional educator, but several years of experience in the X-ray field, coupled with formal training as an instrumental chemist, as well as having taught analytical chemistry for a number of years, provides me the background and opportunity to propose what I consider to be a course outline that would adequately serve to prepare the graduate analytical chemist to assume a highly useful role in an industry using the X-ray spectrometer. The premise on which the outline is based is that the student had no prior exposure to course work in X rays.

It is comprised of a series of ten lectures (50 minutes):

| I and II | Fundamental X-ray Physics |
|----------|---------------------------|
| III | The Origin of the X-ray Spectrum |
| IV and V | Instrumentation and Measurements |
| VI | Specimen Preparation |
| VII | Qualitative Interpretation; Selection of Crystals |
| VIII | Quantitative Measurements |
| IX | Interelement Effects |
| X | Precision vs. Accuracy; Problems of Day-to-Day Analysis on a Routine Basis |

The laboratory work in conjunction with the lectures should begin after lecture VI, and be comprised of five sessions (3 hours):

I    Familiarization with instrumentation. Choice of variables for analysis: X-ray tube target, slits, crystals, counter tubes, exclusion of air.

II   Plotting a counter tube plateau. Familiarization with existing wavelength tables. Qualitative scan and identification of a solid specimen (no specimen preparation required). Qualitative analysis of an unknown.

III Preparation of powdered samples. Preparation of a series of standards. Preparation of a working curve from standards. Quantitative analysis of an unknown.

IV Referral of unknown to same working curve at a later date. Problems of day-to-day analysis.

V Demonstration of interelement effects. Special problems.

In this way, with ten lectures and five laboratory sessions, the work could be incorporated into an instrumental chemistry course for a 20-week semester. It is also suggested that current literature be reviewed in a seminar plan in order to make students aware of current problems and solutions to problems in the emission field. This then serves as one approach to teaching X-ray emission in the university laboratory, to train chemists to keep abreast of the ever-increasing use of X-ray emission in industry.

# X-RAY SPECTROSCOPY IN BIOLOGY AND MEDICINE
# PRELIMINARY REPORT ON THE MICROANALYSIS OF HUMAN TISSUES FOR IRON, ZINC, AND POTASSIUM

## James C. Mathies and Paul K. Lund

Pacific Northwest Research Foundation and
Laboratory of Pathology, The Swedish Hospital
Seattle, Washington

## INTRODUCTION

The reliability of the X-ray spectrograph in biology and medicine as a tool for the analysis of microspecimens of serum for calcium, potassium, bromide, and chloride has been established [1-4]. Extension of this method to the analysis of tissues was therefore both feasible and desirable. The availability of a technique suitable for the analysis of needle biopsy specimens of tissue for a number of important metabolic constituents would allow more accurate diagnosis and therapy of diseases of cellular metabolism. The present preliminary report demonstrates the feasibility of measuring potassium, zinc, and iron using both macro and biopsy specimens of human tissue.

## EXPERIMENTAL

### Spectrometer Operation

A Norelco vacuum path X-ray spectrograph with associated electronic circuit panel and pulse-height analyzer was used. The tungsten target X-ray tube was operated at 50 kv and 40 ma, and all elements were measured using first-order $K_\alpha$ fluorescent radiation. Except where otherwise defined, instrumental operation was as previously described [1, 3, 5]. Specific settings for the various elements are summarized in Table I.

This investigation was supported in part by National Institutes of Health research grant A-3751.

## TABLE I
### Instrumental Settings

| Element | Iron | Zinc | Potassium | Sulfur |
|---|---|---|---|---|
| Analyzing crystal | LiF | LiF | LiF | NaCl |
| X-ray path | Air | Air | Vacuum | Vacuum |
| Goniometer $2\theta$ setting | 57.50° | 41.80° | 136.65° | 144.70° |
| Detector* voltage† | 1650 | 1600 | 1650 | 1725 |
| Counting threshold voltage† | 21 | 12 | 7.5 | 15 |
| Channel width voltage | 30 | 30 | 30 | 30 |
| Count collection $\times 10^{-3}$ | 64 | 32 | 64 | 16 |

*Gas-flow proportional detector with 90% argon-10% methane as the filling gas.
†Infrequent minor readjustments are necessary.

### Tissues and Hydrolysis

Macrospecimens ranged in weight from 1 to 5 g and were hydrolyzed by briefly heating with 2.5 volumes of concentrated nitric acid. The resulting solution was diluted with an equal volume of distilled water, chilled to congeal the supernatant lipid phase, and filtered through a pyrex glass wool plug in a polyethylene funnel into a volumetric flask. The transfer was completed using 0.1 N nitric acid. The hydrolyzate was then diluted to yield a solution containing 100 mg original tissue wet weight per milliliter.

Biopsy specimens were blotted free of excess blood and fluid and placed in preweighed 400-$\mu$l polyethylene centrifuge tubes. The tubes were immediately stoppered and weighed to obtain specimen size. The tissue was then hydrolyzed directly in the centrifuge tube by heating for a few minutes at 100° in the presence of 30 $\mu$l of concentrated nitric acid.

### X-ray Specimen Preparation

For iron and zinc measurements, 1 ml aliquots of macro-hydrolyzate were taken to dryness and redissolved in sufficient 2N $HNO_3$ to yield a final volume of 200 $\mu$l. Duplicate 50 $\mu$l aliquots were dried on 13-mm filter paper disks (25 mg tissue per disk) for insertion in the spectrograph. For the measurement of potassium 100-$\mu$l aliquots were dried and reconstituted in 100 $\mu$l of 1% aqueous methylcellulose in 1N ammonium hydroxide. Specimens were prepared for counting by drying 50-$\mu$l aliquots on aluminum foil planchets [1].

Biopsy specimens were prepared by drying the hydrolyzate

in the original microfuge tube. For zinc and iron measurements, the residue was redissolved in 5 $\mu$l 2N HNO$_3$ per milligram of original tissue wet weight. Suitable aliquots were prepared for counting by drying on filter paper disks as before. Potassium was estimated by dissolving the residue in 10 $\mu$l of ammoniacal methylcellulose per milligram of original tissue wet weight, and preparing planchets using 50-$\mu$l aliquots.

### Reference Procedures

Iron was determined using a wet-ashing procedure with 1,10-phenanthroline [6], zinc by the dithizone method [7], and potassium was measured with the Coleman flame photometer according to a minor modification of the manufacturer's directions.

### Calibration

Values for zinc and iron were calculated using an internal standardization procedure. Separate aliquots of each tissue hydrolyzate were mixed with a multiple iron-zinc standard and processed as usual. The counting rate per microgram of added element was used to calculate concentrations from counting rates of specimens not containing added standard. Generally, 12.5 and 2.5 $\mu$g of iron and zinc, respectively, were added per disk. Potassium counting rates were calibrated by comparison

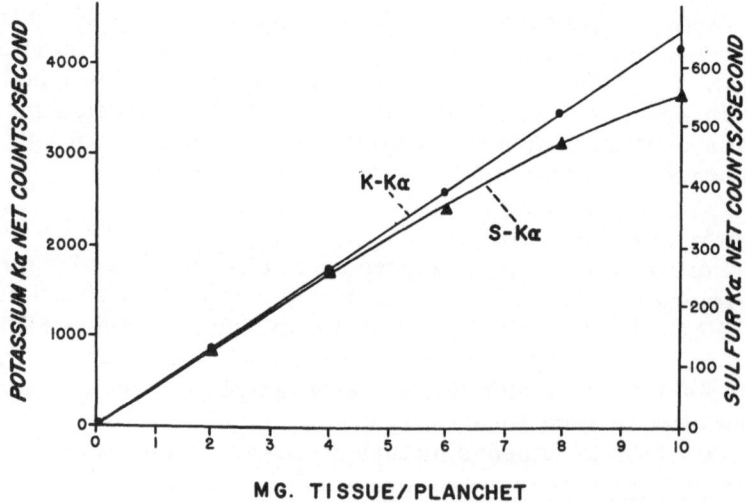

MG. TISSUE/PLANCHET

Fig. 1. Effects of increased tissue hydrolyzate per planchet on potassium and sulfur net counting rates. A myocardial hydrolyzate was employed.

with values obtained using the flame photometer. Ultimately, permanent bracketing standards would be used in making routine assays.

## RESULTS AND DISCUSSION

Quantitation of spectrographic line intensities is not difficult when a linear relation exists between the concentration of the element being measured and instrumental response. That such a relation holds for potassium where the amount of tissue wet weight per planchet is varied from 2 to 8 mg is demonstrated in Fig. 1. Similar experiments, in which the matrix was held constant, yielded a linear response when the potassium was varied from 0 to 40 $\mu$g per planchet. Sulfur counting rates, on the other hand, exhibit a significant decrease in counting rate per milligram of tissue as the amount of tissue per planchet is increased. This effect is regular and not severe, and with suitable controls, measurements of sulfur $K_\alpha$ fluorescent radiation should serve as a suitable monitor as to the amount of tissue (protein) per individual planchet.

Similar data are presented in Fig. 2 for zinc and iron. Results for zinc are linear; however, iron exhibits moderate,

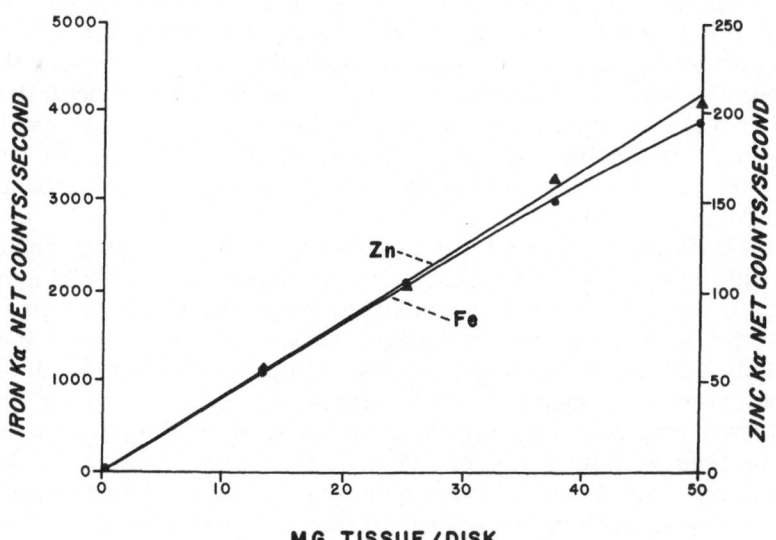

Fig. 2. Response of iron and zinc counting rates to increased amounts of liver hydrolyzate.

## TABLE II
### Specimen Preparation Reproducibility

| Expt. No. | No. Replicates | S-$K_\alpha$ Net c/sec Per 5 mg tissue | % $\sigma$ | K-$K_\alpha$ Net c/sec Per 5 mg tissue | % $\sigma$ | Fe-$K_\alpha$ Net c/sec Per 25 mg tissue | % $\sigma$ | Zn-$K_\alpha$ Net c/sec Per 25 mg tissue | % $\sigma$ |
|---|---|---|---|---|---|---|---|---|---|
| I | 12 | 276±3.8 | 1.4 | ·2727±37 | 1.4 | | | | |
| II | 10 | 300±12 | 4.0 | 2747±131 | 4.7 | | | | |
| III | 10 | | | | | 1072±18 | 1.7 | 435±6 | 1.4 |
| IV | 10 | | | | | 582±27 | 4.7 | 125±11 | 8.8 |

Experiments I and III are replicates from single macrohydrolyzates; experiments II and IV represent biopsy replicates from individual specimens of tissue. Myocardial muscle was used in I and II; liver in III and IV.

but significant, nonlinearity. In view of this latter observation, similar experiments were carried out in which the matrix was held constant at 25 mg per disk. Both zinc and iron were varied from 0 to 100 $\mu$g and the resulting instrumental responses were quite linear.

Having demonstrated that a proportionality exists between the instrumental response and the concentration of iron, zinc, and potassium, it was then necessary to examine the reproducibility of the techniques devised for preparation and mounting of specimens. This aspect is studied in Table II, where the counting rate of replicate specimens is studied. Reproducibility of replicates where a single macrohydrolyzate is analyzed is entirely adequate for the intended application, and in fact closely approaches the statistically predicted accuracy based on count collection alone.

Deviations of analytical results when assaying serial specimens obtained by punch biopsy are significantly higher, and are a reflection of increased sampling error. Although all of the sources of this increased error are not certain, it is likely that tissue heterogeneity is an important factor. Variable inclusion of connective tissue, extracellular fluid, blood, and plasma are certainly involved. Since analytical deviations of this type will increase in magnitude as specimen size is decreased, the level of accuracy obtainable is therefore inherently limited by the degree of tissue inhomogeneity and the size of the sample. The biopsy specimens varied in weight from 3 to 20 mg and were weighed to the nearest 0.1 mg. The

TABLE III
Tissue Potassium Analyses: Correlation of X-Ray
and Reference Procedures

| Specimen | | Milliequivalents/kg tissue wet weight | | |
|---|---|---|---|---|
| No. | Tissue | Reference | X-ray | Deviation |
| 18 | Pooled serum | 5 | 5 | 0 |
| 61-A-16 | Kidney | 46 | 48.5 | +2.5 |
| 61-A-285 | Adrenal | 43 | 43 | 0 |
| 62-A-1 | Heart | 69 | 69.5 | +0.5 |
| " | Spleen | 88 | 88 | 0 |
| " | Brain | 71 | 76 | +5 |
| " | Skeletal muscle | 95 | 98 | +3 |
| 62-A-10 | Kidney | 50 | 50.5 | +0.5 |
| " | Liver | 65 | 65.5 | +0.5 |
| 62-A-12 | Gastric mucosa | 30 | 32.5 | +2.5 |

TABLE IV
Tissue Iron Analyses: Correlation of X-ray and
Reference Procedures

| Specimen | | mg/100 g tissue wet weight* | | |
|---|---|---|---|---|
| No. | Tissue | Reference | X-ray | Deviation |
| 61-A-16 | Pancreas | 3.5 | 3.7 | +0.2 |
| 61-A-16 | Kidney | 7.8 | 8.9 | +1.1 |
| 61-A-31 | Liver | 63.1 | 65.2 | +2.1 |
| 61-A-37 | Liver | 11.5 | 11.6 | +0.1 |
| 61-A-45 | Spleen | 21.3 | 20.6 | -0.7 |
| 61-A-87 | Liver, "normal" | 27.5 | 28.3 | +0.8 |
| 61-A-87 | Liver, "tumor area" | 3.2 | 3.5 | +0.3 |
| 61-A-189 | Liver, 1st lobe | 432 | 464 | +32 |
| 61-A-189 | Liver, 2nd lobe | 653 | 646 | -8 |
| 18 | Serum pool | — | 0.5 | — |

*X-ray results were calculated using an internal standard. The minimum
number of replicates was two for each determination.

## TABLE V
### Tissue Zinc Analyses: Correlation of X-ray and Reference Procedures

| Specimen | | mg/100 g tissue wet weight* | | |
|---|---|---|---|---|
| No. | Tissue | Reference | X-ray | Deviation |
| 61-A-16 | Kidney | 3.4 | 3.3 | -0.1 |
| " | Pancreas | 1.9 | 2.1 | +0.2 |
| 61-A-57 | Liver | 13.8 | 12.6 | -1.2 |
| 61-A-58 | Liver | 10.1 | 10.0 | -0.1 |
| 61-A-60 | Liver | 7.6 | 7.7 | +0.1 |
| 61-A-87 | Liver, "normal" | 4.9 | 4.4 | -0.5 |
| " | Liver, "tumor area" | 2.3 | 2.4 | +0.1 |
| 61-A-189 | Liver, 1st lobe | 7.6 | 7.4 | -0.2 |
| " | Liver, 2nd lobe | 7.0 | 7.0 | 0 |
| 61-1515 | Prostate | - | 35.8 | - |
| 62-A-1 | Skeletal muscle | 4.7 | 4.7 | 0 |

*X-ray results were calculated using internal standardization, and variation in background was compensated for by use of offset $2\theta$ measurements at 40.50°. All determinations were carried out at least in duplicate.

## TABLE VI
### Serial Tissue Potassium Values During Experimental Hypothermia and Heart Surgery in Dogs*

| Time (min) | 5 | 40 | 60 | 90 | 105 |
|---|---|---|---|---|---|
| Body temperature (°C) | 37° | 32° | 27.5° | 33° | 35.5° |

| Tissue | Myocardial muscle | | | | |
|---|---|---|---|---|---|
| Avg. biopsy wt. (mg) | 5.2 | 4.2 | 2.75 | 6.55 | 4.5 |
| meq K/kg tissue | 85 | 88 | 77 | 68 | 67 |

| Tissue | Skeletal muscle | | | | |
|---|---|---|---|---|---|
| Avg. biopsy wt. (mg) | 13.3 | 15.1 | 11.8 | - | 21.9 |
| meq K/kg tissue | 78 | 78 | 79 | | 73 |

*All results represent averaged values of duplicate measurements.

weighing error is therefore significant, but can be readily diminished by use of a more sensitive analytical balance or equating results in terms of a constant sulfur counting rate.

As a final check on the reliability of the analytical procedures devised for the X-ray spectrograph, a series of macrohydrolyzates was prepared. These specimens were then assayed for potassium, iron, and zinc using both the X-ray and reference procedures. The comparative data obtained are presented in Tables III-V. Specimens were selected to provide the widest possible range of values which might be encountered in analyzing human tissue. The absence of biologically significant analytical deviations substantiates the belief that valid practical procedures have been devised. Measurement of calcium and chloride in these same specimens should be possible after minor procedural changes.

No attempt has been made in this investigation to determine minimum sample sizes, due to marked variations in elemental concentrations from tissue to tissue, and microheterogeneity within single tissues. Considerable reductions in sample size are possible, however, with uniform material. The measurement of potassium, for example, is feasible with as little as 200 $\mu$g tissue wet weight.

Development of X-ray spectrographic procedures for the elemental analysis of microspecimens of tissue will allow clinical measurements that have heretofore been quite difficult or virtually impossible. An example of the type of investigation that becomes possible is presented in Table VI. These data were obtained during the course of experimental heart surgery on dogs to determine the effects of inducing hypothermia on myocardial tissue potassium. The small size of the tissue specimens was such that trauma to the myocardium was minimized, thus allowing serial analysis throughout the surgical manipulations. The future usefulness of such *in vivo* tissue analysis in clinical medicine remains to be explored; however, many fruitful diagnostic applications seem likely.

## SUMMARY

X-ray spectrographic procedures for the determination of potassium, iron, and zinc using micro- and macrospecimens of human tissue are presented. Analytical results obtained with these procedures are demonstrated to be in good agreement with those obtained by previously established methods.

Advantages of the X-ray microprocedures are pointed out, and an example of the useful application of such microdeterminations of myocardial potassium during experimental animal surgery is presented.

## ACKNOWLEDGMENT

The technical assistance of Miss Beverly Weber and Miss Elizabeth Giedt is gratefully acknowledged.

### REFERENCES

1. J.C. Mathies and P.K. Lund, Norelco Reporter 7, 130 (1960).
2. S. Natelson et al., Clin. Chem. 5, 519 (1959).
3. J.C. Mathies and P.K. Lund, Norelco Reporter 7, 134 (1960).
4. S. Natelson and B. Sheid, Clin. Chem. 6, 299 (1960).
5. J.C. Mathies, P.K. Lund, and Wilamina Eide, Anal. Biochem. 3, 408 (1962).
6. Colorimetric Determination of Traces of Metals, E.B. Sandell (ed.), 3rd ed., Interscience Publishers, Inc., New York (1959), p. 552.
7. B.L. Vallee and J.G. Gibson, II, J. Biol. Chem. 176, 435 (1948).

# ANALYSIS OF NONMETALLICS BY X-RAY FLUORESCENCE TECHNIQUES

## B. R. Boyd* and H. T. Dryer

Applied Research Laboratories, Inc.
Dearborn, Michigan

Last year at this Symposium, we discussed the use of X-ray fluorescence techniques for analyzing the products and by-products of the iron and steel industry, including some of the primary raw materials such as ores and sinters.

The increased interest in this method during the past year has led to a series of studies into the potential applications of nonmetallic samples, including ores, sinters, slags, process streams, additives in oil, etc. These studies were made not only to evaluate performance data for the equipment but also to determine the effects of sample preparation on the analyses.

An Applied Research Laboratories Vacuum Production X-ray Quantometer (VPXQ) (Fig. 1) was used for these studies. Because of the inherent speed of this polychromator design, a wealth of information is rapidly obtained. Instrument time for the usual sample containing light elements is about 5 min for duplicate analyses. The integration time may be varied as desired to meet the precision requirements.

On this instrument, the X-ray tube is vertically mounted, providing for a 90° incident beam with the individual monochromators arranged radially around the X-ray tube. The outputs from all of the detectors are integrated simultaneously, the integration period being controlled by the measured intensity of an external standard, internal standard, or scattered radiation. After termination, each channel is read out sequentially on a strip chart recorder as a ratio of the intensity of an element to the intensity of the controlling radiation.

The VPXQ, as shown, can accommodate up to 22 elements simultaneously, while the VXQ (Fig. 2), a physically smaller unit, can accommodate up to 9 elements. Where increased flexibility is desired on either instrument, a scanning mono-

*Present address: Angstrom, Inc., 2454 W 38th St., Chicago 32, Ill.

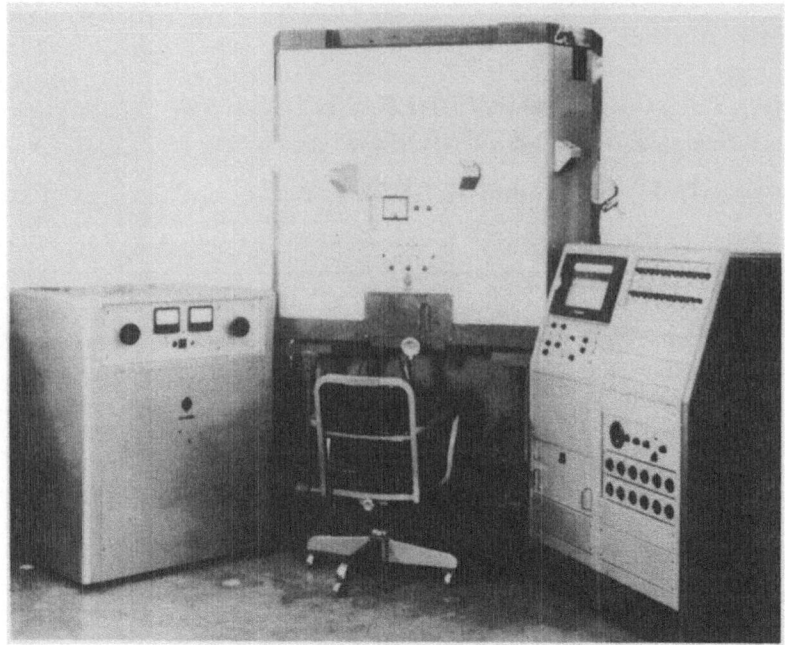

Fig. 1. The ARL Vacuum Production X-ray Quantometer.

chromator may be substituted for a fixed channel to provide speed plus versatility.

Three modes of operation are incorporated into the design of the Vacuum X-ray Quantometers—air, helium and vacuum—and each may be selected as desired by a convenient switching arrangement. The air mode would be used for a program involving only the heavy elements, the helium mode for light elements in volatile liquids, and the vacuum mode for light elements in powders and solids. For our studies, the vacuum mode was used for all of the powdered nonmetallics and solid metal samples, while the helium mode was used for the analysis of process solutions and additives in oils.

## MATERIAL STUDIES

### Iron Ore

Iron ore samples of a variety of types were chosen for this study (Table I). These samples covered a larger variety of grades than would actually be experienced in normal operation. A number of sample preparation methods were investigated,

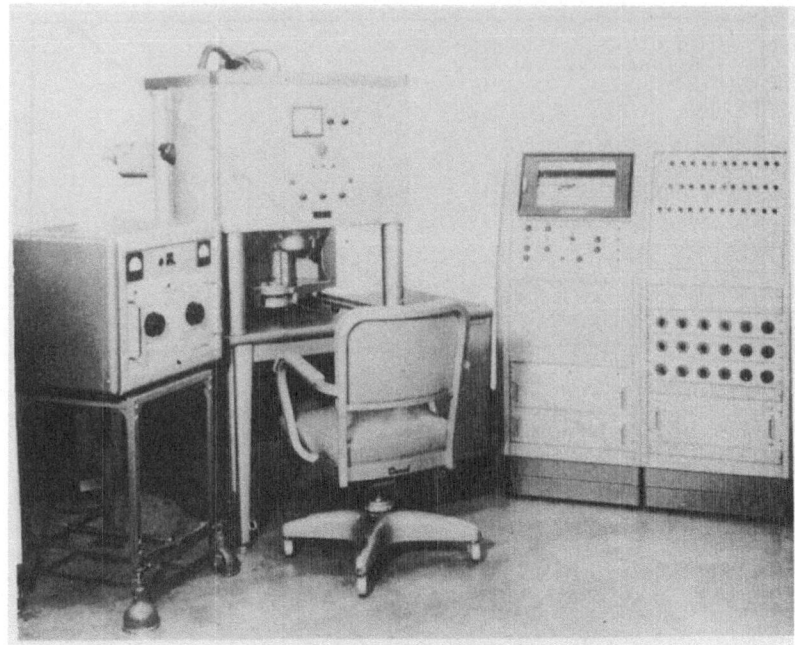

Fig. 2. The ARL Vacuum X-ray Quantometer.

## TABLE I
### Descriptions and Analyses of Samples

| Description | Chemical Analysis | | | | Loss on ignition |
| --- | --- | --- | --- | --- | --- |
| | Fe | $SiO_2$ | $Al_2O_3$ | $TiO_2$ | |
| Low Ignition Fines | 69.73 | 0.30 | 0.18 | 0.01 | 0.24 |
| Fines | 68.75 | 0.28 | 0.24 | 0.01 | 2.32 |
| Siliceous Fines | 67.23 | 3.04 | 0.44 | 0.02 | 0.46 |
| Crust | 64.08 | 0.47 | 1.80 | 0.09 | 6.10 |
| Aluminous Fines | 61.43 | 1.00 | 3.27 | 0.15 | 8.00 |
| Very Siliceous Fines | 62.68 | 9.74 | 0.22 | 0.01 | 0.56 |
| Ocherous Limonite | 57.75 | 0.61 | 4.30 | 0.22 | 12.43 |
| Friable Quartzite | 35.00 | 19.41 | 0.21 | 0.01 | 1.84 |
| Laterite Canga | 53.00 | 6.91 | 7.34 | 3.66 | 9.91 |
| Limonite and Laterite | 48.68 | 2.07 | 13.57 | 0.72 | 14.48 |
| Feruginous Quartzite | 34.28 | 48.90 | 0.39 | 0.02 | 1.71 |
| Laterite | 17.05 | 27.32 | 28.76 | 1.55 | 18.71 |

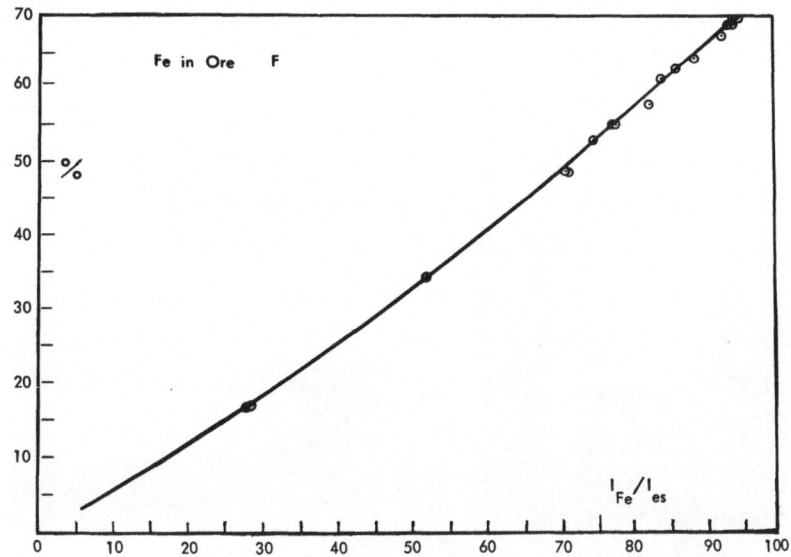

Fig. 3. VPXQ analysis of Fe in ore (fused).

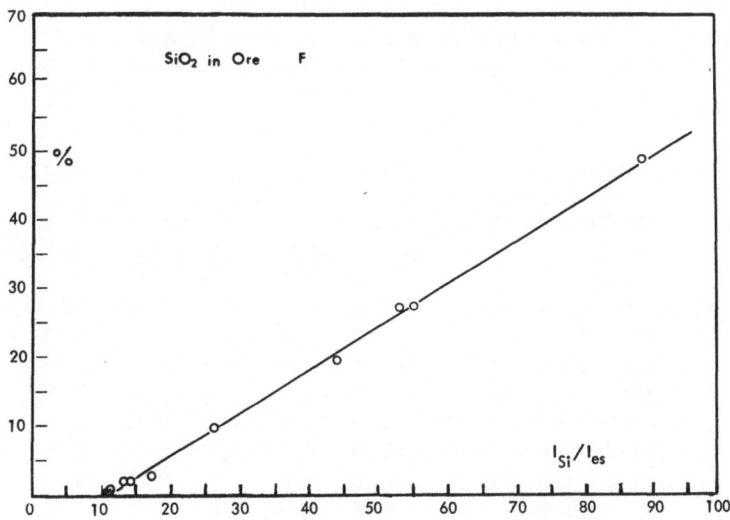

Fig. 4. VPXQ analysis of SiO₂ in ore (fused).

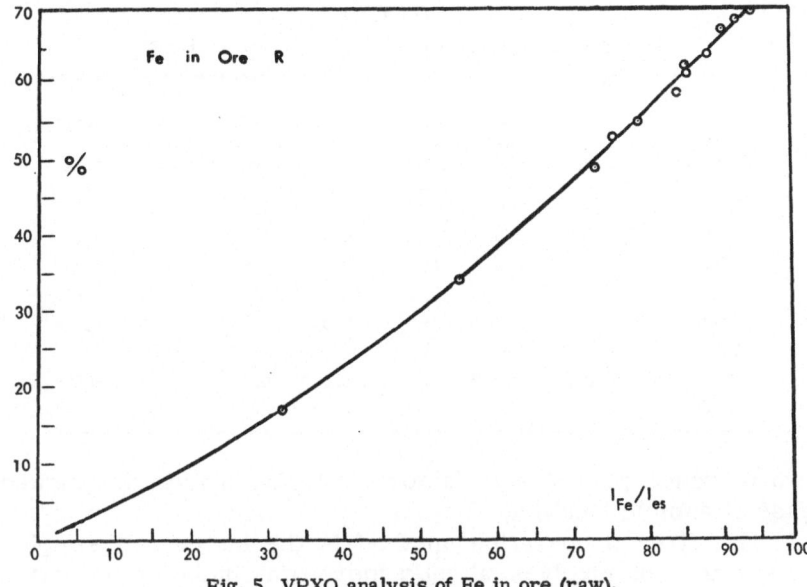

Fig. 5. VPXQ analysis of Fe in ore (raw).

Fig. 6. VPXQ analysis of SiO₂ in ore (raw).

## TABLE II
### Accuracy of Chemical and X-ray Values

| Element | Range | Chemical Aver. | Chemical Max. | X-ray Fused Aver. | X-ray Fused Max. | X-ray Raw Corr. Aver. | X-ray Raw Corr. Max. |
|---------|-------|------|------|------|------|------|------|
| Fe | 17-70 | 0.6 | 2.5 | 0.3 | 1.3 | 0.6 | 1.5 |
| SiO$_2$ | 0-50 | 1.0 | 4.0 | 0.4 | 1.3 | 0.5 | 1.4 |
| Al$_2$O$_3$ | 0-30 | 0.8 | 4.0 | 0.3 | 0.8 | 0.3 | 0.8 |
| Mn | 0-4 | 0.03 | 0.15 | 0.01 | 0.05 | 0.01 | 0.05 |
| TiO$_2$ | 0-4 | – | – | 0.03 | 0.2 | 0.03 | 0.25 |
| CaO | 0-1.5 | – | – | – | – | 0.01 | 0.02 |
| P | 0-0.5 | 0.007 | 0.08 | 0.004 | 0.02 | 0.01 | 0.05 |
| MgO | 0-1 | – | – | – | – | 0.03 | 0.14 |

two of which proved satisfactory, both for analytical data and ease of sample handling.

The first preparation method is the faster and simpler procedure but involves slightly more time in data correlation than the alternate method.

The iron ore sample is ground in a Bleuler Rotary Mill with a binder for 1.5 min; the ground sample is then briquetted on a backing of boric acid, and the resulting briquet is used for analysis.

For the alternate method, the iron ore is fused with borax at a dilution of 1:5; the resulting glass bead is ground in the Bleuler Mill, briquetted as above, and analyzed.

These samples were analyzed for Fe, SiO$_2$, Al$_2$O$_3$, Mn, and TiO$_2$. Other grades and types of ores were also studied and analyzed for P$_2$O$_5$, S, CaO, and MgO in addition to those mentioned for the control samples. Typical analytical curves for these two methods are shown in Figs. 3-6. A comparison of the two X-ray methods and chemical methods is shown in Table II. In general, both of the X-ray methods provide data comparable to those of chemical methods, with the fusion method being somewhat better than the unfused method.

### Slags

Slag samples from several plants and processes were analyzed for CaO, SiO$_2$, MgO, Al$_2$O$_3$, Fe, MnO, TiO$_2$, Cr$_2$O$_3$, and P$_2$O$_5$. These samples included slags from the acid cupola, basic cupola, open hearth, blast furnace, converter, and electric furnace.

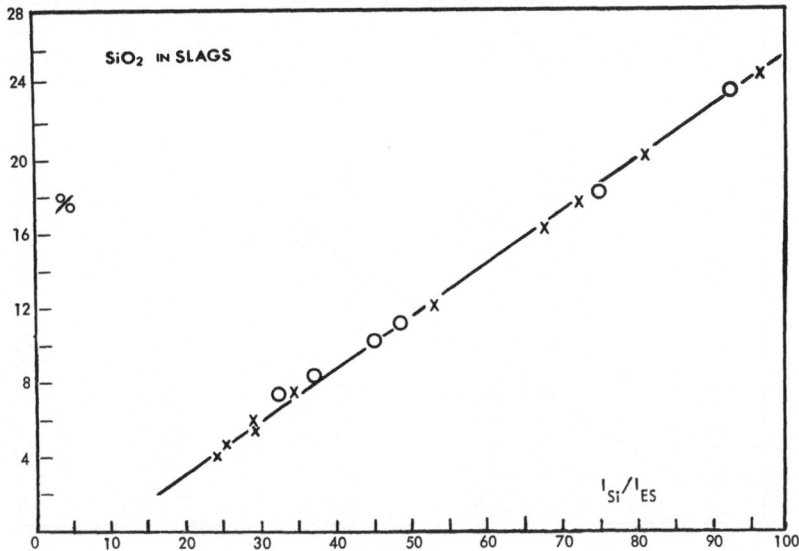

Fig. 7. Analytical data for SiO$_2$ in slags.

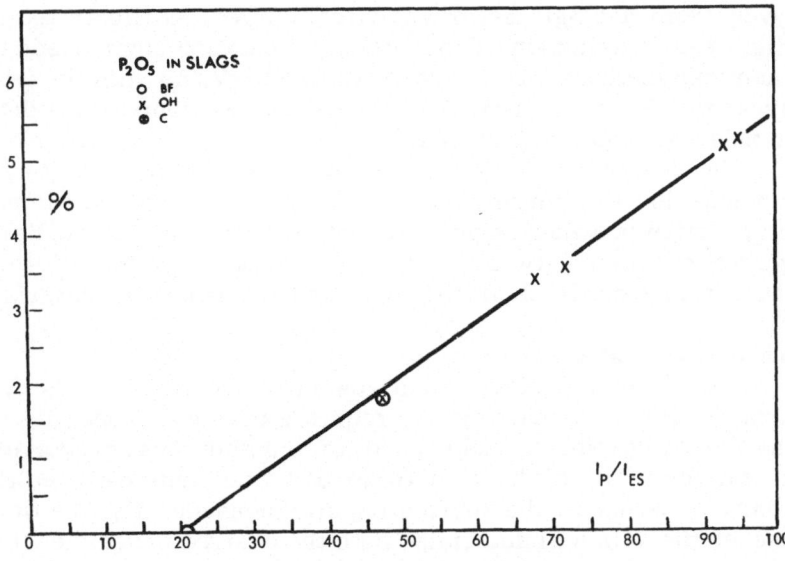

Fig. 8. Analytical data for P$_2$O$_5$ in slags.

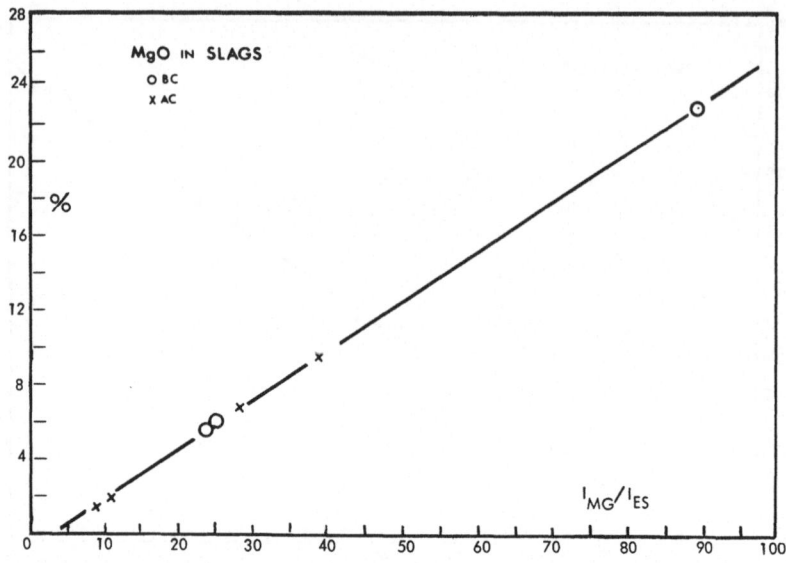

Fig. 9. Analytical data for MgO in slags.

The samples were prepared by grinding in a Bleuler Mill and briquetted as for the iron ores. An evaluation of the analytical data indicates that samples from different laboratories and for many elements from different processes may be analyzed from a single set of analytical curves, as illustrated in Figs. 7-9. A summary of the analytical data and comparison to chemical methods are shown in Tables III-V. As with the iron ores, the X-ray analytical method provides comparable data to those of chemical methods.

The analysis of slag samples by either chemical or X-ray methods is very much dependent on the sampling and sample preparation procedures prior to receipt in the laboratory. Very good correlation between the two methods and between different laboratories can be obtained from adequately prepared material.

### Additives in Lubricating Oils

Control of petroleum additives in lubricating oils is desirable, first, to maintain the required physical properties of the finished products, and second, to maintain these properties at minimum cost. Each of these additives imparts specific characteristics to the lubricating oils; consequently, the final use of the oils will determine both the desired properties and the types and amount of additives. Since each additive may con-

## TABLE III
### Comparative Data on the Precision of X-ray Analysis of Slags

| Compound | Conc. Range | Precision | At Conc. |
|----------|-------------|-----------|----------|
| CaO | 14-54 % | ± 0.07 % | 35.0 % |
| $SiO_2$ | 5-56 % | ± 0.15 % | 24.0 % |
| MgO | 1-24 % | ± 0.21 % | 6.82% |
| $Al_2O_3$ | 5-24 % | ± 0.13 % | 11.59% |
| MnO | 0.2-20 % | ± 0.012% | 4.04% |
| Fe | 0.2-48 % | ± 0.030% | 9.0 % |
| Fe | 0.2-48 % | ± 0.088% | 37.0 % |
| $TiO_2$ | 0.2-2.5% | ± 0.012% | 1.64% |
| $P_2O_5$ | 0.02-6.0% | ± 0.084% | 3.30% |
| $Cr_2O_3$ | 0.01-6.5% | ± 0.009% | 0.72% |
| FeO | 0.5-9.0% | ± 0.008% | 1.73% |

tain more than one essential ingredient and the final oils may contain more than one additive, the control of these materials must be accomplished by the determination of a number of elements. Typical information is shown in Table VI, which lists the additive types of interest.

The samples used for this study were prepared by the addition of the additives to base stocks of different viscosities in order to match the types of oils which might be encountered in routine processing. Five sets of standards were analyzed and covered an equal number of oil types.

The VPXQ was used in the helium mode of operation and was programed as shown in Table VII. For this mode of operation the light element spectrometers are filled with helium and separated from the sample chamber during loading. After loading, the sample chamber is flushed for a short period of time, the shutter separating the spectrometers and sample chamber is opened, and the entire unit is flushed with helium at a low flow rate.

Two types of liquid sample holders were used for these studies (Fig. 10). The first of these was made in the laboratory from a polystyrene vial and the second is a production ARL holder. There are several advantages to the ARL holder: (1) elimination of air bubbles at the analytical surface; (2) reproducible sample surface; (3) ease of loading; (4) minimizes effects of analyst's technique; and (5) sample may be used for an extended period of time. The 0.25 mil Mylar film which is

## TABLE IV
### Comparative Data on the Accuracy of X-ray Analysis of Slags

| Compound | Conc. Range | Aver. Dev.* | Aver. Dev.[†] |
|----------|-------------|-------------|---------------|
| CaO | 14-54 % | ± 0.15 % | ± 0.10 % |
| $SiO_2$ | 5-56 % | ± 0.28 % | ± 0.20 % |
| MgO | 1-24 % | ± 0.17 % | ± 0.10 % |
| $Al_2O_3$ | 0.5-24 % | ± 0.22 % | ± 0.15 % |
| MnO | 0.2-9 % | ± 0.07 % | ± 0.05 % |
| MnO | 4-20 % | ± 0.40 % | |
| Fe | 0.2-1.5% | ± 0.04 % | |
| Fe | 6-48 % | ± 0.30 % | |
| $TiO_2$ | 0.2-2.5% | ± 0.032% | ± 0.015% |
| $P_2O_5$ | 0.1-6.0% | ± 0.16 % | ± 0.08 % |
| $Cr_2O_3$ | 0.01-6.5% | ± 0.03 % | ± 0.01 % |
| FeO | 0.5-9.0% | ± 0.10 % | ± 0.10 % |

*Average for all companies and types of slags, including reference standard and samples with routine chemical values. Duplicate X-ray analyses.
†Data evaluation on reference standards, one company only. Duplicate X-ray analyses.


## TABLE V
### Comparative Data on the Accuracy of Chemical Analysis of Slags

| Compound | Conc. Range | Max. Spread* |
|----------|-------------|--------------|
| CaO | 33-54 % | 2.0% |
| $SiO_2$ | 8-34 % | 1.9% |
| MgO | 7-12 % | 0.6% |
| $Al_2O_3$ | 2-16 % | 2.3% |
| MnO | 0.5-10 % | 1.0% |
| Fe | 0.2-22 % | 0.9% |
| $TiO_2$ | 0.2-3 % | 0.5% |
| $P_2O_5$ | 0.008-6 % | 0.6% |
| S | 0.05-1.5% | 0.2% |
| $Cr_2O_3$ | 0.007-1.5% | 0.3% |

*Difference between high and low results for eight laboratories using good chemical methods. The averages of the eight results were used as standard values for the X-ray analyses.

## TABLE VI
### Oil Additives

| Type | P | Ba | Ca | Zn | S | Pb | Cl |
|------|---|----|----|----|---|----|----|
| A    | X | X  |    | X  | X |    |    |
| B    | X |    |    | X  | X |    |    |
| C    | X | X  |    | X  | X |    |    |
| D    |   |    |    |    |   | X  |    |
| E    |   |    |    |    | X |    |    |
| F    |   |    |    |    | X |    | X  |
| G    |   |    | X  |    | X |    |    |

used on both holders reduces the intensities from the light elements by a factor of about two.

As described previously, the information provided by the Quantometer is the ratios of element intensities to control intensities. An external standard, internal standard, or scattered radiation may be used for control purposes. For this study, the external standard and two scattered wavelengths were used for control. The use of scattered radiation to compensate for various sample effects has been described by Kemp and Andermann (Anal. Chem., Aug., 1958) and numerous cases have been investigated to demonstrate its applicability. The two scattered wavelengths used for these studies were 0.6 A and 1.51 A (Compton scatter of the $WL_\alpha$ line).

Pronounced effects of mixed additives and viscosity are readily apparent on plots of concentration vs. $l_{el}/l_{es}$; however, satisfactory data can be obtained by using families of curves for the various types of oils. Both scattered wavelengths im-

## TABLE VII
### VPXQ Programing for Analysis of Oil Additives

| Element | Line | Crystal | Detector |
|---------|------|---------|----------|
| P       | $K_\alpha$ | 4 in. $SiO_2$ | 2 cm Minitron |
| Ba      | $L_\alpha$ | 4 in. EDT | 2 cm Minitron |
| Ca      | $K_\alpha$ | 4 in. LiF | Neon Multitron |
| Zr      | $K_\alpha$ | 11 in. LiF | Argon Multitron |
| S       | $K_\alpha$ | 4 in. NaCl | 2 cm Minitron |
| Pb      | $L_\alpha$ | 11 in. LiF | Krypton Multitron |
| Cl      | $K_\alpha$ | 4 in. EDT | 2 cm Minitron |
| SR      |      | 11 in. LiF | Krypton Multitron |
| Ext. Std. |    |         | Argon Multitron |

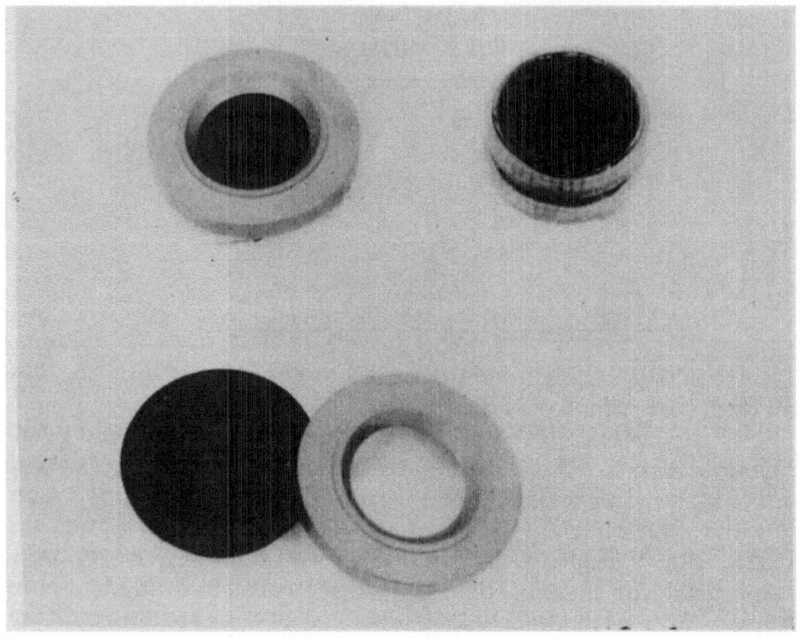

Fig. 10. Sample holders: Made from a polystyrene vial (top) and produced by
ARL (bottom).

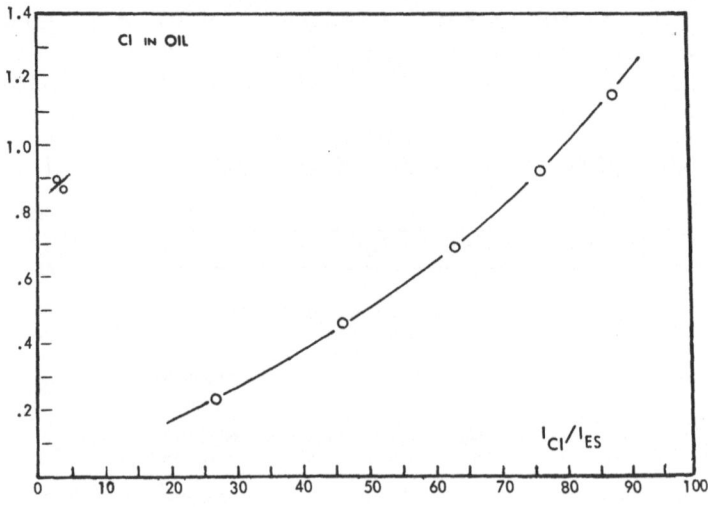

Fig. 11. VPXQ analysis of Cl in oil.

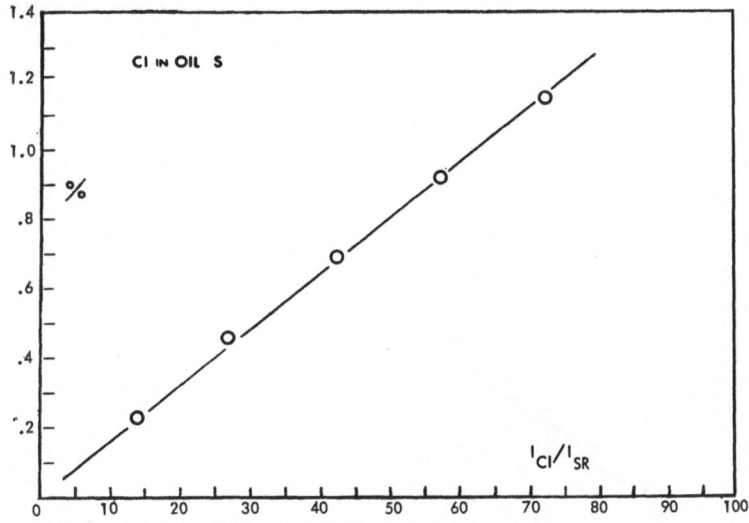

Fig. 12. VPXQ analysis of Cl in oil.

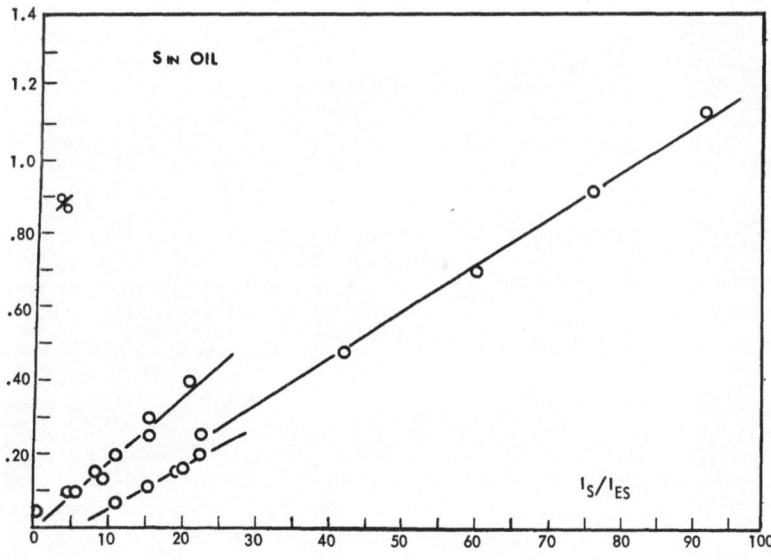

Fig. 13. VPXQ analysis of S in oil.

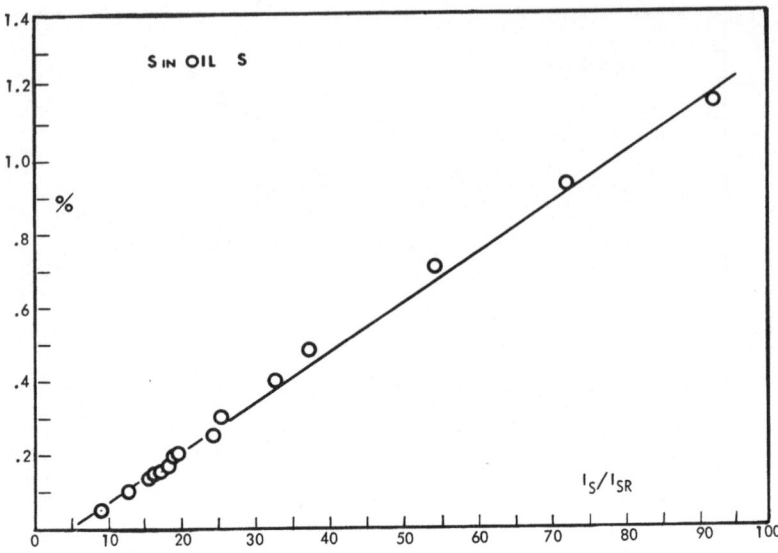

Fig. 14. VPXQ analysis of S in oil.

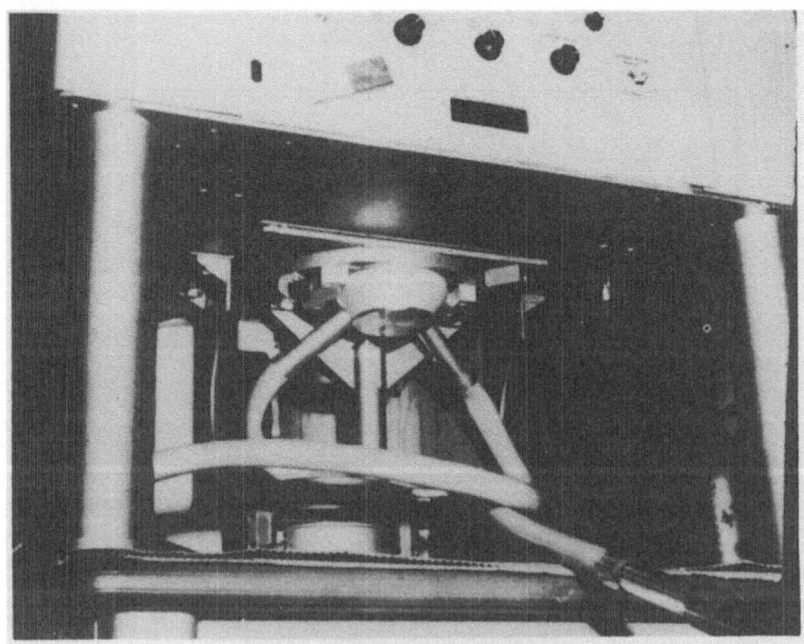

Fig. 15. The ARL slurry chamber.

prove the correlation and accuracy of the analytical data, with the Compton scattered $WL_\alpha$ radiation proving much superior to the other SR.

As shown in Figs. 11-14, which are representative analytical curves, straight intensity plots require families of curves, 0.6 A SR reduces the number of curves, while $WL\alpha(C)$ SR almost eliminates the need for additional curves.

The small error shown in the analytical curves using Compton SR would in all probability be eliminated by using chemically analyzed standards. The standards used for these were prepared by using nominal compositions and densities for the various additives and the calculated values used for correlation. For this study, all errors were attributed to the X-ray method; however, for a process control application more reliable standards would be required for standardization.

### Slurry Analysis

While there have been a number of applications using the single element Quantrol for process streams or slurries, multielement analysis, including the light elements in the same manner, has received little attention.

ARL has designed a slurry chamber (Fig. 15), which is adaptable to the Multichannel VXQ to handle process streams and to permit the detection and analysis of elements down to magnesium. Acceptable intensities for both magnesium and aluminum in cement slurries have been obtained.

### CONCLUSION

These studies have covered a variety of material types and demonstrated the applicability of the X-ray emission method of analysis. The X-ray method is ideally suited to the control of many other nonmetallic processes. The successful application of the available X-ray equipment to a control program is dependent on the development of adequate sampling and sample preparation procedures.

# SOME SPECIAL PROBLEMS IN THE X-RAY SPECTROGRAPHIC ANALYSIS OF PLATE AND CONTAINER GLASS

Stephen H. Laning

Pittsburgh Plate Glass Company
Chemical Division
Barberton, Ohio

## ABSTRACT

The X-ray spectrographic analysis of glass is straight-forward for most of the elements with atomic numbers greater than 11. However, three of the elements require special attention, since matrix effects are serious enough to require correction. These three elements are calcium, silicon, and sulfur.

The X-ray spectrographic analysis of calcium in glass is somewhat more difficult since any potassium in the glass reduces the intensity of the Ca $K_\alpha$ line by absorption. A correction factor has been found which when applied to the X-ray data gives results in excellent agreement with wet chemical analysis. The evaluation of the factor is discussed.

The analysis of silica in glass is quite difficult for glasses from different sources. It is less difficult when analyzing glass from the same tank for quality control. Calcium is the interfering element in the analysis. The nature of its interference and a procedure for overcoming it are discussed.

The analysis of sulfur in glass is made difficult by two interferences—a change in background with calcium content and the presence of the $WL_{\gamma 6}$ fifth-order line very near the S $K_\alpha$ line. Both interferences must be evaluated. A procedure for this analysis is presented.

## INTRODUCTION

The analysis of glass is a time-consuming operation requiring many hours by conventional chemical methods. To reduce the time needed for an analysis, X-ray methods have been developed for the elements with atomic numbers greater than 11 commonly associated with plate and container glass. In a few cases such methods have encountered special problems which warrant description and our approach to their solutions.

Our investigations have shown a linear relationship between concentration and intensity for all the elements studied within the limits of wet chemical methods of analysis. Where interelement effects have been observed, corrections have been applied to the linear X-ray data. Once suitable standards have been procured, glass can be analyzed without much difficulty with the exception of the elements calcium, silicon, and sulfur and to a lesser extent, barium and titanium. These elements will be discussed individually to point out the particular problem and our solution to it.

## INSTRUMENTATION

In our analysis of glass two X-ray spectrographs are employed. A standard Norelco spectrograph is equipped with helium path, FA-60 tungsten target X-ray tube, 4 × 0.005 in. plate collimator, gas-flow proportional counter with P-10 gas, and various analyzing crystals. This instrument is used for a preliminary semiquantitative analysis of unknown glasses to identify the elements to be analyzed in the glass. This instrument is also used for quantitative analysis of the elements Ba, Ti, Sr, and Zr.

The second instrument is a Norelco 3-position spectrograph capable of high intensity employing an FA-60 tungsten target X-ray tube, hydrogen atmosphere, wide-window gas-flow proportion counter using P-10 gas, 1 × 0.005 in. plate collimator before the detector, and a 4 × 0.125 in. plate collimator between the sample and analyzing crystal. An RIDL pulse-height analyzer (PHA) is used when necessary to reduce background. Analyzing crystals recommended for glass analysis include ammonium dihydrogen phosphate (ADP), ethylenediamine d-tartrate (EDDT), germanium (111), sodium chloride, lithium fluoride, and topaz.

## SAMPLE PREPARATION

A single piece of glass is preferred to a compacted powder for glass analysis. A disc $1\frac{3}{8}$ in. in diameter by at least $\frac{1}{8}$ in. thick is cut from plate glass or the bottom of the container and

## TABLE I
### Glass Standards

National Bureau of Standards
      No. 80: soda-lime-silica glass
      No. 89: lead-barium-silica glass
      No. 91: opal glass
      No. 93: borosilicate glass
Society of Glass Technology*
      Standard No. 1: soda-lime-alumina-magnesia-silica glass
      Standard No. 2: borosilicate glass
      Standard No. 3: potassium-lead-silica glass
      Standard No. 4: opal glass

*These standards may be obtained by writing to: University of Sheffield; Department of Glass Technology; Elmfield, Northumberland Road; Sheffield 10, England.

is given a standard finish by wet grinding with 180 grit alumina to make it flat and finished with 302 grit alumina. Both operations are performed on flat glass polishing plates and take about five minutes to complete.

## REFERENCE STANDARDS

Glass standards are few in number at the present time; the National Bureau of Standards offers four different types of glass in powder form only, and the Society of Glass Technology offers four different types in plate or rod form. A list of these standards is presented in Table I.

Additional standards for the various elements of interest must be obtained by individual laboratory chemical analyses performed by experienced analysts.

## PROCEDURES

Several different procedures are possible, depending on available instrumentation, the degree of accuracy desired, the concentration range of the element sought, and the presence or absence of interferences. An unknown glass that is being analyzed for the first time should be given a general X-ray spectrographic scan using a sodium chloride analyzing crystal to determine the elements present and their approximate concentrations. Following the general scan, one is ready to begin the analysis. High and low standards for an element of interest

## TABLE II
### General Instrumental Conditions for Analysis of Various Elements in Glass

| Component | Technique | Analyzing Crystals | Angular Position | PHA |
|-----------|-----------|--------------------|------------------|-----|
| $Al_2O_3$ | Chart * | EDDT | Al $K_\alpha$ | Yes |
| $As_2O_5$ | Chart | LiF | As $K_\alpha$ | — |
| BaO | Scan† 1°/min | LiF | 89-84° | No |
| CaO | Chart | LiF | Ca $K_\alpha$ | — |
| $Fe_2O_3$ | Chart | LiF or Topaz | Fe $K_\alpha$ | — |
| $K_2O$ | Chart | LiF | K $K_\alpha$ | — |
| MgO | Chart | ADP | Mg $K_\alpha$ | Yes |
| $SO_3$ | Chart | NaCl | S $K_\alpha$ and 148° | Yes |
| $SiO_2$ | Chart | EDDT | Si $K_\alpha$ | No |
| SrO | Scan 1°/min | LiF | 26-21° | No |
| $TiO_2$ | Scan 1°/min | LiF | 89-84° | No |
| $ZrO_2$ | Scan 1°/min | LiF | 26-21° | No |

Special Analyses

| Component | Technique | Analyzing Crystals | Angular Position | PHA |
|-----------|-----------|--------------------|------------------|-----|
| Cl | Chart | Ge | Cl $K_\alpha$ | Yes |
| CoO | Chart | Topaz | Co $K_\alpha$ | — |
| $Cr_2O_3$ | Chart | Topaz | Cr $K_\alpha$ | — |
| MnO | Chart | Topaz | Mn $K_\alpha$ | — |
| $P_2O_5$ | Chart | Ge | P $K_\alpha$ | Yes |
| Se | Chart | Topaz | Se $K_\alpha$ | Yes |
| ZnO | Chart | LiF | Zn $K_\alpha$ | — |

*This technique consists of operating the recorder while the goniometer remains in a fixed angular position, such as the peak position for an element.
†This technique consists of recording intensity while the goniometer traverses an angular range of an element of interest.

are run, followed by the unknowns. One may measure counting rates on the standards and unknowns or chart the intensities. In some instances calibration curves can be obtained from scanning data where one measures peak height above background. A summary of the instrumental conditions frequently employed in our laboratory for the various elements is presented in Table II.

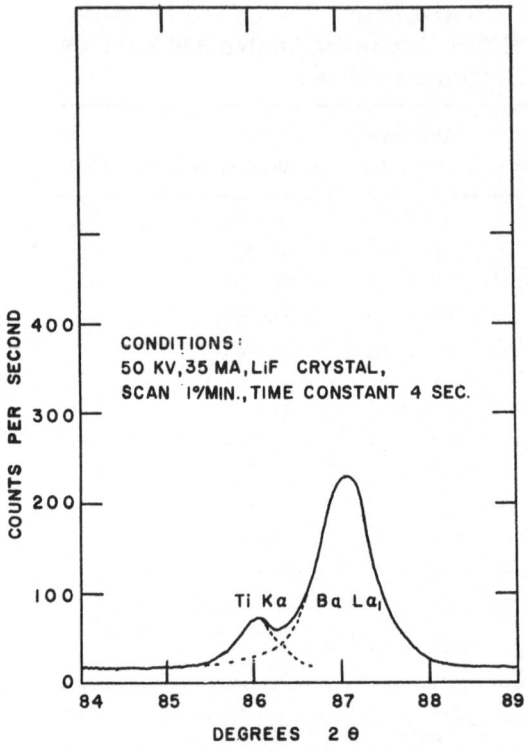

CONDITIONS:
50 KV, 35 MA, LiF CRYSTAL,
SCAN 1°/MIN., TIME CONSTANT 4 SEC.

Ti Kα    Ba Lα₁

Fig. 1. Scanning technique for determining barium and titanium in glass. The diagram shows a situation frequently encountered in glass analysis when using our standard spectrograph to scan the angular range from 89° to 84° for determining barium and titanium. This determination would be extremely difficult on an instrument with less resolution, if it could be done at all.

## GENERAL PROCEDURES

When an analysis for an element is not complicated by interferences, the procedure is to chart a high and low standard for the particular element and then the unknowns. A linear equation is derived from the data for the knowns and the unknowns are then evaluated. The recorder is run continuously during the period of the determination in order to keep a record of the time and to correct for any drift. Correction for drift is accomplished by repeating the high and low standards following the unknowns (as illustrated in Figs. 3 and 5).

### Determination of BaO and TiO$_2$

These two elements are determined together by a scanning technique. Figure 1 illustrates a frequent situation in glass composition with from 0 to 0.8% BaO and traces of TiO$_2$. With a lithium fluoride analyzing crystal the peak positions of the Ba$L_{\alpha1}$ and Ti$K_\alpha$ lines are only 1° apart. Frequently there is enough barium in glasses to give sufficient background under

## TABLE III
### Comparison of Chemical and X-ray Data for CaO in Glass

| CaO by Wet Analysis, % | $K_2O$, % | CaO by X-ray Uncorrected, % | CaO by X-ray Corrected for $K_2O$, % |
|---|---|---|---|
| 12.51 | 0.08 | 12.49 | 12.51 |
| 6.35 | 0.31 | 6.29 | 6.36 |
| 6.77 | 0.50 | 6.65 | 6.76 |
| 8.49 * | 0.72 | 8.32 | 8.48 |
| 7.17 | 1.09 | 6.96 | 7.20 |
| 7.05 | 1.12 | 6.78 | 7.03 |
| 10.87 | 2.23 | 10.37 | 10.87 |

*Society of Glass Technology Standard No. 1.

the Ti$K_\alpha$peak so that it must be subtracted from the Ti$K_\alpha$peak intensity. This can best be done graphically, as in Fig. 1. The angular scan from 89° to 84° $2\theta$ is made at 1°/min and peak height above background corrected for overlap is measured against a standard that has been run under the same conditions of X-ray intensity.

### Determination of CaO

The determination of CaO is only slightly more involved than the elements that have no interferences. A noticeable matrix effect caused by potassium in the glass must be compensated for in order to have agreement with wet chemical values.

The determination is performed following the general procedures by charting the peak intensity of Ca$K_\alpha$ for the high and low standards and the unknown. An X-ray determined CaO value is obtained, using a linear equation derived from the standards. To this value is a d d e d the potassium correction, which is found by multiplying the percent $K_2O$ found in the glass by (0.225% CaO/1% $K_2O$). Table III presents a comparison between wet chemical and X-ray data for CaO in glasses having a range of over 2% $K_2O$. These data are shown graphically in Fig. 2.

### Determination of SO₃

The determination of $SO_3$ in glass is made somewhat more complex than that of the previously mentioned elements for two reasons: (1) The general background intensity at the sulfur

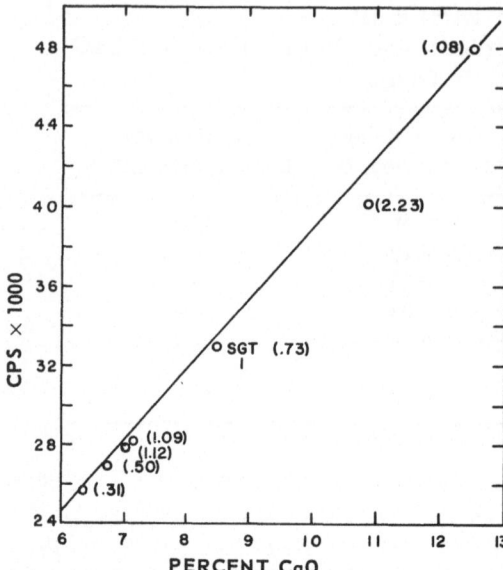

Fig. 2. Diagram presenting the data of Table III in graphic form. The curve represents the value for CaO derived from X-ray data; the circles represent the values obtained by chemical methods. The corrections applied to the X-ray data place the determined values within the circles.

$K_\alpha$ line position is influenced by the calcium content of the glass; and (2) a fifth-order tungsten $L_{\gamma 6}$ line falls very near the sulfur peak. The latter situation results in a higher background under the sulfur peak than on either side of it. A high and low sulfur standard are necessary to evaluate the contribution of the tungsten line to the peak intensity. The general background intensity is measured a few degrees from the sulfur peak at 148° $2\theta$. The peak height above the general background represents the sum of the two lines $S K_\alpha$ and $W L_{\gamma 6}$, $n=5$. The latter must be subtracted from the peak height to get the net intensity due to sulfur alone.

Figure 3 illustrates the problem in the determination of $SO_3$. It can be seen that the background intensity at 148° varies with glasses of different CaO content. The low standard contains no $SO_3$ and therefore represents the $W L_{\gamma 6}$, $n=5$ intensity, which has to be subtracted from the peak intensity to evaluate the sulfur peak alone. Figure 4 shows our calibration curve, including the data for the Society of Glass Technology Standard No. 1.

### Determination of $SiO_2$

The determination of $SiO_2$ in glass is by far the most complex of the elements determined by X ray. There is no single calibration curve, but rather a family of parallel curves de-

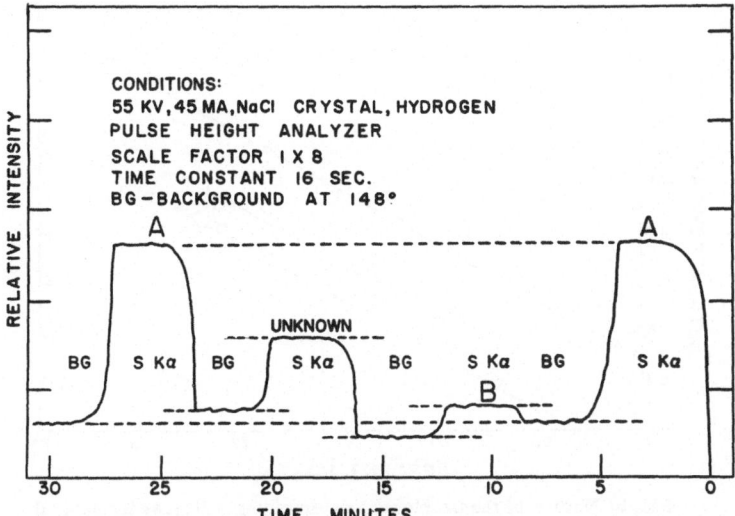

Fig. 3. Charting technique for determining sulfur in glass. The high standard "A" is charted first on the $SK_\alpha$ peak and then on background at 148°. Next, the low standard "B" is similarly charted, then the unknowns, followed again by the high standard to check on stability of the instrument.

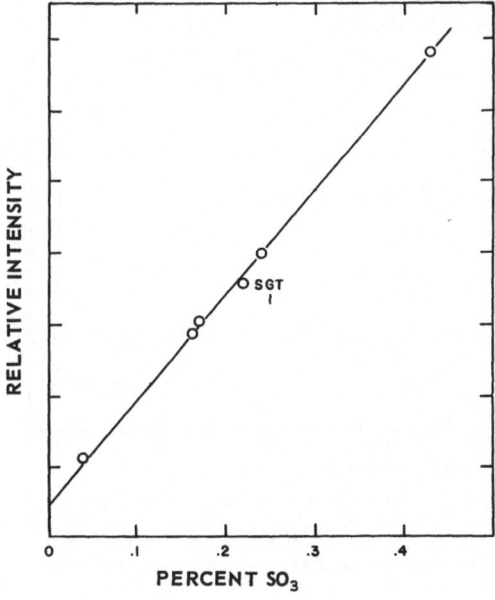

Fig. 4. Results of our procedure as outlined in Fig. 3. The Society of Glass Technology Standard No. 1 was evaluated and is indicated on the drawing. The data indicate a linear relationship between intensity and concentration.

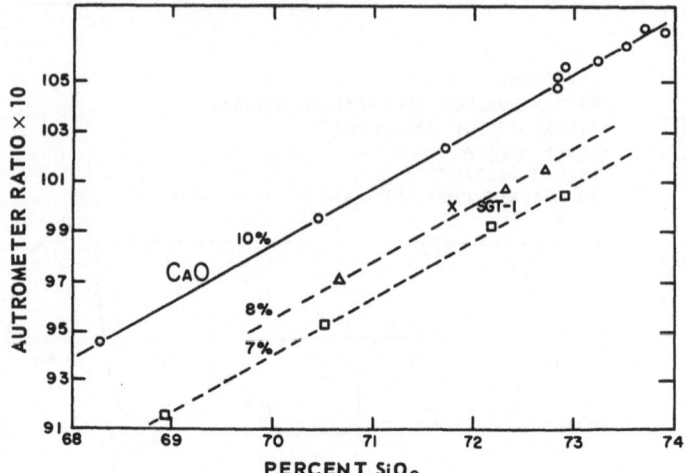

Fig. 5. Nature of the problem in determining silica, as illustrated by data obtained using an Autrometer with the Society of Glass Technology Standard No. 1 as reference. It is evident that calcium has a significant influence on the silica intensity which outweighs the influences of other elements, although they may contribute to the general inaccuracies in the determination. To overcome the latter problem, a standard having a composition close to that of the unknown is used for comparison.

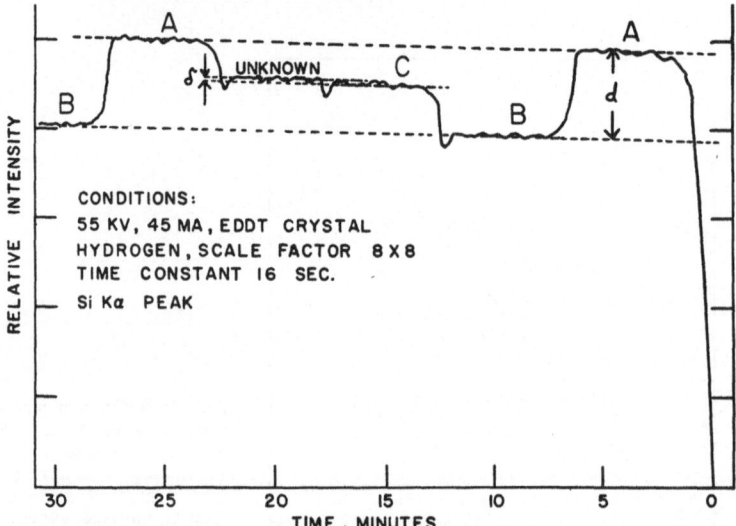

Fig. 6. Charting technique for determining silica in glass, using the three-position spectrograph. Three standards ("A", "B", and "C") are generally required for this determination. Note that the intensity on the standards increases with time, a circumstance which is quite difficult to compensate for when using counting techniques rather than charting.

pendent on the CaO content of the glass. Some of our data are shown in Fig. 5 to illustrate the nature of the problem.

After the slope has been established, a glass standard having the same CaO content as the unknown is run, followed by the unknown. The difference in intensity between the known and unknown when divided by the slope gives the percent of $SiO_2$ to be added to or subtracted from the $SiO_2$ value of the known. A chart of this determination is shown in Fig. 6 to illustrate the procedure.

## CONCLUSIONS

With the procedures outlined here, it has been possible for us to analyze container and plate glass having a wide range of composition. Definite improvement in our silica procedure is needed to shorten the time of analysis and increase its accuracy.

Additional studies are needed to evaluate matrix changes on the silica intensity, particularly by the elements aluminum, magnesium, iron, and potassium. These studies are being made as time permits.

## ACKNOWLEDGMENT

Table III and Figs. 2 and 5 were previously published in slightly different form in my paper entitled "The application of X rays to the analysis of container glass" [William M. Mueller (ed.): Advances in X-ray Analysis, Vol. 5, Plenum Press, Inc., New York (1962), pp. 457-463] as Table I and Figs. 4 and 9, respectively, and are reproduced here by permission of the University of Denver.

# A RAPID X-RAY FLUORESCENCE METHOD FOR THE DETERMINATION OF V, Cu, Mo, Ti, Co, AND Ni

## E. D. Pierron and R. H. Munch

Monsanto Chemical Company,
St. Louis, Missouri

## ABSTRACT

The development of a rapid and quantitative X-ray fluorescence method for the determination of vanadium, copper, molybdenum, titanium, cobalt, and nickel present in mixed oxides is discussed. Each element is successively determined from the same solution. Instrument settings, matrix effects, and accuracy of the method are given.

## INTRODUCTION

Rapid and precise procedures were needed to study the composition of mixed oxides at various steps during their preparation and utilization. No simple wet chemical or colorimetric method exists for the quantitative determination of vanadium in the presence of molybdenum. The use of X-ray fluorescence as an analytical tool has advanced rapidly since 1948, when equipment became available commercially. It permits the general analysis of multiple components present in a sample without separation of the interfering elements prior to measurement. In addition, it offers advantages of speed and accuracy seldom found in other analytical methods.

## ANALYTICAL METHOD

### Apparatus

1. Norelco Universal Vacuum X-ray Spectrograph equipped with an FA-60 tungsten target X-ray tube, a lithium fluoride analyzing crystal, and a flow proportional counter.

2. Laboratory centrifuge equipped with 15-ml centrifuge tubes and capable of 1500 rpm.

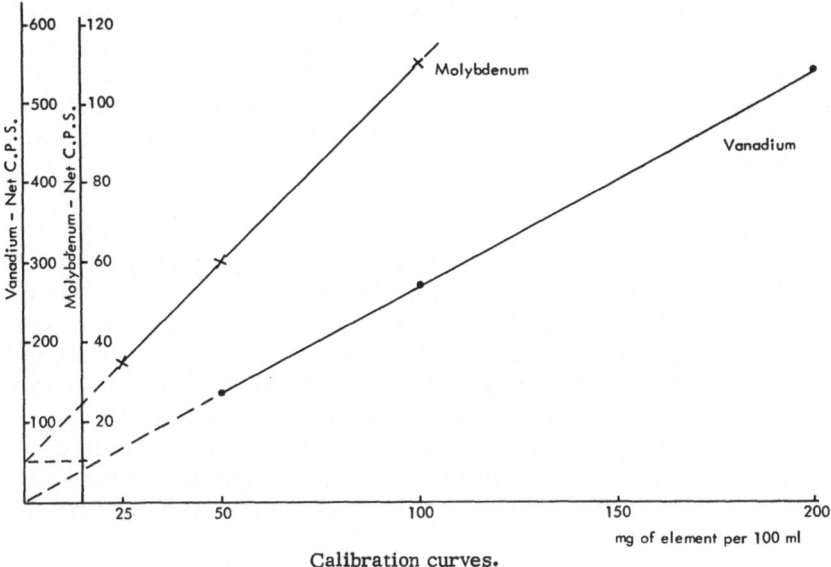

Calibration curves.

### Reagents

1. Sulfuric acid, concentrated—Reagent grade.
2. Ammonium sulfate—Reagent grade.
3. Ammonium metavanadate—Reagent grade.
4. Cupric sulfate—$CuSO_4 \cdot 5H_2O$.
5. Ammonium molybdate, $(NH_4)_6Mo_7O_{24} \cdot 4H_2O$—Reagent grade.
6. Titanium dioxide, rutile.
7. Cobalt sulfate, $CoSO_4 \cdot 7H_2O$—Reagent grade.
8. Nickel sulfate, $NiSO_4 \cdot 6H_2O$—Reagent grade.

### Calibration

Stock standard solutions of concentration equal to 1 g element/100 ml are made up by dissolving the appropriate amount of reagent and 5 g of ammonium sulfate in 20 ml of concentrated sulfuric acid and diluting this solution to 100 ml with water. The concentration of the titanium stock standard solution equals 50 mg/100 ml. A 100-ml sample of a composite solution containing 150 mg of vanadium and 10 mg of each of the other elements is used to determine the instrument panel settings. Table I shows the settings necessary to obtain maximum resolution for each element considered in a sulfuric acid—ammonium sulfate solution.

The calibration curves shown in the figure were established by determining the net counts per second (cps at $K_\alpha$ peak minus cps at chosen background) for each element present in three standard solutions. The curves show a straight line relationship between the net counts per second and concentration within the range to be used.

### Procedure

Transfer approximately 0.2 g of finely ground sample (-100 mesh), weighed to the nearest milligram, into a 100-ml volumetric flask. Add 20 ml of concentrated sulfuric acid and 5 g of ammonium sulfate. After 2 to 3 min of gentle boiling, the solution is cooled to room temperature and diluted to 100 ml. Prior to analysis a 25-ml aliquot portion of the solution is centrifuged at 1500 rpm for 10 min. This operation permits separation of the undissolved silica which is present in certain samples and which would interfere during X-ray measurement.

Transfer approximately 8 ml of the clear supernatant solution into a plastic cup and cover the cup with a 0.00025 in. Mylar film window. Place the sample container in the spectrograph and set the instrument panel in accordance with Table I for each element to be analyzed. For each element the counts per second are determined at the corresponding $2\theta$ angle for $K_\alpha$ peak and background. To increase the accuracy of the measurement a minimum counting time of 3 min is used and all measurements are made in duplicate.

## DISCUSSION

The use of the pulse-height analyzer permits measuring radiation which is characteristic of the element being determined and discriminates against foreign radiations. In addition, this electronic component of the instrument increases the peak-to-background ratio and thereby increases the sensitivity.

Preliminary investigation showed that serious matrix effects were obtained when solid powder samples were used. To minimize such difficulties and facilitate the preparation of standard mixtures, the solution technique of sample handling was adopted.

The presence of relatively insoluble oxides such as $TiO_2$ requires the use of boiling concentrated sulfuric acid in the presence of ammonium sulfate to achieve complete solubility

## TABLE I
### X-ray Spectrograph Settings (W tube; 40 kv, 25 ma; LiF crystal; air path; flow proportional counter)

| Element | Pulse-Height Analyzer | | Detector High Voltage | $2\theta$ | |
|---------|-----------------------|----------------|-----------------------|---------|---------|
|         | Base Line, Volts | Window, Volts |                       | BG *    | $K_\alpha$ |
| V       | 12                | 18            | 1675                  | 75.00   | 77.05   |
| Cu      | 18                | 15            | 1675                  | 46.50   | 44.95   |
| Mo      | 9                 | 12            | 1600                  | 23.00   | 20.50   |
| Ti      | 9                 | 6             | 1675                  | 88.00   | 86.20   |
| Co      | 21                | 15            | 1650                  | 54.50   | 52.80   |
| Ni      | 21                | 15            | 1650                  | 23.00   | 48.80   |

*Background.

of the sample. A plastic cup sample container and a piece of Mylar film in contact with 20% sulfuric acid solution showed no deterioration over a period of 48 hr, thus demonstrating the safe use of such solution media in the instrument chamber.

The matrix effects were studied by determining in duplicate and taking the average values of the net counts per second of each element present in six series of standard solution. The series of solutions were prepared to contain (per 100 ml): 200 mg of vanadium + 20 mg of each element in binary mixtures; 200 mg of vanadium + 20 mg of each of the other 5 elements in composite mixtures. In addition, for each minor component series, standard solutions were prepared to contain (per 100 ml): 20 mg of the element + 20 mg of each of the other minor elements in binary mixtures. Each element series was composed of a minimum of six different standard solutions.

## TABLE II
### Matrix Effects

| Series | Counts Per Second | | Solution No. Counted | % Std. Deviation |
|--------|-------------------|-----------|----------------------|------------------|
|        | Mean Net Count    | St. Dev.  |                      |                  |
| V      | 1024.6            | 5.1       | 7                    | 0.5              |
| Cu     | 293.5             | 7.6       | 9                    | 2.6              |
| Mo     | 84.3              | 2.4       | 7                    | 2.8              |
| Ti     | 25.3              | 0.8       | 6                    | 3.0              |
| Co     | 371.7             | 1.0       | 6                    | 0.3              |
| Ni     | 520.9             | 11.8      | 8                    | 2.3              |

TABLE III
Recovery Study

| Elements | Solution No. Studied | Recovery (in %) | | Mean Dev. (in %) |
|---|---|---|---|---|
| | | Minimum | Maximum | |
| V | 5 | 99.1 | 102.9 | 0.8 |
| Ni | 3 | 98.3 | 101.8 | 1.5 |
| Mo | 4 | 99.0 | 100.8 | 0.6 |
| Co | 2 | 100.3 | 101.7 | 0.7 |
| Cu | 2 | 100.1 | 104.2 | 1.5 |

The mean net count, the standard deviation, and the percent standard deviation for each series of element as shown in Table II were computed from the net counts of that element obtained for each standard solution series. In certain cases, as for the copper series, composite mixtures with various concentrations of copper were included in the study. The relatively low percent standard deviations shown in Table II indicate that no significant matrix effect exists in the composition range investigated.

To save time the background counts are made on the same solution for each $K_\alpha 2\theta$ angle at adjacent angles where the background count is low. Except for the vanadium calibration curve, which passes by the coordinate origin, all others intercept the cps axis. However, it has been found that the ordinates of the calibration curves at zero concentration are reproducible and constant. If background corrections are made by counting a blank solution at the $K_\alpha 2\theta$ angles, all the curves intercept the coordinate origin.

To compensate for day-to-day variations in instrument performance, a standard composite solution is counted before each series of determinations. It was found that in most cases the variations which occur are less than 1%, and therefore the original calibration curves are used for calculation. If the variations are greater than 1%, new calibration curves are established by counting a standard composite solution in its original concentration and in one dilution.

A recovery study of elements from various synthetic mixtures (Table III), shows a minimum mean deviation of 0.6% for molybdenum and a maximum mean deviation of 1.5% for copper.

The reproducibility appears to vary from a minimum of 2%

relative for vanadium to a maximum of 5% relative for copper and nickel; other elements are between these limits.

Preliminary studies for the determination of phosphorus in a concentration equal to 5 mg/100 ml and in the same matrix as described therein indicated that only very low sensitivity could be obtained. For this purpose an EDDT crystal was used.

In conclusion, we may state that the aqueous solution method has several advantages over the solid sample method:

1. No significant matrix effect.
2. No specific surface area or finish on the sample is necessary.
3. There is uniform distribution of each element within the sample.
4. The concentration of the sample can be reduced accurately several hundred times, if necessary.
5. The means of addition of internal standards to eliminate any interelement effect or instrument variations is simplified.
6. One standard sample with several of its dilutions is sufficient to establish working curves.

### REFERENCES

1. J. E. Fagel, Jr., H. A. Liebhafsky, and P. D. Zemany, Anal. Chem. 30, 1918 (1958).
2. W. W. Houk and L. Silverman, Anal. Chem. 31, 1069 (1959).
3. R. W. Jones and R. W. Ashley, Anal. Chem. 31, 1629 (1959).

# GAS CHROMATOGRAPHY

# A REVIEW OF PREPARATIVE
# GAS CHROMATOGRAPHY

## John A. Perry

Sinclair Research, Inc.
Harvey, Illinois

Preparative gas chromatography can be used to prepare fractions for further study, pure-grade reactants, or highly purified reference standards. In this paper some current methods and problems in processing and collecting micro- and milliliter quantities of liquids are reviewed.

## A REVIEW OF PREPARATIVE GAS CHROMATOGRAPHY

Analytical gas chromatography once followed but now has gone far beyond emission spectrophotometry, infrared spectroscopy, and mass spectrometry in revolutionizing analytical chemistry. Compared to these other instrumental methods, the equipment and methods of analytical gas chromatography are simpler and more available, the results generally quicker and more obvious, and the need greater, being based on the previously unsatisfied requirements of almost the entire realm of organic chemistry. Part of this need has led to preparative gas chromatography, wherein components are separated by gas chromatography and then collected for further use or study. In this paper, the status of and some criteria for the practice of preparative gas chromatography as reported in the literature are summarized in two sections: (1) The use of normal analytical equipment for preparative work; and (2) the design and use of scaled-up equipment. In each section, the problems of sample introduction, column design and use, and component collection are considered.

## PREPARATIVE GAS CHROMATOGRAPHY
## WITH ANALYTICAL EQUIPMENT

By "analytical equipment" we mean gas chromatographs designed primarily to accommodate microliter-sized liquid samples and packed columns of $\frac{1}{4}$-in. outside diameter and up to about 40 ft in length.

### Sample Introduction—Analytical Equipment

In any type of gas chromatography, sample introduction is the first operation: It can determine the minimum peak width and thus either permit or vitiate the desired separation. As Martin and Synge [21] pointed out, the charge should be instantaneously contained within the first theoretical plate. Thus, the sample must be introduced as a plug rather than by logarithmic introduction from a chamber in which more or less complete mixing with the carrier gas takes place. This "plug" introduction requires immediate sample volatilization as well as a continuing smooth diameter and the elimination of effective "tees" from the complete gas chromatographic system from introduction point to detector. If we can assume that the normal gas chromatograph and accompanying sampling systems—such as the syringe and the Mikrotek dipper, which by now are both very well known—satisfy these requirements for instantaneous vaporization and "plug" injection, then we can proceed to consideration of the $\frac{1}{4}$-in.-OD preparative column.

### Column Design and Use—Analytical Equipment

In this section the different natures, functions, and uses of $\frac{1}{4}$-in.-OD analytical—i.e., yielding minimum HETP—and $\frac{1}{4}$-in.-OD preparative columns are outlined. About the only features the two types of $\frac{1}{4}$-in. column have in common are acceptability of normal bending or coiling [9] and usability with the average gas chromatograph. However, if the differences between the two types of columns are to be delineated and established, the recommendations for analytical columns must first be reviewed.

A $\frac{1}{4}$-in. packed column designed primarily to achieve a high separation factor [25] should contain fine (e.g., 100 mesh) support particles having a minimum adsorptivity [6] and size distribution [7] and supporting a minimum weight, approximately 2%, of stationary phase that will avoid tailing. A minimum—as near zero as possible—sample volume should be charged to the column [7, 24]. Finally, smooth, if not linear, temperature programing [20] saves analysis time with materials having an appreciable boiling range and improves sensitivity with these materials. Every one of these points is different for a $\frac{1}{4}$-in. preparative column because the criterion of serviceability is different.

## TABLE I
### Comparison of Analytical and Preparative 1/4-in. Columns

|  | Analytical (Minimum HETP) | Preparative (Maximum Throughput) |
|---|---|---|
| Sample size | Minimum; usually 1-5 µl | Maximum permitted by column and mode of operation; 50-100 µl or more |
| Support size | 80-100 mesh | 20-30 mesh |
| Support Loading | 1-5% | 20-30% |
| Length | X | 2X-4X |
| Temperature Programing | Linear | Stepwise |

Although efficiency falls off rapidly with increasing sample size [7, 24], Golay [12] has suggested that throughput per unit time rather than separation factor or HETP is the prime criterion in preparative work. Thus, he further suggests that preparative columns are likely to be longer than analytical columns and to be used at an appreciably higher flow rate. To allow a higher flow rate and to permit reasonable ease of sample injection against column head pressure, a coarser support is indicated. More stationary phase, probably 20-30% by weight, is required for the much greater sample quantities injected. Finally, using a segmented, long, narrow ($\frac{3}{8}$ in.) preparative column, and samples of up to 10 ml, Perry [22] demonstrated that stepwise temperature programing of narrow columns with one step per peak after collection of each peak-component affords great sample-loading capacity per column area and maximum time-separation of peaks consistent with minimum time per sample.

The contrast between analytical $\frac{1}{4}$-in. columns prepared primarily for minimum HETP and preparative $\frac{1}{4}$-in. columns prepared primarily for maximum sample throughput per unit time is summarized in Table I.

### Component Collection—Analytical Equipment

Although the preceding distinctions between the two extreme types of $\frac{1}{4}$-in. columns, minimum HETP and preparative, are valid enough and may also emphasize the usability of $\frac{1}{4}$-in. preparative columns with adequate sample vaporizers for samples up to 1 ml in size, nevertheless a standing problem in the gas chromatographic laboratory is to identify a given peak using the column currently in the instrument. In this section a few recent simple, ingenious, and effective microtraps for use with run-of-the-mill $\frac{1}{4}$-in. columns are reviewed.

Before any peak-component can be trapped, it must be brought to the trap; thus, a requirement for successful peak identification is that the detector and detector-exit lines be free of voids or tee's and be at the highest temperature reached in the column during the run. In other words, the component must be conveyed to the trap as carefully as it was to the detector.

The idea of using a simple 2- or 3-mm thin-wall glass tube 9 or 10 in. long comes from Lohr and Kaier [19]. Such a tube can be partially inserted into a heated exit port, if there is a septum around the tube which fits the port so that the exit gas is forced through the tube. Alternatively, the glass tube could be firmly butt-ended against the port by a Teflon sleeve. Again, the port could terminate in a heated syringe needle onto which the tube, containing or terminated by a septum, could be impaled. With any such arrangement, the desired peak-component condenses within the air- or ice-cooled section of the tube and can subsequently be washed down. One tube may then be readily used to collect each peak-component. The tubes can be conveniently stored in test tubes held in a test-tube rack.

Several variations add utility to the basic tube idea. For example, Mr. D. Ford in our laboratory has used a tube which terminates in a fritted disc about 5 mm in diameter; the tube rests in a test tube in perhaps 2-3 ml of solvent. The tube then condenses the peak-component partly within the tube and absorbs nearly all of the remainder in the solvent. Later the solvent can be drawn by vacuum back through the disc into the tube to dissolve and thus recover the condensate. Finally, if 25-30 ml of solvent are used, recovery is better than 99%, according to measurements with radioactive materials.

In all these cases, the resulting solution is usually turned over as such to the infrared spectroscopist. The advent of commercially available microcells and micro-gas cells specifically designed to be used first with the gas chromatograph and then in the infrared spectrometer should further simplify and strengthen peak identification. In the microcell, the exit gas passes through a glass tube-condenser; the condensed film is then centrifuged off the tube into the microcell, both meanwhile being held in a special jig. The microcell containing the condensate is then used in conjunction with beam-condensing optics. With volatile components, the micro-gas cell is either equipped with Irtran (water-insoluble, infrared-transmitting) windows and directly immersed in the chilling bath, or if equipped with rock salt windows it may be encased in a polyethylene bag before immersion; the infrared cell thus becomes the trap within which the condensate appears.

Another ingenious method developed by an infrared spectroscopist is the KBr powder trap [18]. In this, the peak-component in the exit gas is passed through a plug of tamped KBr powder on which the component is deposited. This KBr is then formed by conventional techniques into a disc for examination by infrared spectrophotometry.

## PREPARATIVE GAS CHROMATOGRAPHY WITH SCALED-UP EQUIPMENT

By "scaled-up" equipment we mean any gas chromatograph designed to accommodate or produce samples or products of $1~\mu l$ or larger size, and packed columns larger than $\frac{1}{4}$ in. in diameter. However, before proceeding with a discussion of scaled-up equipment, let us review the criteria for judging gas chromatographic separations with $\frac{1}{4}$-in. columns and develop from these a target criterion for judging preparative gas chromatographic separations. Trying for maximum column efficiency, Bohemen and Purnell [7] obtained 1280 theoretical plates/ft (HETP, 0.24 mm) and Scott [26] obtained 600 theoretical plates/ft (HETP, 0.5 mm). Perry [22] reported 300 plates/ft (HETP, 1 mm) with a $\frac{1}{4}$-in. column used on his analytical preparative gas chromatograph. Most workers, it seems, are satisfied to achieve about 200 theoretical plates/ft (HETP, 1.5 mm). Johns et al. [15] obtained 117 plates/ft (HETP, 2.6 mm) with a $\frac{1}{4}$-in. column containing 30% Apiezon J and used as a standard of comparison for their preparative gas chromatograph containing similar $\frac{5}{8}$-in.-OD columns.

We apparently can conclude that the average worker, moderately interested in column efficiency and injecting 1- to 5-$\mu l$ samples into a typical $\frac{1}{4}$-in. column gas chromatograph, usually can expect separations corresponding to about 200 theoretical plates/ft (HETP, 1.5 mm). Because efficiency tends to decrease with increasing sample size, with which preparative gas chromatography is primarily concerned, approximately 200 theoretical plates/ft (HETP, 1.5 mm) may be taken as a reasonable target efficiency of separation to be expected of a scaled-up gas chromatograph. We shall refer to this target efficiency from time to time in the following review, and indeed shall see that in one case it has already been exceeded.

### Sample Introduction—Scaled-Up Equipment

The requirements stated previously for sample introduction as a vapor "plug" become more difficult to meet with increasing sample size, but they still apply. The sample must be completely vaporized without mixing with the carrier gas, then transferred as a plug onto the first "plate" of the column. Thus, the use of a low-mass, high heat-input vaporizer requires that the heat transfer from the heating element into the sample and consequent vaporization of the sample must be accomplished within a certain period which is short compared to the peak width which would be obtained with plug introduction. (Such a peak width can be determined by air injection into the top of the column; it must be determined separately for each component of the chromatograph.) In the same way, the unpacked space preceding the column, into which the sample vapor is introduced, must be swept out by the carrier gas also within this certain period. However, temperature programing can alleviate both these requirements under certain conditions to be described.

If the sample could be deposited uniformly on the first "plate" of the column while the plate is at a temperature sufficiently low to act as a perfect collector or reservoir, and to prevent its releasing any part of a volatile component into the gas phase to move along the column while the remainder of the component is still arriving on the column from the vaporizer or from the space preceding the column, then the sample would not have to arrive as a plug. Once the vapor had been deposited uniformly on the cold first plate, the plate could be warmed and flow then recommenced at a given, controlled rate. This approach would make the results of preparative gas chromatography independent of the designs of the vaporizer and vapor-distributor, inappropriate design of either of which can undermine the effectiveness of preparative separations.

Temperature programing has another theoretically favorable effect which will be mentioned under "Column Design and Use."

An alternative and apparently feasible approach which has often been mentioned but not reported would be to dissolve the sample in the stationary phase of just the first plate, place this first plate physically in position on top of the column, bring the plate to temperature, and then start the flow at a given, controlled rate.

The problem of introducing milliliters of sample against column head pressure can be approached in three ways: (1) The sample can be manually injected with some alleviation of the head- plus sample evaporation-pressure effect, such as by the use of a check valve or a valve to turn off the carrier gas flow entirely; this method is limited to 1- to 2-ml volumes and to fairly short columns, e.g., 10-15 ft. (2) The injection can be done automatically, in which case the operator is protected from the effect of sudden pressure increases; however, precautions must be taken against tee's in the gas stream. (3) The sample can be put into some sort of container which is then incorporated into the chromatograph by valve action, and presumably swept out at once by the carrier gas.

We will now consider some experimental arrangements for sample introduction in scaled-up preparative gas chromatography.

Kirkland [16] used manual syringe injection onto heated $\frac{1}{8}$-in.-diameter steel balls held in a 12-mm-OD × 12.5-cm section of glass tubing. The carrier gas was heated upstream from the vaporizer. The 10-mm-ID vaporizer opened through a ball joint directly into a 31-mm-ID column. Similarly, Atkinson and Tuey [2] used automatic injection onto heated steel balls held in a $\frac{1}{2}$-in.-ID × 3-in. steel vaporizer placed just above a 20-mm-ID (0.8-in.) column. The peak shape issuing from the vaporizer was not tested in either case, but ideally should have been.

Alternatively, Johns et al. [15] used a thermostatted, low-mass, high heat-input vaporizer which apparently is also effective. Such an arrangement should allow faster clean-out by the carrier gas. Again, no data on the peak shape issuing from this type of vaporizer operating on different quantities of injected liquids have been made available.

Huyten et al. [14], in a very important paper, reported studies of chromatographing n-pentane over a silicone oil. They experienced no trouble with vaporization samples up to 10 ml, but did not describe their vaporizer. However, they found that with the 3-in.-ID column the inlet cone with which they topped their column should be 80% filled, although with the 10-in.-ID column the degree of filling of the inlet cone was immaterial. With either column, the exit cone should be completely filled. Bayer [3] found diffusers more efficient than cones for capping or terminating columns.

Perry [22] arranged for manual injection into an isolated reservoir against atmospheric pressure only. The reservoir was then incorporated into the chromatograph by valves. The vaporizer was a hot, initial 3-ft section of the 30-ft, $\frac{3}{8}$-in.-OD column; a better arrangement would have been to pass the sample leaving the reservoir through an adequate length of $\frac{1}{8}$-in. steel tubing cast in aluminum and held 20-30°C above the boiling point of the highest-boiling sample component, thus assuring vaporization without mixing with the carrier gas and with quick clean-out of the vaporizer.

Bayer et al. [4] have most recently reported using pump injection against column head pressure into a vaporizer packed with steel wool held 30-60°C above the boiling point of the l o w-e s t-boiling component. During the vaporization the vaporizer is closed off from the column. The carrier gas issues uniformly from a perforated plate at the top of the vaporizer interior; the completely vaporized contents are swept out through the cone-shaped bottom of the vaporizer into the filled, inverted-cone top of the column.

To summarize: Few data and no comparisons have been reported at this time concerning the peak shape of vapor "plugs" issuing from the vaporizer- plus inlet-space onto the first plate of the column, so that vaporizer design is at the moment a matter for personal preference. A narrow tube embedded in a mass of metal such that the sample can be vaporized without cracking and swept from the vaporizer without mixing with the carrier gas would seem to be an efficient and easily used arrangement. With scaled-up columns, diffusers such as sintered steel or glass disc may perhaps be preferred to cones for terminating columns at either end.

### Column Design and Use—Scaled-Up Equipment

Several problems arise in designing scaled-up gas chromatographic columns: Are U-shaped columns allowable, or should only straight column sections be used? What is the optimum stationary phase-to-support ratio? What can be said about linear temperature programing with thick columns? Most important: Why are scaled-up columns so much less efficient than $\frac{1}{4}$-in. columns, area for area?

Consider the U-shaped preparative column. What limitation on resolution does the 180° bend impose? Giddings [9] derived an expression giving the plate height contribution of one 360° bend; in general, it may be concluded that large-diameter preparative columns designed for minimum HETP should not be bent.

## TABLE II
Liquid Sample Quantities for Injection into Various Columns without Causing Overloading

| Diameter | Area Ratio | Liquid Charged |
|----------|-----------|----------------|
| 1/4 in. OD | 1.0 | 5 μl |
| 1/2 in. ID | 6.9 | 35 μl |
| 1 in. ID | 27.7 | 140 μl |
| 2 in. ID | 111.0 | 550 μl |
| 3 in. ID | 249.3 | 1.2 ml |
| 5 in. ID | 692.5 | 3.5 ml |
| 10 in. ID | 2770 | 14 ml |

Does an optimum stationary phase-to-support ratio for preparative work exist? Or, if the art of preparative gas chromatography were better understood, could the ratio be lowered to gain resolution [26, 27]? Bayer and Witsch [5] have shown that for a 50-μl sample injected onto a 1.77-cm-diameter column, a 10% ratio gives a lower HETP than a 30% ratio. On the other hand, the 10%-ratio support overloads more quickly. Their conclusion is that a 20% ratio gains the low HETP of the 10% ratio plus the higher capacity of the 30% ratio; the 20%-ratio support exhibits performance superior to each of the other ratios at all but the smallest sample sizes. And what are the sample sizes one should expect to handle in preparative gas chromatography?

In Table II are presented some liquid sample quantities, based on ratios of cross-sectional area, which could probably be injected onto columns of various diameters and stationary phase-to-support ratios without causing overloading, i.e., an increase in HETP. Thus, if a $\frac{1}{4}$-in. column with a given packing can take no more than 5 μl without overloading, then a 1-in.-ID preparative column with the same packing should have no more than about 150 μl charged to it. To charge 1 ml, given the same column packing, at least a 3-in.-ID column would be needed.

Although HETP seems to decrease uniformly with decreasing sample size with $\frac{1}{4}$-in. columns [7, 15], this does not always hold with thicker columns. In 1959, Mr. J. Krupowicz of our laboratory established that a $\frac{3}{8}$-in.-OD column shows the same resolution, length-for-length, as a $\frac{1}{4}$-in. column; and moreover that it can handle up to 20 μl of sample without losing resolution. Bayer et al. [5] and Huyten et al. [14] found that, with their columns from 0.8-cm to 4.09-cm-ID, increasing diameter gives decreasing resolution at minimum sample sizes but maintains resolution over a wider sample-size range before overloading. In contrast, Peurifoy et al. [23] found a uniform decrease in resolution for all increases in sample size. Thus, we have so far considered the variables in preparative gas chromatography of bending the column (undesirable), of optimum stationary phase-to-support ratio (about 20%), and of quantity of sample to be charged (less than 1 ml with columns up to about 3 in. ID). Another variable is temperature programing, after which we will take up the fundamental problem of maintaining separation efficiency while increasing column diameter.

Temperature programing offers, in addition to its other advantages, another type of benefit to preparative gas chromatography with thick columns. As will be pointed out in the following paragraph, gas tends to travel faster down the walls of the thicker columns. Also, with thicker columns subjected to linear temperature programing, the walls are likely to be considerably hotter than the center; indeed, the thicker the column, the more likely there is to be a steep temperature gradient restricted to the wall region. Therefore, because the viscosity of gases increases with temperature, a viscosity gradient offsetting the usual velocity profile exists across linear temperature-programed thick columns. In other words, the phenomena associated with linear temperature programing of thick columns comprise a specific answer to the boundary-layer effect (the velocity profile), which will be described later, which so plagues preparative gas chromatography. On the other hand, stationary phases are required which offer partition coefficients of relatively low temperature sensitivity, but this is a requirement in any case for linear temperature programing of thick columns. (My colleague, Mr. F. L. Boys, has pointed out in personal discussions that the same type of advantageous effect can be brought about in an isothermal run with a thick, relatively short column if, just before the sample is introduced, the wall temperature is suddenly increased, thus tending to create the desired viscosity gradient without substantially affecting the column temperature during the run.) Thus, linear temperature-programed preparative gas chromatography with thick columns is an area which needs more development of theory and more detailed measurements of temperature and velocity profiles within the column. The final area to be considered here, that of maintaining separation efficiency in scaled-up columns, seems to be yielding to study.

The fundamental problem in scaling-up columns seems to be a "boundary-layer effect," i.e., the presence of interparticle passages near the column wall which are larger than those in the column center. Consequently, the carrier gas moves faster near the wall; in moving faster, it distorts the spatial concentration gradients of the component vapor bands. Because the vapor bands are usually collected as if they were still planar and had not been distorted, the separation of components may deteriorate with large-diameter columns. The problem has received considerable effective attention and will no doubt receive a great deal more—witness the sudden proliferation of commercial preparative gas chromatographs.

Golay [11] stated the theory of the problem and suggested insertion into the column of "mixing washers" which would periodically remix the gas crossing planes normal to the column axis. Frisone [8], observing the boundary-layer effect in a 9-ft, 2-in.-ID column, inserted solvent-saturated, column-fitting, filter-paper rings as mixing washers at 1-ft intervals. The rings had concentric holes decreasing with ring height from $1\frac{7}{8}$ in. to $1\frac{3}{8}$ in. ID. As a result, after numerous and painstaking failures with the column before trying mixing washers, Frisone now reported "very good" peak shape and resolution from a chromatogram which was "very gratifying indeed." Further work showed 68 theoretical plates/ft (HETP, 4.7 mm), or a total of 615 theoretical plates available with a sample charge of 0.2 ml.

We now refer again to the important paper of Huyten et al. [14]: Using a 3-in.-ID column, these authors studied velocity profiles and radial and axial diffusivities as a function of column length, method of packing, and mesh size. They found a "pronounced wall effect" with a velocity 12 times greater than average within $\frac{1}{2}$ particle diameter of the wall. They also found radial symmetry in gas movement within the column.

The effects of the velocity profile are compensated by radial diffusion in small columns; in large columns, the wall-located velocity becomes less important. Thus, the worst effects of the higher wall velocity are believed to be found with a column of about 1 cm ID.

The manner of packing has a pronounced effect on the density profile of the packing: the more agitation, the denser the packing in the center of the column. The velocity profile, the radial diffusion, and the axial diffusion are all affected in a complex way by the manner of packing. Separation efficiency is increased by a low velocity profile and low axial diffusivity, and by a high radial diffusivity.

Experimentally, Huyten et al., found packing at a low filling rate and with radial beating gave the most favorable results. Equally important is limiting individual column lengths to 1 m; such 1-m columns, connected by narrow tubes, can be used in series with no HETP increases.

Bayer [3] confirmed the low-filling rate, radial-beating packing practice, and also improved over-all efficiency by enameling the column wall with sintered firebrick in order to lower gas wall velocities. The firebrick was then coated with the stationary phase. This column yielded 170 theoretical plates/ft, an HETP of 1.8 mm.

Finally, Huyten et al. [14], injecting a 1-ml n-pentane charge onto a 1-m, 3-in.-ID column, achieved 153 theoretical plates/ft (HETP, 2 mm); and injecting a 10-ml n-pentane charge onto a 1-m, 10-in.-ID column, 122 theoretical plates/ft (HETP, 2.5 mm). Further, when concentric outlet cones were used to collect only the center half of the total flow from the 10-in.-ID column, 260 theoretical plates/ft (HETP, 1.25 mm) were observed. Thus, the "target efficiency" proposed earlier can be and has been surpassed.

Looking at the boundary-layer effect from an entirely different angle, Giddings [10] has recently proposed, with supporting evidence, that the normal process of packing concentrates fines in the center of the column and coarser particles nearer the wall. He suggests that this maldistribution of particles causes the dome-shaped velocity profile. He further suggests an annular column geometry as the most promising solution; his work on his approach was in press in February.

To summarize: Definite predictions can be made concerning the allowable bending radius for U-shaped preparative columns of given ID and desired resolution; in general, a column with an internal radius larger than about 1 cm should not be bent.

In the present state of the art, a 20%-by-weight stationary phase-to-support ratio seems optimum for preparative packings.

Increasing sample size increases HETP; some workers report a continuous increase, others report a minimum HETP maintained over a range of minimum sample sizes.

Scaled-up columns tend to be less efficient, even given perfect sample injection, because the carrier gas moves faster near the column walls. The effect is presumably most noticeable and damaging to resolution with 1-cm-ID columns. The most efficient preparative separations to date have been obtained with columns not exceeding 1 m in length, which have roughened walls, which have been slowly filled with beating, and which are terminated by diffusers; if longer columns are required, the shorter sections are combined by narrow tubing. With these columns, at least with isothermal runs, gases eluted from the axial region should for most efficient separations be collected separately from gases from the wall region. With these practices, resolution of over 150 theoretical plates/ft (HETP, 2 mm) is readily obtainable; with only axial gases being collected, resolution of 260 theoretical plates/ft (HETP, 1.25 mm) has been achieved.

Linear temperature programing of thick columns offers two advantages which have not been brought out by its proponents: Given an adequately low column temperature during sample injection, the chromatographic development can start with the solute mixture completely and ideally distributed throughout the first plate. Also given that a stationary phase having a partition coefficient of low temperature sensitivity is used, then the unfavorable velocity profile found with thick columns can be offset by a viscosity gradient derived from the temperature gradient and primarily restricted to the wall region.

### Detection—Scaled-Up Equipment

With preparative gas chromatographs, thermal conductivity detectors are generally used. If the carrier gas flow is above about 200 ml/min, certain workers [14, 15, 22] take only a small part of the effluent through the detector in order to improve sensitivity and to eliminate instability caused by turbulence. The reference side of the thermal conductivity detector is generally either sealed or else supplied with a separated stream of the carrier gas. Finally, the improvement in baseline stability gained by shielding the entire detector bridge very carefully from any air movement is becoming increasingly realized.

### Component Collection—Scaled-Up Equipment

When a hot stream of a separated component and its carrier gas enters a cold trap, the component vapor very frequently forms tiny solid particles, or f o g [17]. This fog, which is the desired component, moves straight through the trap, inhibiting or preventing component recovery. Despite the large number of varied and ingenious solutions to the problem of fog which have been reported, the problem remains with us. We can, however, at least review some of what has been done.

Kirkland [16] devised an "extremely effective" glass trap consisting of an indented glass tube followed by a narrower helical tube. The glass receiver of Perry [22] "eliminates" fogging by passing the gas first through a narrow helical tube, then against a cold finger within the trap.

Atkinson and Tuey [2] devised a "completely effective" fog precipitator placed within a trap which condensed vapors "reasonably well." Their precipitator consisted of a 10-cm-long steel cylinder of 16-mm diameter with a center steel rod-electrode; 7-8 kv DC was applied across the electrodes. The trap consisted of about 16 in. of narrow tubing producing turbulent gas flow and therefore fairly effective condensation.

Wehrli and Kovats [28] reported a device in which the cooled receiving-tube is rotated on its axis to become an in situ centrifuge, thus throwing the fog particles onto the receiving-tube walls. This device appears workable enough and less complicated than it sounds. It allows only a "relatively low loss" of desired components.

Peurifoy et al. [23], using traps packed with glass wool and with copper turnings, reported 97% or better recovery with the copper-packed trap. They also used successively cooler traps in series.

Ambrose [1] packed traps with activated alumina, which not only adsorbed the fog but also allowed its recovery. He also mentioned the value of traps in series with intervening connectors in which the fog particles can revaporize.

Bayer et al. [4] reported 80% to 98% collection from a condenser-separator in which the component-laden carrier gas enters a sharply curving helical space which both condenses vapor on the walls and centrifuges the component fog particles onto the walls. The many-turned helix surrounds the carrier gas exit tube; the condensed vapor droplets gather, run off into a reservoir, and can be drawn off through a valve.

Hajra and Radin [13] use the "Millipore" plastic filter, which has very uniform small holes, as trap packing. Measurements based on radioactivity show nearly complete recovery.

# CONCLUSION

The scope of preparative gas chromatography is quite as broad as that of its parent, but it will not be applied so widely: Information is more useful than matter. However, preparative gas chromatography will grow rapidly and supplant or complement present distillation practices to supply fractions for identification and analysis, pure-grade materials as reactants for study, and highly purified materials as reference standards.

## REFERENCES

1. D. Ambrose, in D. H. Desty (ed.): Vapour Phase Chromatography, Butterworths, London (1957), p. 209.
2. E. P. Atkinson and G. A. P. Tuey, in D. H. Desty (ed.): Gas Chromatography 1958, Academic Press, New York (1958), p. 270.
3. E. Bayer, in R. P. W. Scott (ed.): Gas Chromatography 1960, Butterworths, Washington (1960), p. 236.
4. E. Bayer, K. P. Hupe, and H. G. Witsch, Angew. Chem. 73, 525 (1961).
5. E. Bayer and H. G. Witsch, Z. anal. Chimie 170, 278 (1959).
6. J. Bohemen et al., J. Chem. Soc. 1960, 2444.
7. J. Bohemen and J. H. Purnell, in D. H. Desty (ed.): Gas Chromatography 1958, Academic Press, New York (1958), p. 6.
8. G. J. Frisone, J. Chrom. 6, 97 (1961).
9. J. C. Giddings, J. Chrom. 3, 520 (1960).
10. J. C. Giddings, J. Chrom. 7, 255 (1962).
11. M. J. E. Golay, in Noebels, Wall, and Brenner (eds.): Gas Chromatography 1959, Academic Press, New York (1961), p. 11.
12. M. J. E. Golay, in R. P. W. Scott (ed.): Gas Chromatography 1960, Butterworths, Washington (1960), p. 139.
13. A. K. Hajra and N. S. Radin, J. Lipid Research 3, 131 (1962).
14. F. H. Huyten, W. van Beersum, and G. W. A. Rijnders, in R. P. W. Scott (ed.): Gas Chromatography 1960, Butterworths, Washington (1960), p. 224.
15. T. Johns, M. R. Burnell, and D. W. Cable, in Noebels, Wall, and Brenner (eds.): Gas Chromatography 1959, Academic Press, New York (1961), p. 207.
16. J. J. Kirkland, in Coates, Noebels, and Fagerson (eds.): Gas Chromatography 1957, Academic Press, New York (1958), p. 203.
17. S. Klinkenberg, in D. H. Desty (ed.): Vapour Phase Chromatography, Butterworths, London (1957), p. 211.
18. H. W. Leggon, Anal. Chem. 33, 1295 (1961).
19. L. T. Lohr and R. J. Kaier, Facts and Methods 2, 2, 1 (1961), F & M Scientific Corp., Avondale, Pennsylvania.
20. A. J. Martin, C. E. Bennett, and F. W. Martinez, in Noebels, Wall, and Brenner (eds.): Gas Chromatography 1959, Academic Press, New York (1961), p. 363.
21. H. J. P. Martin and R. L. M. Synge, Biochem. J. 35, 1358 (1941).
22. J. A. Perry, in Noebels, Wall, and Brenner (eds.): Gas Chromatography 1959, Academic Press, New York (1961), p. 183.
23. P. V. Peurifoy, J. L. Ogilvie, and I. Dvoretzky, J. Chrom. 5, 418 (1961).
24. F. H. Pollard and C. J. Hardy, in D. H. Desty (ed.): Vapour Phase Chromatography, Butterworths, London (1957), p. 115.
25. J. H. Purnell, J. Chem. Soc. 1960, 1268.
26. R. P. W. Scott, in D. H. Desty (ed.): Gas Chromatography 1958, Academic Press, New York (1958), p. 189.
27. R. P. W. Scott and J. D. Cheshire, Nature 180, 702 (1957).
28. A. Wehrli and E. Kovats, J. Chrom. 3, 313 (1960).

# A COMPLETELY AUTOMATIC PREPARATIVE SCALE GAS CHROMATOGRAPH

## J. M. Kauss, J. Peters, and C. B. Euston

F & M Scientific Corporation
Avondale, Pennsylvania

## INTRODUCTION

During recent years the use of gas chromatography for the separation and isolation of relatively large quantities of material has become increasingly popular. Although preparative scale chromatography is still used in conjunction with other techniques as an aid in positive qualitative identification of components, this application is becoming less important. Other applications, such as the preparation of spectroscopic solvents, pure materials for reaction studies, and perhaps even the isolation of valuable pharmaceuticals, require even higher sample capacities.

Attaining the desired capacities by direct scale-up is limited by the rapid decrease in separating efficiency as the column diameter is increased. One approach to this problem is to limit the scale-up to the point where resolution is not seriously affected and then to obtain high capacity by repetitive sampling.

A lab bench model was constructed to study large columns and large sample injections. It was found that separating efficiencies comparable to analytical columns could be obtained with $3/4$-in. diameter columns. The final step in such an approach is to increase daily yields by making the chromatograph fully automatic. This makes possible unattended overnight operation. This paper describes a unit which utilizes this approach.

## DESCRIPTION

The instrument is shown in Fig. 1.

The sample mixture is stored in the 300-cc pressurized

Fig. 1. The Model 770 Automatic Preparative Gas Chromatograph.

reservoir which feeds the automatic injector. A standard injection port is provided for manual syringe injections.

The columns are located inside the circulating air oven.

After passing through the detector, components emerge
through the heated manifold and are collected in glass traps
which are immersed in an insulated coolant bath. The automatic
fraction collection valve may also be operated manually at any
time. Column head and back pressures are indicated by the
dual pointer gauge. A high-temperature back-pressure control
valve is located between the column exit and the detector.
Other panel-mounted controls are for various heaters and carrier gas and reference flows.

The unit was originally designed for isothermal operation
only, but it was found that programmed temperature operation is
just as desirable in prep scale work as it is in analytical work.

In order to obtain maximum versatility it was felt that the automatic features of the unit should depend solely on the conditions of the separation being performed rather than on preset timing devices. This requires coordination of four major systems: the injection system, the fraction-collection system, the recorder and, in the case of programmed operation, the oven-cooling system.

A flow schematic of the instrument is shown in Fig. 2.

The injection is basically a pneumatic-hydraulic intensifier. Force is supplied by carrier gas pressure which enters the pneumatic side of a differential area stepped piston. The pneumatic piston area is 2.5 times the area of the hydraulic liquid sample piston area. Thus, when a sample injection begins, there is sufficient mechanical advantage to overcome any back-pressure caused when the sample flash vaporizes in the hot injection chamber. The injector will inject from 0.25 to 12 cc. This volume is infinitely variable by a simple screw adjustment which determines the sample volume by limiting the length of the refill stroke. Refilling force is provided by the sample itself because the reservoir is under carrier gas pressure at all times. A 12-cc sample is completely injected in approximately five seconds. The injection port is a heated cylinder having an empty volume of 35 cc. It can be heated to 400°C and is filled with stainless steel shot which acts as a heat sink. Immediately preceding the injection port is an identical cylinder which serves as a carrier gas preheater. The heat capacity of this injection system is sufficient to rapidly vaporize more than 12 cc of sample.

The vaporized sample then passes through the column or columns. The oven chamber accommodates up to six columns. They are $3/4$ in. in diameter and 8 ft long and may be manifolded for any combination of series or parallel configurations.

At the exit of the columns, a high-temperature needle valve provides a means of varying the back pressure on the system. The proper head-to-back pressure ratio is often important in large separations. This valve is kept isothermally at or near the temperature of the thermal conductivity type detector cell. This can be controlled from 100 to 375°C. The entire flow passes through the detector. Since the carrier gas flow is usually more than 1 liter/min, a specially designed detector cell is required for noiseless operation.

The reference side of the detector is supplied by a complete analytical system which is an integral part of the unit.

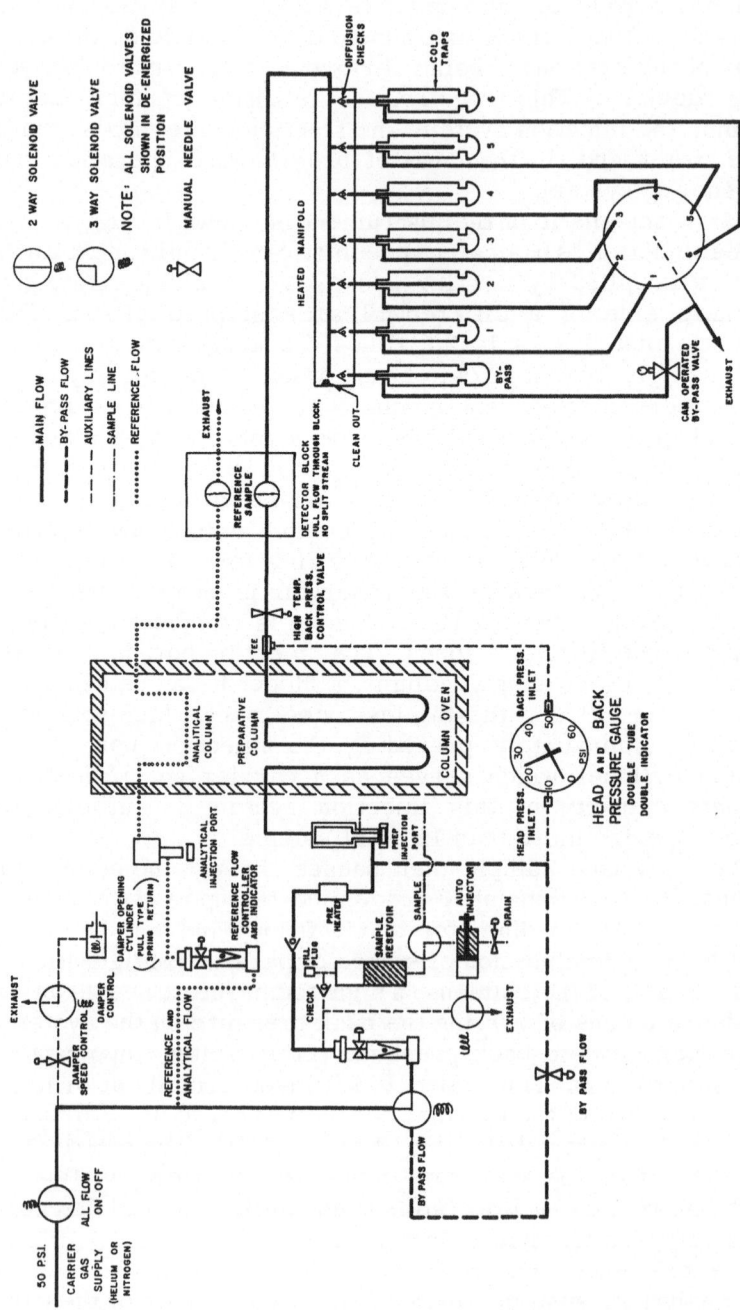

Fig. 2. Flow schematic of Model 770 Automatic Preparative Gas Chromatograph.

This system has its own flow controller, injection port, and $\frac{1}{4}$-in. column. It is very useful for column scouting work. Merely by reversing the detector leads to the recorder with a panel mounted switch, the preparative system provides the reference flow for the small analytical column. This eliminates the labor involved in packing and repacking the large columns while determining the best column packing for a given sample. From the detector the component enters a heated manifold which can be heated to 400°C.

Before the first fraction is eluted, all flow passes through a by-pass clean-up trap. When a peak rises on the recorder, a recorder pen activated switch energizes the fraction collection valve. This switch can be preset to any position from zero to full scale wherever the operator desires. The valve then indexes to the first collection position. This closes the by-pass valve and opens the exit of trap no. 1 to the atmosphere. All flow then passes through trap no. 1 and the first component is condensed out of the carrier gas. After the peak passes the set point on the way back downscale, the valve indexes to the by-pass position, again closing the component trap exit and again opening the by-pass valve. The second component causes the valve to index to the second collection

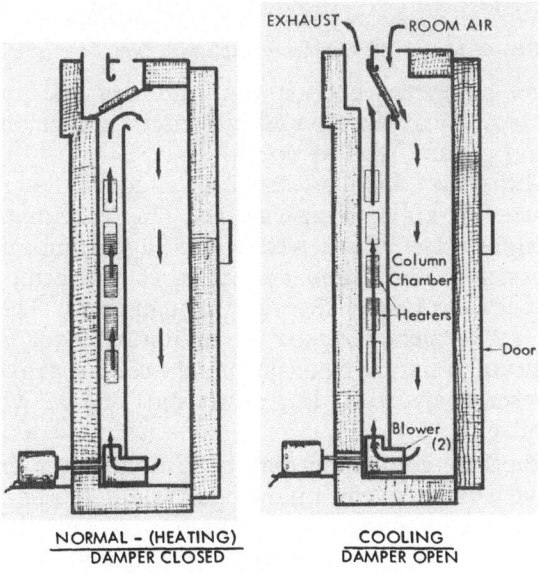

NORMAL - (HEATING)
DAMPER CLOSED

COOLING
DAMPER OPEN

Fig. 3. Schematic of Model 770 column oven.

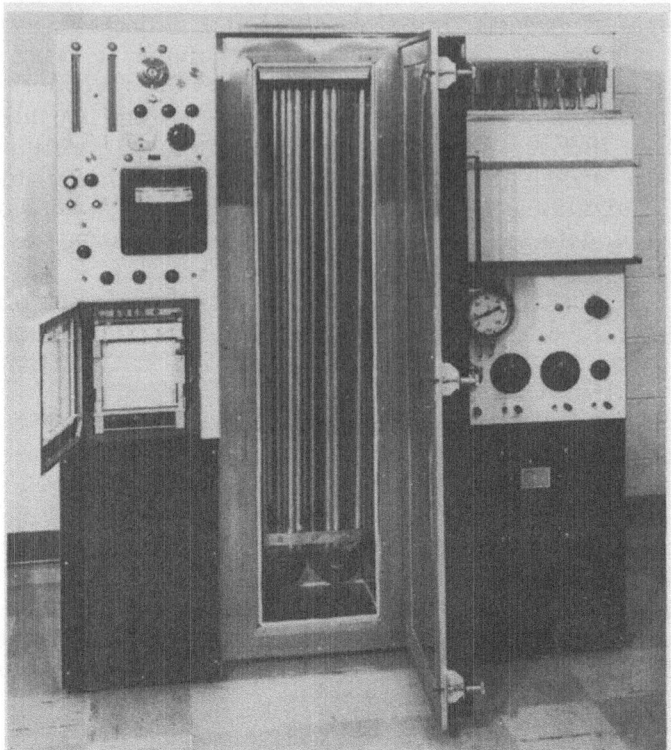

Fig. 4. Model 770 Chromatograph with oven door open.

position and so forth, until a maximum of six fractions have been collected. Thus, the flow is by-pass, collect one, by-pass, collect two, by-pass, and so forth.

All valving is done on the exit side of the traps and the valve operates at room temperature. This eliminates the problems of high-temperature seals and high-temperature valve body corrosion. The traps are open to the atmosphere only while the carrier gas is flowing through them. The trap fractions are not exposed to air or moisture because the traps always contain a carrier gas atmosphere. To prevent any diffusion between traps, simple gravity ball checks are located in the hot manifold.

After the last component has been eluted, the fraction collection valve resets to the inject position. In isothermal operation, this resetting operation activates the automatic injector which causes the next sample to be introduced.

For programmed temperature operation, however, the se-

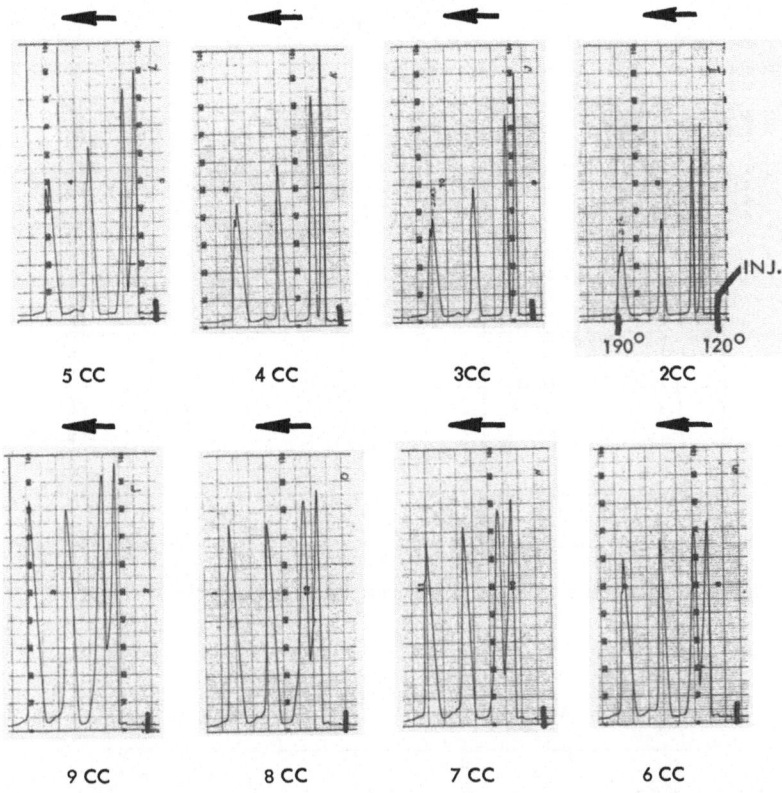

5 CC   4 CC   3CC   2CC

9 CC   8 CC   7 CC   6 CC

Fig. 5. Effect of sample size on separation. $C_7$, $C_8$, $C_{10}$, $C_{12}$ programmed 120° to 195°.

quence of events is slightly different. The elution of the last peak still causes the fraction collection valve to index to the inject position, but this does not activate the automatic injector. Instead, the programmer continues upscale until an upper temperature limit has been reached. This limit can be preset from 50 to 325°C. At this point, a cooling damper is opened and the programmer reverses and drives the controller set point back downscale. With the damper open, the blowers draw room air into the oven chamber, where it circulates over the columns and heaters and passes out an exhaust duct. When a preset lower limit has been reached, the damper closes and the injector introduces another sample. The programmer then starts upscale again and the cycle is repeated. The oven damper operation is shown in Fig. 3.

The unit is shown with the oven door open in Fig. 4. Il-

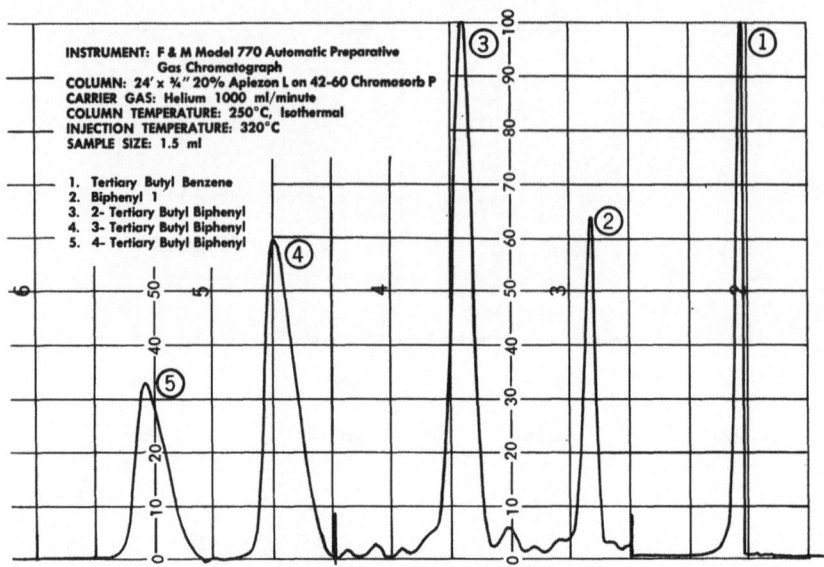

Fig. 6. Preparative separation of peroxide decomposition reaction product.

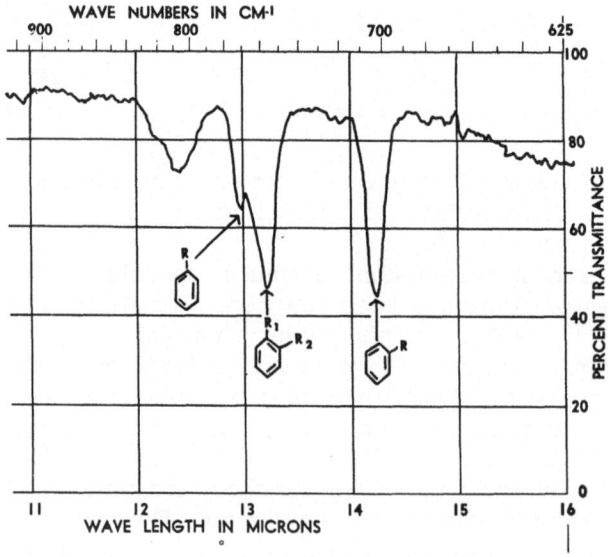

Fig. 7. Partial infrared spectrum of material collected in peak no. 3 (2-tertiary butyl biphenyl).

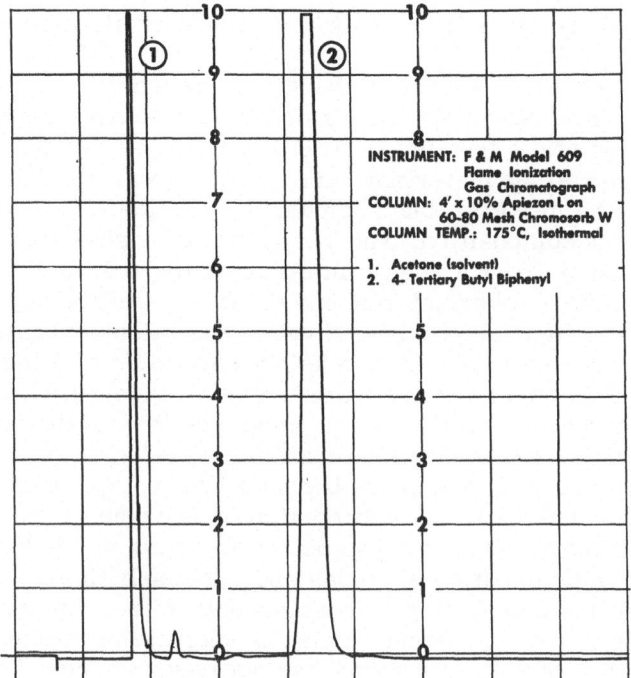

INSTRUMENT: F & M Model 609
            Flame Ionization
            Gas Chromatograph
COLUMN: 4' x 10% Apiezon L on
            60-80 Mesh Chromosorb W
COLUMN TEMP.: 175°C, Isothermal

1. Acetone (solvent)
2. 4- Tertiary Butyl Biphenyl

Fig. 8. Analytical chromatogram of trapped peak (no. 5) from
preparative separation.

lustrated are the two circulating blowers, the damper, the
$\frac{1}{4}$-in. analytical column, and 48 ft of prep columns manifolded in series. The performance of the unit has been very
satisfactory and unattended operation of four days has been
successfully carried out.

The type of chromatograms obtained are shown in Fig. 5.
The chromatograms show the effect of sample size on resolution. These runs were made with three 8-ft columns connected
in parallel; good resolution was obtained up to about 4 cc of
sample. Above this value the first two components are not sufficiently resolved for effective collection, although separation
of the latter two components is adequate up to values of 9 cc.
The extraneous peaks on the latter curves do not represent
impurities in the sample, but are caused by imperfect matching
of the pressure drop across the three parallel columns. However, even with appreciably mismatched columns the collection
of the component is in no way impaired, since this function is

controlled by the recorder switch, which in this case was set at 25% of scale.

The F & M Research Department recently worked with Jack Cazes of New York University on an interesting practical application of the instrument which illustrates its performance with high-boiling materials. The problem was to isolate and identify the tertiary butyl biphenyl reaction products of a peroxide decomposition. The chromatogram shown in Fig. 6 was obtained from a 1.5-cc injection of the reaction mixture. This was an isothermal run and the column efficiency, based on the 4-tertiary butyl biphenyl peak, was calculated as a value of 2300 theoretical plates or a height equivalent to 3 mm.

The collected fractions were tentatively identified by infrared spectroscopy. Perhaps the most reliable of the correlations between the absorption maximum wavelengths and molecular structure are the bands between 12 and 15$\mu$, which relate to the type and degree of substitution on benzene rings. Figure 7 shows this region of the IR spectrum of peak no. 3. From the fact that a combination of mono- and 1,2-disubstituted benzene rings is indicated, it was concluded that the compound was 2-tertiary butyl biphenyl. This is interesting because the preparation of this compound has not been reported except in very low yields and attempts to duplicate the published syntheses have been unsuccessful.

An analytical chromatogram of the 4-tertiary butyl biphenyl collected as peak no. 5 is shown in Fig. 8. The sample was run on a flame ionization chromatogram. The first peak and the second small peak are caused by the acetone used as solvent for the solid biphenyl in the collection trap. This chromatogram indicates that the collected biphenyl has a purity of better than 99%.

## SUMMARY

A preparative scale gas chromatograph featuring completely automatic programmed temperature or isothermal operation has been developed. Column efficiencies approaching those of $1/4$-in.-analytical columns have been obtained and efficient collection of very pure components has been demonstrated. The application of this instrument to the study of a complex reaction mixture has provided a quick, effective solution to a separation problem which would have been extremely tedious and time consuming by conventional methods.

## ACKNOWLEDGMENT

The authors of this paper are grateful to Mr. Louis Mikkelsen of F & M Scientific Corporation and Mr. Jack Cazes of New York University for their assistance in preparing data for this paper.

# PROGRAMED-TEMPERATURE PREPARATIVE CHROMATOGRAPHY

## Nathaniel Brenner and Donald R. Bresky

The Perkin-Elmer Corporation
Norwalk, Connecticut

When we speak of preparative chromatographs we may mean either one or both of two distinctly different types of devices. The first type is a small production plant, a device intended to produce a large quantity of a known component of interest and generally of high value to the user. The second type is more closely related to analytical purposes. In this case, the user will place the sample into a preparative chromatograph for the primary purpose of obtaining a reasonably small but very useful pure fraction for further analysis by analytical chromatography; for use in a mass spectrometer, infrared spectrometer, or other analytical instrument; or for standard means of semimicroanalysis which are commonly available in organic and analytical laboratories. In designing a preparative chromatograph, an instrument company must first decide which type it is to be. If it is to be the first type, then such features as automatic sampling and automatic collection for repetitive handling on an unattended basis must be given primary consideration. On the other hand, if it is to be the second type of instrument, automation of the device is not necessary; since it is probable that the quantity of component which must be prepared will be small, the type of problem to which it will be applied will vary from day to day. The user who is interested in the device on this basis would consider it well worth his time and attention to watch the chromatograph while it is in operation. This fundamental decision will affect the scale and dimensions, and to a large extent the cost, of the equipment.

At the Perkin-Elmer Corporation it was considered that the primary effort in preparative chromatography should be directed to the second purpose, i.e., to constructing a device intended largely for the convenience of analytical and organic chemists who wish to prepare a pure fraction of a component of interest for further analysis by other means available to the user. Implicit here are some compromises: For example, we estimated that a 1-ml sample should be reasonably manageable in the chromatographic column. By allowing for the introduction of 1 ml, we are effectively saying that sample components of 0.1% or more would be obtained in sufficient quantity to make possible convenient handling by the analyst who wishes to fill an infrared cell, ultraviolet cell, or some other equivalent requirement. We also, however, wish to maximize the throughput per unit time, since in many cases the analyst will be dealing either with sample components of such low concentration that repetitive introduction of the materials is required to collect sufficient sample for further use, or with a large number of samples which must be treated so that the chromatograph will be made ready for subsequent sampling on a fairly repetitive basis. The third requirement, of course, is that the low cost and convenience associated with chromatographic separation be maintained. A fourth requirement was noted in our study of those users who are presently doing preparative chromatography as an analytical adjunct. This requirement was that the instrument be usable for analytical as well as preparative purposes. Although a preparative chromatograph would be considered most useful in the average analytical laboratory, there are very few laboratories in which the requirement for such instruments is so great that the cost of this instrument and the space required could be justified by the amount of time devoted to its use. For example, we estimate that the average laboratory might use a preparative chromatograph for 10% to 20% of the usual work week. The device would remain idle the remaining 80% or 90% of the time. However, if the chromatograph were sufficiently flexible in design so that it could be employed as a reasonably efficient analytical instrument, the amount of useful time could be substantially increased.

In order to meet these requirements, it was decided to build a column system which would operate on a direct resistance heating principle, employing temperature programing as the key instrumental feature in the preparative process. The temperature-programing feature was considered most useful in preparative chromatography for several reasons: Figure 1 illustrates the separation of some common chlorinated hydrocarbon components on a 10 ft. x 1 in. column run iso-

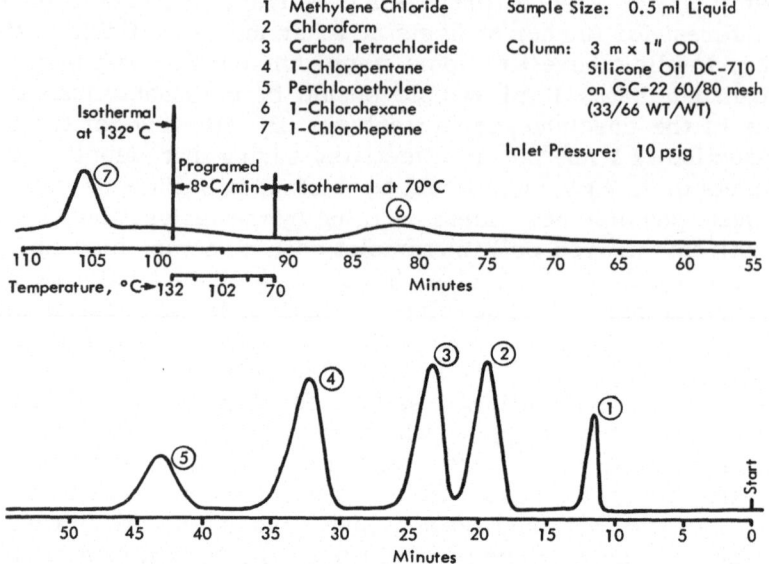

Fig. 1. Isothermal separation (70°C) of 0.5 ml of mixed chlorinated hydrocarbons.

thermally. Since this is not a real problem, of course, a certain arbitrary situation was considered; specifically, that the separation between components 2 and 3 be sufficient for preparative purposes—that is, that they be sufficiently resolved so that a reasonably pure fraction of each of these materials could have been collected. In order to do this a certain temperature limit was obtained, which happened to be 70°C. Note that components 2 and 3 are adequately resolved, but that at the end of 90 min component 7 still has not appeared at the effluent end of the column. What this means in practical terms is that a new sample of either the same material or a new sample could not have been introduced into this chromatograph until well over 90 min had elapsed. Therefore, the potential throughput of this column is low indeed. We must recognize, of course, that no new samples can be run in a preparative chromatograph if there is danger of contamination of these new samples by coincidence of elution of a heavy component from a previous run. Therefore, in preparative chromatography more than in analytical chromatography, one must be absolutely certain that all components of a previous sample have been removed from the colum prior to introduction of a succeeding sample. In the case shown, the temperature of the column was raised after 90

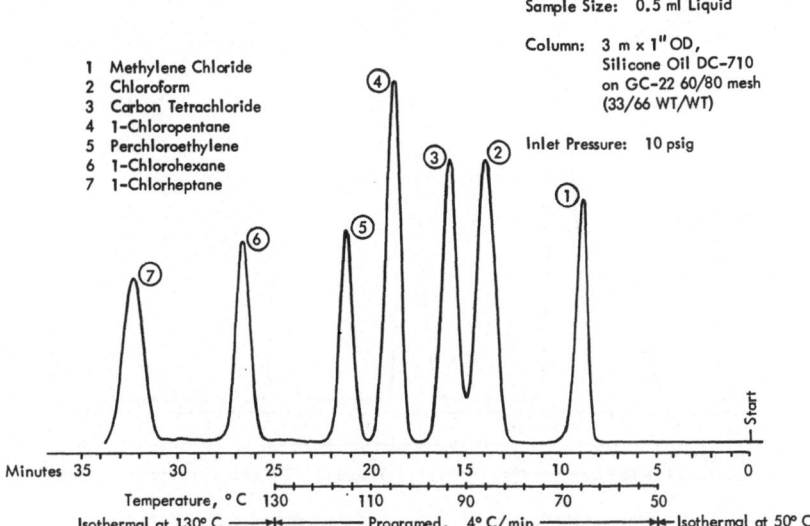

Sample Size:  0.5 ml Liquid

Column:  3 m x 1" OD,
Silicone Oil DC-710
on GC-22 60/80 mesh
(33/66 WT/WT)

Inlet Pressure:  10 psig

1  Methylene Chloride
2  Chloroform
3  Carbon Tetrachloride
4  1-Chloropentane
5  Perchloroethylene
6  1-Chlorohexane
7  1-Chlorheptane

Minutes  35        30        25        20        15        10        5        0
Temperature, °C  130        110        90        70        50
Isothermal at 130° C ——►|◄——— Programed, 4° C/min ————►|◄—Isothermal at 50° C

Fig. 2. Programed-temperature separation of the same mixture as in Fig. 1.

min and, as shown, after 105 min the last component (1-chloro-heptane) is eluted. Had isothermal conditions been maintained, it would probably have taken nearly two full hours to remove that last component.

Contrast this run with that shown in Fig. 2, in which the temperature-programing feature was employed to increase throughput. Here we have exactly the same sample, run on the same column and under the same conditions, with a single exception: The temperature was programed from 50°C up to 130°C at a rate of 4°C/min after a 5-min isothermal start. Note that components 2 and 3 are still just barely resolved, so that the same situation with respect to these components exists as in Fig. 1. In this case, however, 1-chloroheptane had been eluted at a sharp peak around the 32-min mark, so that after 35 min the column is essentially clean and is ready for introduction of the next sample. We can therefore say that the incorporation of the programed temperature feature has in-creased the throughput of this device by a factor of at least 3. Note also that the band sharpness has increased indicating that the average concentration of the later components as they are eluted is higher and therefore that collection efficiency will be increased.

Figure 3 illustrates a second sample run on the same type

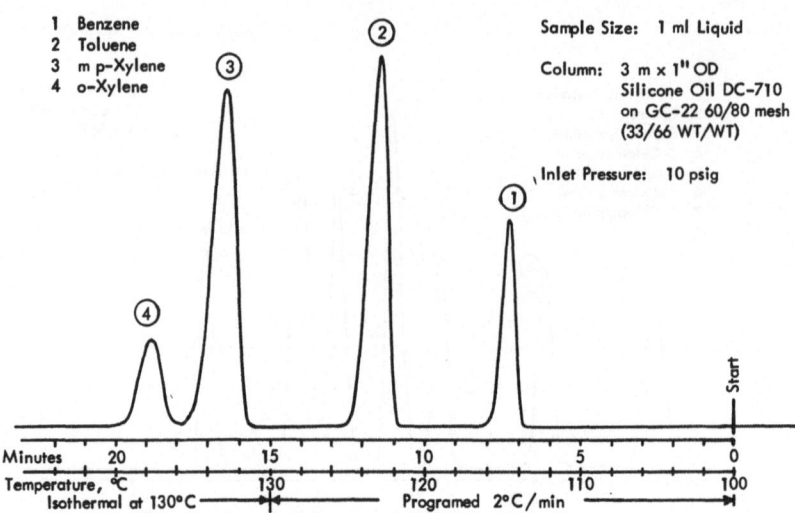

Fig. 3. Programed–temperature separation of pure fractions of an aromatic mixture.

of column, but with the sample load increased to 1 ml of liquid. The peak shapes still indicate that the column is not overloaded to any great extent and the separation is very fine. Once again, through use of a temperature program we obtain complete elution of all components within 20 min of the start and the column is ready to go again.

The problem of column capacity is constantly brought up in preparative chromatography, of course, and the question is always asked, "How much sample can I put into your preparative chromatograph?" This is a question which is not easily answered except with another question, which is generally "Well, exactly what is your problem?" since the amount of sample that can be put into any chromatograph is largely dependent on the degree of separation required for the particular problem to be handled. Obviously, if the separation involves something as simple as a two-component mixture of normal butane and toluene, extremely large samples could be placed in even a standard analytical chromatograph. Therefore, it is well to examine the special properties of a particular column in terms of pure resolving power in order to be able to anticipate just how the column will react to a given problem.

Figure 4 illustrates the resolution in terms of band shape as the sample size is increased from very low to high levels. In this figure we see succeeding injections of samples varying

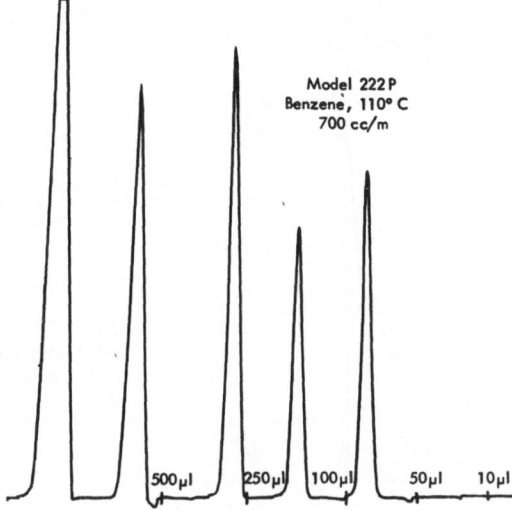

Fig. 4. Band shape vs. sample size.

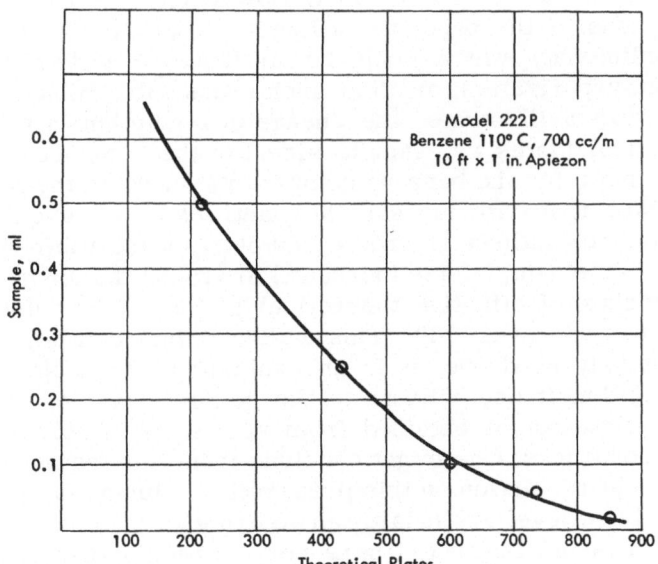

Fig. 5. Effect of sample size on theoretical plates.

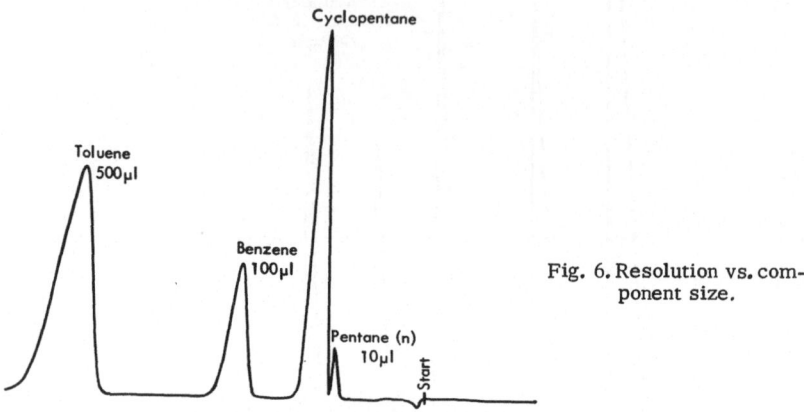

Fig. 6. Resolution vs. component size.

from $10\,\mu$l to $500\,\mu$l. As can be seen, at the 10- and 50-$\mu$l levels the bands are quite symmetrical and show no evidence of over-loading. However, starting with $100\,\mu$l of pure component it can be noted that the bands have an extremely sharp front and a somewhat gentle slope on the back, which is generally indicative of an overload of the column. However, there are no other distortions of the band, no tailing of peaks, or any other type of nonlinearity which could be attributed to either poor in-jection, vaporization, or other such undesirable behavior.

Figure 5 illustrates the change in resolution as plotted in theoretical plates vs. sample size for the components which were shown for the benzene components shown in the previous run. Note that with low sample quantities (10-$\mu$l level) on this column, resolutions of around 850 plates were attained. How-ever, as the sample size increased to around the 0.5-ml level, the number of effective theoretical plates of the column was down to just over 200. This column, incidentally, is not a particularly good one; it is an ordinary 10 ft. × 1 in. Apiezon column, but it does illustrate the general level of resolving power that can be obtained from this scale of column under these conditions. One might conclude from this that in order to get decent resolution on this preparative column, sample sizes of the 10-$\mu$l level are as large as we can expect to handle. How-ever, it should be noted here very carefully that this is not true. It should be understood that the 10-, 50-, and 100-$\mu$l levels shown in Fig. 4 were loadings for a single pure component.

Figure 6 shows the resolution vs. component size for a mixture of components in which 10 $\mu$l of pentane, 100 $\mu$l of benzene, and 500 $\mu$l of toluene, in addition to some cyclopentane, were run as a 1-ml mixed sample into the same chromatograph with the same column. As can be seen from the graphic illustration, the peak shapes are quite different for the different components, depending on their individual concentration level in the mixture. The peak for benzene is not of as nearly ideal appearance as the small pentane peak, and the toluene peak is rather more distorted than that of the benzene peak. This illustrates that the capacity of the column is related to the individual amount of sample present as well as to the total amount of sample injected.

The table shows some measurements that were made on these peaks. On the first line are the resolutions obtained with a pure benzene sample injected as the pure compound. On the bottom line are the resolutions obtained from components in the mixture which were present at the same levels as the corresponding pure benzene samples plotted above. Note that there is reasonable agreement in resolution between components injected, either as pure or as parts of samples, depending merely on the number of microliters of that component injected and not on the total amount of sample injected. So, while we have essentially begged the question of how large a sample a chromatograph can conveniently handle, we do point out that, given a reasonable number of components, 1-ml samples are just about of the proper scale for this chromatograph. The larger the number of components, the larger the sample that can be conveniently handled by the chromatograph. In any case, the scale is reasonably well adapted to the applications for which the instrument was designed: running samples large enough to obtain pure components suitable for analytical purposes by other forms of instrumentation. We stated as one of our first objectives the possibility of conveniently running the chromatograph with no more trouble than is generally associated with analytical chromatography, and also of maximizing the throughput per unit time for the preparative chromatograph. The system as shown in Fig. 7 is the chromatograph itself. Note the control panel (at left), which contains the injection block heating facility, temperature read-out, flow controller, and programer controls. Above this panel is an injection block through which samples are injected by the standard hypodermic syringe method. The samples, which are vaporized by the in-

Resolution vs. Component Size for Pure Benzene
and Mixtures

| Benzene | 10 $\mu$l | 100 $\mu$l | 500 $\mu$l |
|---|---|---|---|
| As pure compound | 1150 | 510 | 450 |
| As part of 1-ml mix | 850 | 600 | 225 |

dependently heated injection block heater, pass into a column which is ordinarily fitted in the open cage-like structure on top of the instrument. The important feature of this system is that the column is a direct resistance-heated element, with current being supplied through a transformer directly to the two ends of the column and the resistance of the column itself determining the temperature characteristics that will be obtained. The column is made in two parts which may be of any suitable length provided they are made of the proper type of stainless steel. This is simply to keep the total resistance of the column within the proper limit for the particular transformer which is employed. The steel itself is easily obtainable

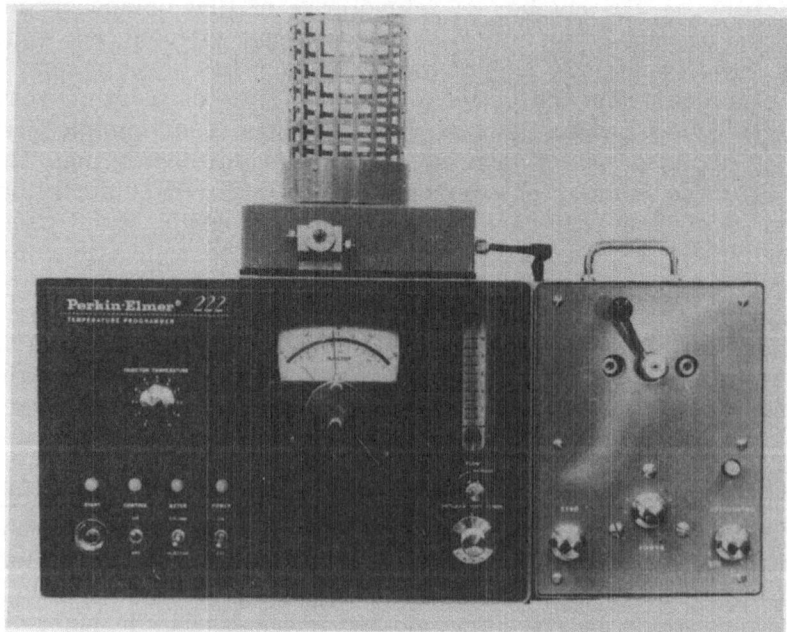

Fig. 7. The Perkin-Elmer Model 222 Preparative Chromatograph.

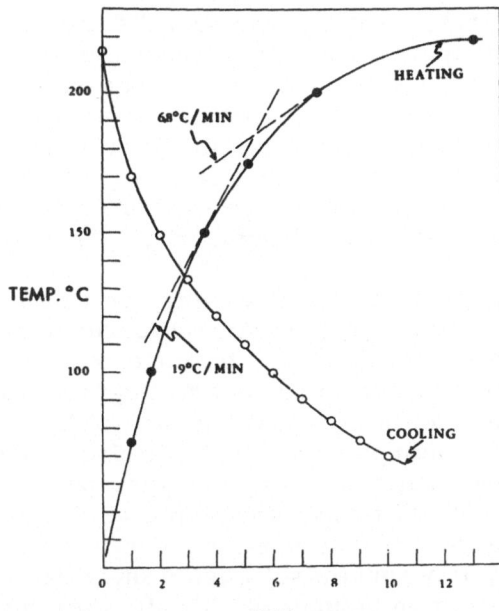

200

68°C/MIN

HEATING

150

TEMP. °C

100

19°C/MIN

COOLING

0    2    4    6    8    10    12

TIME  MINUTES

Fig. 8. Preparative column
heating-cooling rate.

from any standard steel supply house. The two column lengths
are joined at the top by a U-shaped fitting which is also made
of properly specified steel tubing. Since it is not necessary to
purchase additional connectors, but merely to transfer the connectors from one set of columns to the other when changing
columns, this is a problem not to the user but only to the
manufacturer. The ends of the column in all four cases consist
of Swagelok reducer fittings: at the lower ends, 1 to $\frac{1}{4}$ in.
reducers; at the upper ends, 1 to $\frac{3}{8}$ in. reducers. The connector between the two column paths is not packed with any
support material or substrate, but since it is of narrow diameter no measurable losses in resolving power occur. All
fittings have been machined to a funnel shape and are fitted
with micrometallic diffuser ends so that good flow profiles
are produced and maintained throughout the column. We have
found that this serves to increase the resolving power of the
column by a demonstrably large amount. The direct resistance
feature makes possible certain inherent advantages in this type
of construction. For one, the column may be made as long as
possible without difficulty. The tube is restricted in length
essentially only by the height of the ceiling. Since it can be
run up as high as desired, connected to the return tube, and

then run down to the instrument again provided the proper materials are employed, there is no restriction on length or in fact, in realistic terms, on diameter. A second advantage is that the tube does not have to be placed in an oven, which means that it can be heated and cooled, installed and removed very conveniently. Most important to performance, however, is the fact that the elimination of the large oven space which would ordinarily be required makes possible very rapid temperature programing.

Figure 8 illustrates the heating and cooling curves for a 10 ft. × 1 in. column installed in the preparative chromatograph. You will note that with open loop heating a maximum temperature of 215°C is obtained after about 12 min of heating, and that cooling from 215°C down to 75°C is accomplished within 10 min. The cooling is very simple: By simply turning off the power the convection and radiation loss to the atmosphere is sufficient to bring the column temperature down at the rate shown. The system is provided with linear temperature programer controls which will permit the selection of any program rate from 2°C/min to 26°C/min in steps of 2°C/min. However, a 26°C programing rate cannot be maintained over the entire heating range. The illustration shows, by means of tangents drawn to the curve, that up to a temperature of 150°C any program rate up to 19°C/min could be maintained on a linear basis, and up to a temperature of 200°C program rates of up to 6.8°C/min could be maintained. Therefore, at the lower temperatures virtually any program rate can be run and followed by the column, while at higher temperatures, of course, the maximum heating rates are restricted. The extreme adaptability of the instrument from one problem to another and also the ability to remove unwanted components at the end of the run are, of course, considerably enhanced by this rapid temperature change ability.

One of the problems that was foreseen in using direct heating of a column was that of achieving uniformity of temperature throughout the column, both longitudinally and radially. Longitudinal gradients are negligible, since they are determined by the uniformity of the steel tubing. Since steel tubing is made in extremely precise grades and to excellent specifications, we found no difficulty in controlling the longitudinal gradients in the column. Gradients in the fittings were a little more difficult; since the fittings and connector tubings are of vastly different dimensions from the column tubes, they had to be carefully selected so that their resistivity and dissipation quotients would be as close as possible to those of the column tubing. Much of this was done quite empirically simply by testing the fittings and columns in actual operation with thermocouples and selecting those materials in dimensions which best created a thermal match throughout.

The problem of radial gradients was a somewhat more serious one. It was assumed quite logically that, since we are going to rapidly temperature-program this instrument but that we are effectively applying the heat to the outside of the column and not to the inside, severe gradients might be present between the outside steel jacket of the column and the center of the chromosorb or celite packing. While no efforts were made to measure these gradients exactly, tests were run to show on an empirical basis whether there was a severe enough radial gradient to effect the resolving power of the column. It was assumed that if these gradients were truly serious, severe degradation of the wave front of the component as it passed through the column would result and therefore a severe decrease in the number of theoretical plates would be obtained when an isothermal chromatogram was compared with a temperature-programed chromatogram. A series of tests was run, using pure benzene, in which isothermal runs made on carefully stabilized columns were compared with runs obtained on rather fast step-function programs. In the most characteristic case, a 10°C/min program was impressed as a step-function on a column through which a 50-µl sample of benzene was being run. The elution temperature under programed conditions was 110°C. The band widths obtained here were compared with those obtained on chromatograms which had been run isothermally at 110°C with the same size and type of sample. The band widths were found to be virtually identical and, indeed, the band widths and therefore the resolution of the programed-temperature run was found to be somewhat superior to that of the isothermal run. The actual figures were approximately 1200 theoretical plates obtained under programed conditions, and somewhat fewer than 1100 obtained under isothermal conditions. While it certainly cannot be surmised that there is less of a radial gradient under programed conditions than under isothermal conditions, it is quite possible that the additional efficiency obtained through low-temperature operation of the columns during the injection period resulted in an improvement in performance which overrode any degradation of performance due to gradients obtained during programing.

The matter of injection of samples is a most important one

in preparative chromatography. In most systems used to date, the method used for efficient injection of samples has been that of providing a high-mass vaporizing system—that is, an injection system large enough to supply sufficient calories to vaporize the entire sample prior to passage onto the column. This was done by packing a large mass with aluminum or steel beads or similar material to supply a large number of calories for this purpose. In the system under discussion a different approach was used. In this device a low-mass but fast-response injection system was provided. Injection is done through a septum in the usual way into a very light but easily heated tube where the injection will take place on the way to the column. While the mass of the tube and therefore its function as a calorie reservoir is extremely limited, the tube can be supplied with calories at a very rapid rate from the heating system, resulting in immediate resupply of calories, i.e., reheating of the tube, as fast as it is cooled by the heat of vaporization of the sample. The sample can therefore be efficiently vaporized on a continuous basis without difficulty. A very important feature of the device is the flow restrictor, which can be seen in the lower right-hand corner of the main housing (Fig. 7). This flow restrictor, when provided with adequate input pressure from the carrier gas tank, serves to maintain a constant flow into the injection system and column regardless of the pressure within the system. What this does effectively is to resist the blowback due to the pressure surge crated by the rapidly vaporizing sample which has been injected. It forces the sample onto the column and prevents it from expanding back through the first part of the system and getting to any of the cold areas of the gas-handling system. In effect, as we inject the sample the only way the vapor is permitted to travel is toward the column. In a temperature-programed system this is most effective, because the column is usually quite cool at the start of the run. Therefore, as the vapor is injected it is impelled into the cool part of the column, collects there (in what we might call the first theoretical plate), and does not start to move until such time as the programer is started. Vaporization can occur over a fairly long period without destroying the apparent resolution of the system. In fact, we have obtained better results by injecting samples into the column reasonably slowly and minimizing the surge than by injecting very rapidly.

Figure 9 shows the collection system, which is at the right

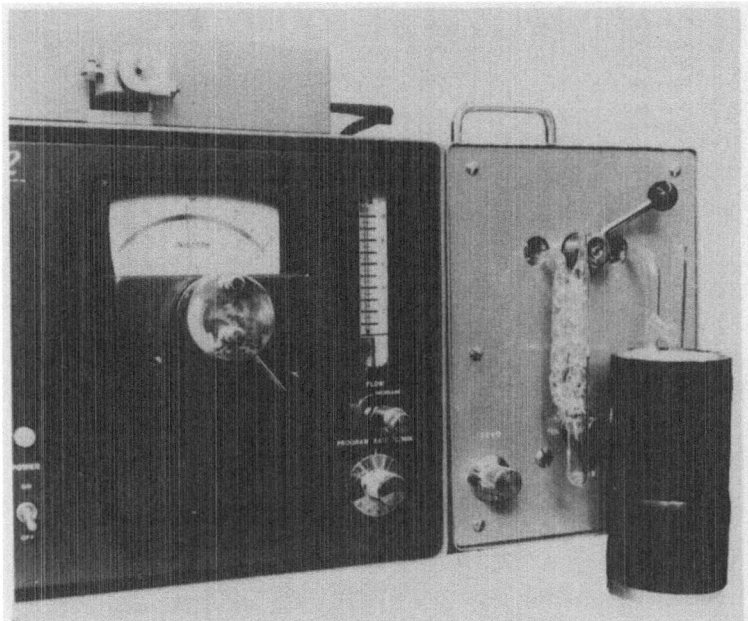

Fig. 9. Model 222-P collection system.

of the instrument. Effluent from the column passes into the collector detector system, in which is located a thermal conductivity cell to monitor the effluent and produce a record on a strip chart recorder, and a two-port collection system. The black knob at the top of the device is a valve handle which can be shifted to either of two positions to direct flow into one or the other of the collection outlets. The two collector systems shown in the illustration are typical of the types used with this device. The one on the left is a Vigreaux type condenser, designed by personnel of the Continental Oil Company which has been found very efficient. The one on the right is another design, developed by Dr. Bombough of the Spencer Chemical Company, which has been found most effective for other types of collection problems. As illustrated, the general technique is simply to suspend the collectors in Dewar flasks, wait for the components of interest, and trap them in either of the respective collection devices.

Finally, let us return to one of the requirements set for this instrument in the original objectives—that of constructing it so that it is useful not only for preparative chromatography but also for analytical chromatography. Since a direct re-

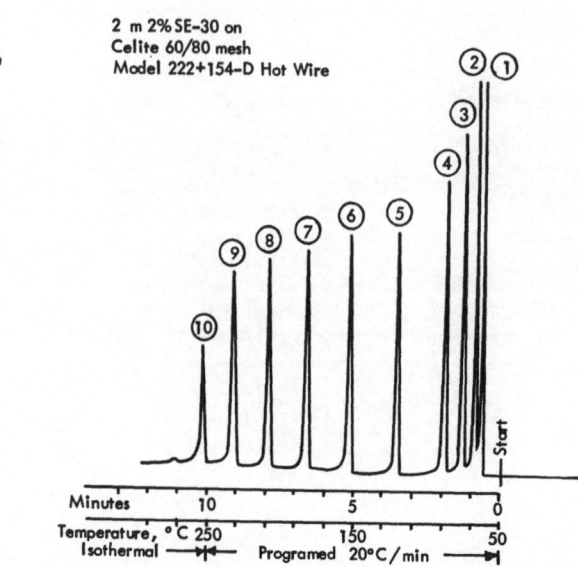

1 Formate
2 Acetate
3 Propionate
4 Butyrate
5 Caproate
6 Caprylate
7 Caprate
8 Laurate
9 Myristate
10 Palmitate

2 m 2% SE-30 on
Celite 60/80 mesh
Model 222+154-D Hot Wire

Fig. 10. Analysis of fatty acid esters using Model 222 temperature
programer accessory with Model 154-D.

sistance-heated column is employed, and since the injection
system has not been compromised with large volume and large
mass as is done in most preparative systems, the preparative
scale column can be replaced with a standard analytical scale
column of similar resistivity and the device immediately converted to analytical purposes. Figure 10 illustrates a chromatogram obtained on this chromatograph in this manner: A
standard Perkin-Elmer 2-m silicone column made with the
usual $1/4$-in. dimension was substituted for the usual 10 ft. $\bar{x}$ 1 in.
diameter column, a mixture of fatty acid esters was run
programed at 20°C/min, and an excellent separation of esters
from C1 to C16 was obtained. While we do not claim that the
precision of the chromatograph is equivalent to that which could
be obtained on the best analytical temperature-programed
chromatograph, it is certainly highly adequate for survey work
and for reasonably good quantitative analysis. We therefore
feel that the unit is sufficiently versatile to provide 100% full-
time utility either as a preparative chromatograph employed
to obtain pure fractions from large samples for other analytical
purposes, or as a standard analytical chromatograph to be
used as a primary quantitative or qualitative tool.

# A REMOTE INJECTION AND FRACTION COLLECTION APPARATUS FOR PREPARATIVE CHROMATOGRAPHY

## W. H. Penney and J. P. Windey

Minnesota Mining & Manufacturing Company
Central Research Laboratories
St. Paul, Minnesota

## INTRODUCTION

A remote injection and fraction collection apparatus for handling explosive or hazardous samples during separation on a Beckman Megachrom Preparative Gas Chromatograph is described. The apparatus could be used with very little modification for other commercially designed preparative gas chromatography units.

All components are operated remotely behind plexiglass and/or steel plates of adequate thickness to stop all shrapnel formed from an explosion in any part of the system. The

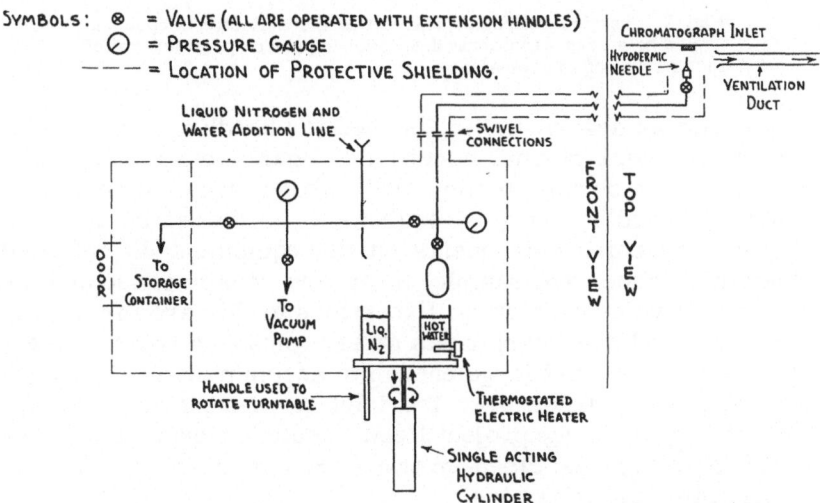

SYMBOLS:  ⊗ = VALVE (ALL ARE OPERATED WITH EXTENSION HANDLES)
Ⓞ = PRESSURE GAUGE
--- = LOCATION OF PROTECTIVE SHIELDING.

Fig. 1. Injector. The back side of the injector is not shielded and should be placed close to a wall to provide pressure relief in the event of an explosion.

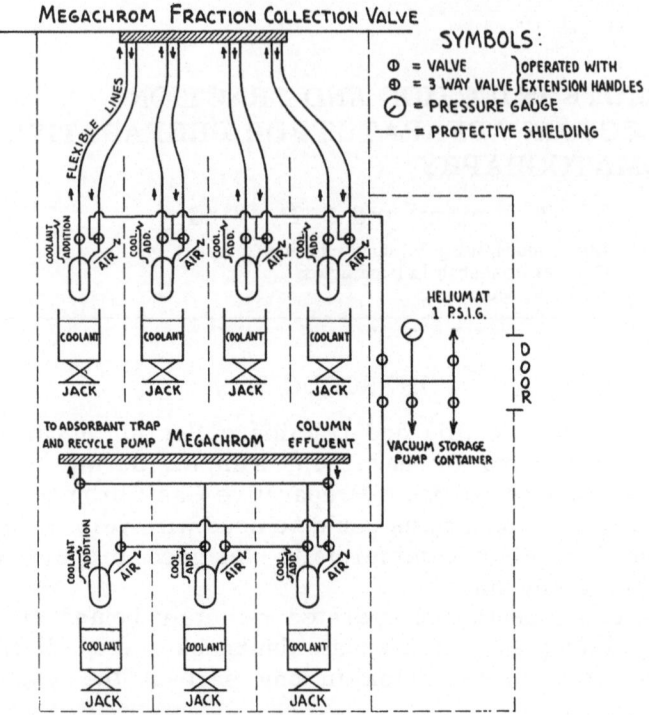

Fig. 2. Fraction collector. The top of the collector is open and should be at least a foot above head level to provide pressure relief in the event of an explosion.

equipment, as described, is designed for handling samples that are gases under normal conditions. Again, with minor modifications, compounds boiling well above room temperature could be handled on this equipment in the same manner.

Most machine components on the equipment do not need shielding, since the sample is in the vapor phase and the cabinet provides protection. Exceptions to this are the sample inlet port and the carrier gas stream clean-up traps. A steel plate with a hole just large enough to accommodate a hypodermic needle is placed over the inlet to prevent the inlet seal retainer from being projectilized should an explosion occur while injecting. The clean-up traps are contained and shielded in the collection apparatus.

The injector and recovery apparatus has been used continuously for over six months and has given very satisfactory service.

Fig. 3

## INJECTOR

Sample injection into the chromatograph is accomplished by differential pressure. The sample is condensed into a chilled cylinder of about 30 to 100 cc volume (depending on sample size) under vacuum. The cylinder is then heated until the pressure reaches about 100 to 200 psig. Alternate heating and cooling of the injector cylinder is accomplished by a hot water bath and a liquid nitrogen bath mounted on a turntable which in turn is mounted on a hydraulic cylinder. The injection cylinder can then be immersed in either bath (see Fig. 1). A thermostatted electric heater is used to keep the water hot and an addition line makes it possible to fill the liquid nitrogen Dewar and replenish the hot water bath remotely.

To inject a sample, pressure is relieved into the chromatograph inlet via a flexible arm extending from the injector. Since injection is against the chromatograph column inlet

Fig. 4

pressure, a fraction of the sample remains in the injector each time so that every sample (except the first) is somewhat diluted with the preceding one. This must be kept in mind if quantitative data are to be obtained from chromatograms. Usually, the amount of sample that can be charged to the injector is several times the amount that can be injected into the chromatograph, making several chromatograph injections possible from one injector charge.

## FRACTION COLLECTOR

The remote fraction collector (Fig. 2) does essentially the same thing the chromatograph operator would do manually if the sample were not hazardous. Fractions are condensed from the column effluent in glass traps chilled to dry ice or liquid nitrogen temperature. Conventional traps, with stopcocks and appropriate connectors added, work very well. The lines be-

Fig. 5

tween the fraction collection valve and the fraction collector
must be flexible to accommodate the rotation of the valve. The
coolant for each of the four glass traps used is contained in a
Dewar mounted on a jack which is raised or lowered remotely
with an extension handle. Addition lines are used to replenish
the coolant in each Dewar. Four traps are used in this design
since the rotary collection valve has four trap positions. Each
collection position is operated in exactly the same way.

After the sample is condensed in a trap, it is removed by
vacuum transfer to a storage container. The trap must be isolated from the collection valve to do this, since the valve could
leak under this vacuum. When the transfer system and storage
container are evacuated, the contents of the trap are warmed
by lowering the coolant and applying a stream of air to the trap.
Normally, the storage container is chilled to effect a complete
transfer.

After transferring, the trap must be pressured to 1 psig

with helium before it can be opened to the valve. Also, the helium pressure must be maintained while the trap is being chilled, since a negative pressure at the valve will result in air leaking into the system.

Incorporated into the lower part of the fraction collector are the clean-up traps that remove sample and other impurities from the column effluent. An extra trap is provided to allow continuous operation with the manifold designed so that two traps can be in the chromatograph stream while the third is being emptied. Except for this feature, the clean-up traps are used in the same way as the collection traps.

Three general views of the entire apparatus are shown in Figs. 3-5.

## ACKNOWLEDGMENTS

The authors wish to acknowledge the efforts of F. S. Schroder and P. M. Ista, both of Minnesota Mining & Manufacturing Company, for their very prominent contributions to the design of this equipment.

# TEMPERATURE-PROGRAMED CAPILLARY COLUMNS

## F. L. Boys

Sinclair Research, Inc.
Harvey, Illinois

---

Capillary columns and programed-temperature columns have shown a roughly parallel and rapid increase in usage. These two types of chromatographs may readily be incorporated into a single instrument to combine their advantages. A satisfactory design is presented.

## INTRODUCTION

For wide boiling range samples, temperature programing of conventional packed columns has demonstrated numerous advantages. The use of a more nearly optimum column temperature for each component offers increased resolution of lower boiling components, increased peak height sensitivity of higher boiling compounds, possible elimination of prior fractionation, and a generally more satisfactory analysis. The use of capillary columns offers resolution not available with packed columns, resulting in a more rapid and complete analysis.

Considering the impact that the separate techniques of capillary column, or Golay, chromatography and programed-column-temperature chromatography have had for several years, it is surprising that so little interest has been shown in combining these techniques. Teranishi et al. [1] had described this combination in 1960, using an argon detector, but gave few details. Little further mention of programed capillary columns was noted until the March, 1962, announcement by Perkin-Elmer of the first commercial instrument. During this time interval we felt the need for such an instrument, for study of wide boiling petroleum fractions, and developed the instrument to be described.

## GENERAL DESIGN

In designing the programed-temperature capillary column gas chromatograph we have aimed for maximum simplicity. It

415

is usually found that the more complex an instrument, the more down time required for servicing. Frequently, the simple instrument is found to be more versatile than the complex apparatus.

The instrument should be designed to be convenient to operate and service. The column must be thermally isolated from the inlet, splitter, and detector, without inclusion of dead spaces or cold zones. It is necessary to provide means for rapid heat transfer to the column during the sample run, and desirable to allow even more rapid column cooling after the run is completed.

The hydrogen flame detector was chosen for this instrument because of its wide dynamic range, sensitivity, predictable response, usable temperature range, and ease of construction.

The electrometer design is not considered to be part of the basic chromatograph design; any electrometer suitable for use with a flame detector is satisfactory, whether the capillary column is operated programed or isothermally.

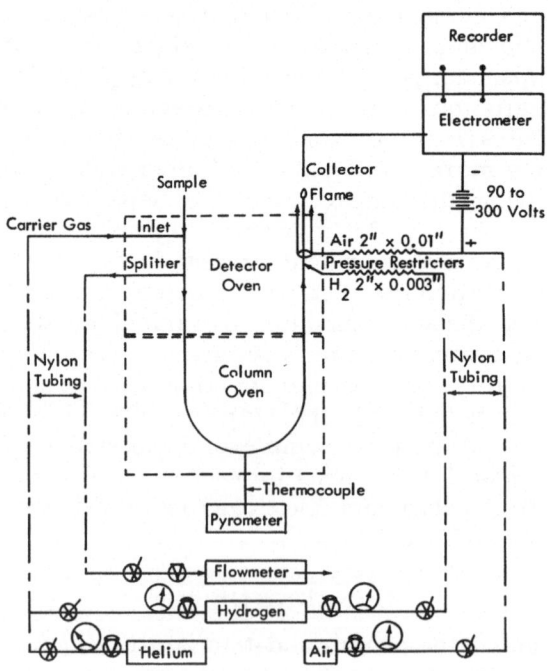

Fig. 1. Chromatograph schematic.

## DESIGN DETAILS

The flow system is conventional and is shown in Fig. 1. The sample is injected, from a microliter syringe, through a silicone rubber septum; the vaporized sample is swept directly to the splitter. Less than 1% of the charged sample is usually allowed to enter the capillary column, the rest of the sample being vented to the atmosphere through a back-pressure needle valve. The sample entering the column is resolved and the separated components are eluted into the flame, where ionization occurs, enabling a measure of component concentration to be made.

### Inlet

The inlet shown in Fig. 2 has no dead spaces and no changes in cross section from the time the sample is injected until the sample enters the capillary column. This feature is of special importance if extreme speed or resolution is desired. The carrier gas is preheated, moves through the annular space of the inlet past the septum, and then sweeps the injected sample cleanly to the splitter.

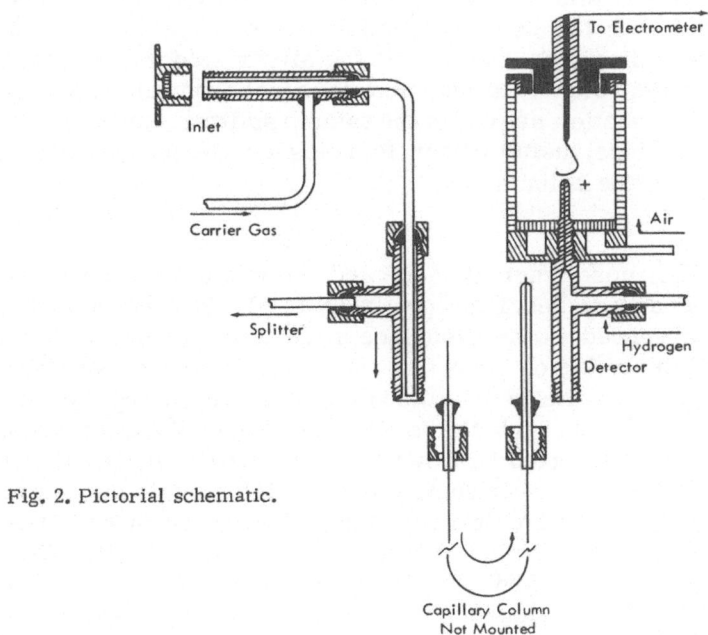

Fig. 2. Pictorial schematic.

## Splitter

- The splitter (Fig. 2) is not greatly unlike other systems, but was designed specifically for programed-temperature operation. The end of the capillary column slips up into the $\frac{1}{8}$-in. tube that is integral with the inlet. The vaporized sample slug sweeps the end of the column; that portion of the sample not entering the column is immediately swept up around the outside of the $\frac{1}{8}$-in. tube and out of the instrument to a back-pressure needle valve. The needle valve is adjusted to give the desired split ratio. Since the coated column is a part of the splitter, and extends into the splitter body, the splitter temperature is usually set for the maximum allowable column temperature. The body of the splitter is mounted on a $\frac{1}{2}$-in.-thick steel plate; this plate acts as a heat source so that heat is always available to be dissipated to the column oven to avoid cold zones.

## Column

The column, usually 100 ft. × 0.01-in.-ID stainless steel tubing, is wound on an aluminum cylinder $4\frac{1}{2}$ in. in diameter up to $4\frac{1}{2}$ in. long. To provide for mounting of the column, short lengths of $\frac{1}{8}$-in. tubing are silver soldered to each end of the capillary column as shown in Fig. 2. After the column has been coated with the desired liquid phase, one end of the column is broken off nearly flush with the silver solder joint; the other end of the column extends about one inch from the silver solder joint to function as part of the sample splitter. Imperial Hi-Seal tube fittings, using either ferrules or O-rings, are used for securing the column.

## Detector

The most generally accepted version of the flame detector is the design described by Desty et al. [2]. This model is especially good in the avoidance of dead spaces and in minimum electron emission from the flame jet. Hydrogen for the flame sweeps past the end of the column to help sweep the resolved components through the jet into the flame. We did find, in line with work reported by Dewar [3], that with this detector design ion collection is not complete with high sample concentration. Addition of a wire spiral to the collector, so that the flame is encircled, appears to result in complete ion collection at all sample levels tried. The detector, shown in Fig. 2, was provided with an extended rigid collector lead to provide thermal isolation of the electrometer pick-up cable.

Fig. 3. Detector oven.

## Detector Oven

The oven containing the inlet, splitter, and detector (Fig. 3) is constructed of $\frac{1}{2}$-in. Transite. The splitter and detector base are mounted on a piece of $\frac{1}{2}$-in. steel plate; this plate provides for component mounting and helps maintain the splitter and detector at the desired temperature. A 1000-w Calrod heater is held in direct contact with the inlet, and approximately $\frac{1}{8}$ in. off the steel plate. The oven detector is filled with gravel to provide thermal ballast during the time the column oven is changing temperature. This arrangement maintains the inlet approximately 50°C hotter than the bases of the splitter and detector. The top of the detector extends through the lid of the oven. The hydrogen flame detector is insensitive to reasonable temperature changes, so that no thermostatting is needed; oven heat is merely set with an autotransformer. Thermocouples are provided to measure temperatures of the inlet and the bases of the splitter and detector. The hydrogen and air pass through pressure restrictors, enter the oven, and then are preheated by passing through about 18 in. of $\frac{1}{8}$-in. tubing so that neither the end of the column nor the jet will be cooled. Carrier gas is preheated in the same manner before entering the inlet.

**Column Oven**

The column is maintained in a separate oven located below the detector oven. This arrangement allows the ends of the capillary column to be placed directly in the splitter and detector; in this way no connecting tubing or dead spaces need be incorporated. The column, wound on the aluminum cylinder, surrounds a 650-w Glocoil heater; air is circulated over the heater and around the column by a fan.

Walls of the column oven are of $\frac{1}{4}$-in. Transite, as a compromise, to allow a rapid temperature rise in the oven, and also to have low thermal capacity to allow rapid cooling. Three walls of the oven slide out to allow column cooling or column change. Column temperature is measured with a calibrated pyrometer; the thermocouple is fastened to the aluminum cylinder.

## TEMPERATURE PROGRAMING

Most of our temperature programing has been by the simplest effective means. The column oven is closed and a preselected voltage is placed on the column heater with an autotransformer. In general this method of programing is satisfactory; however, occasional changes of line voltage will result in retention time shifts. To protect the column, the heater voltage is set to a value that will give a safe final column temperature; this usually creates no inconvenience except that it limits the rate at which the column may be heated. The instrument has

Fig. 4. Assembled chromatographs.

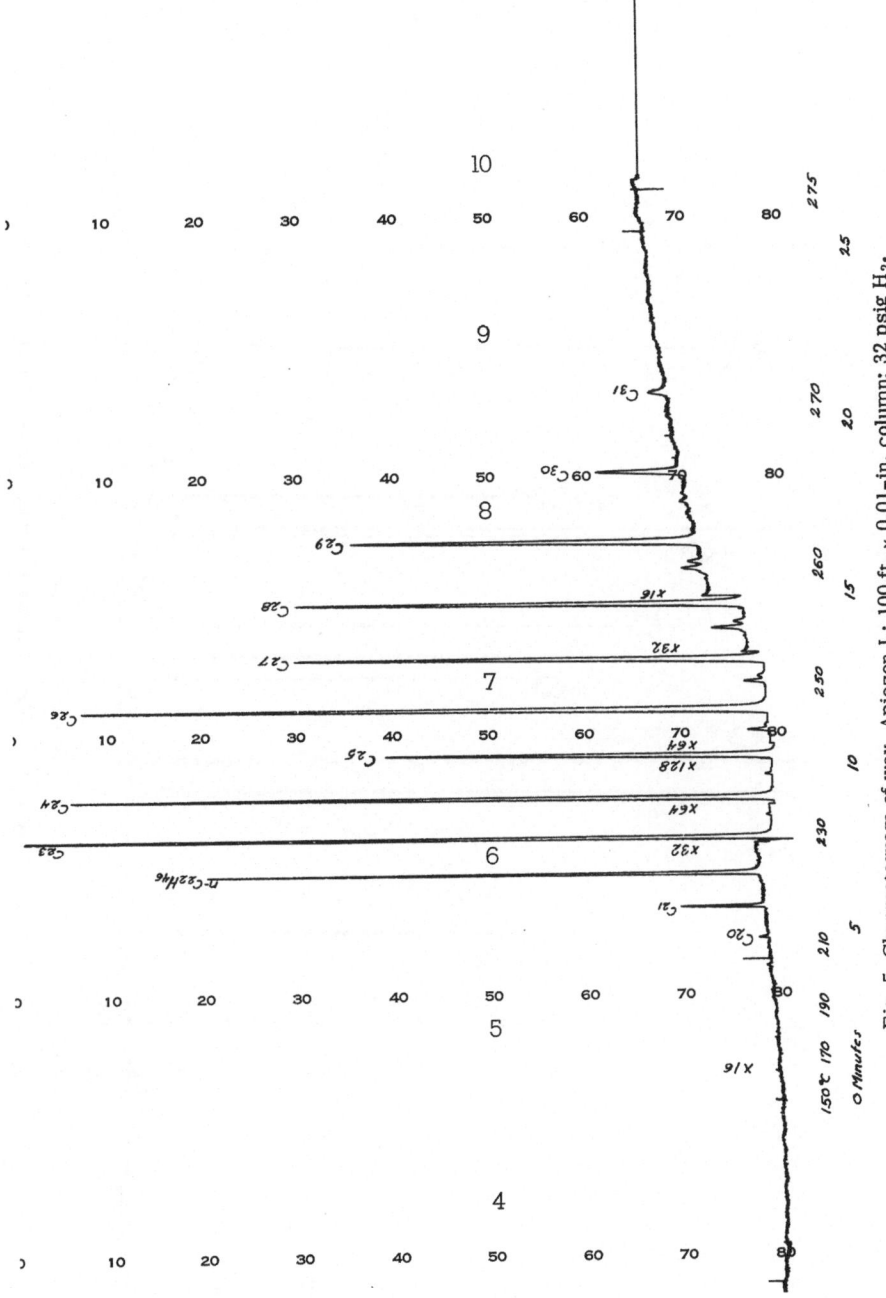

Fig. 5. Chromatogram of wax. Apiezon L; 100 ft. × 0.01-in. column; 32 psig H₂.

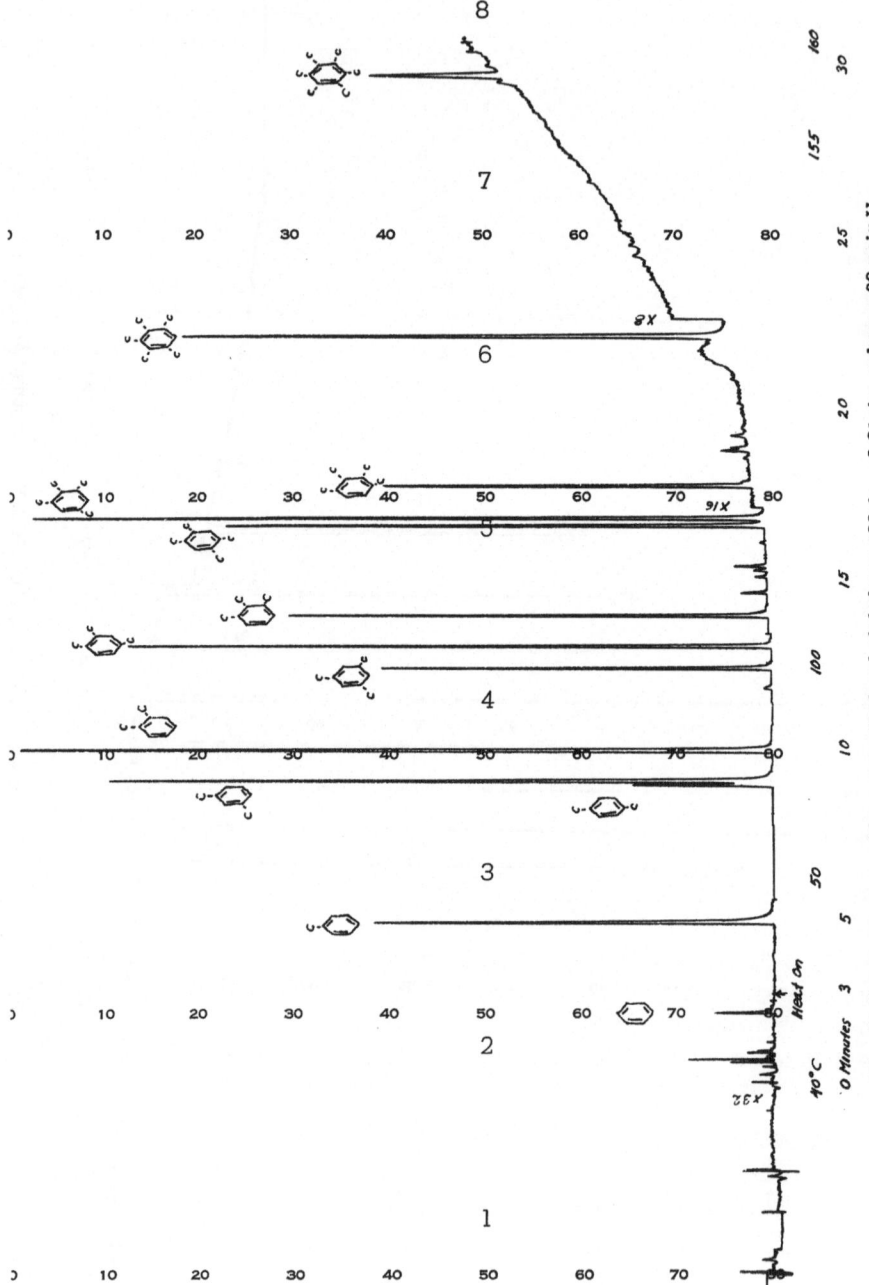

Fig. 6. Chromatogram of benzenes. Didecyl phthalate; 100 ft. × 0.01-in. column; 30 psig H₂.

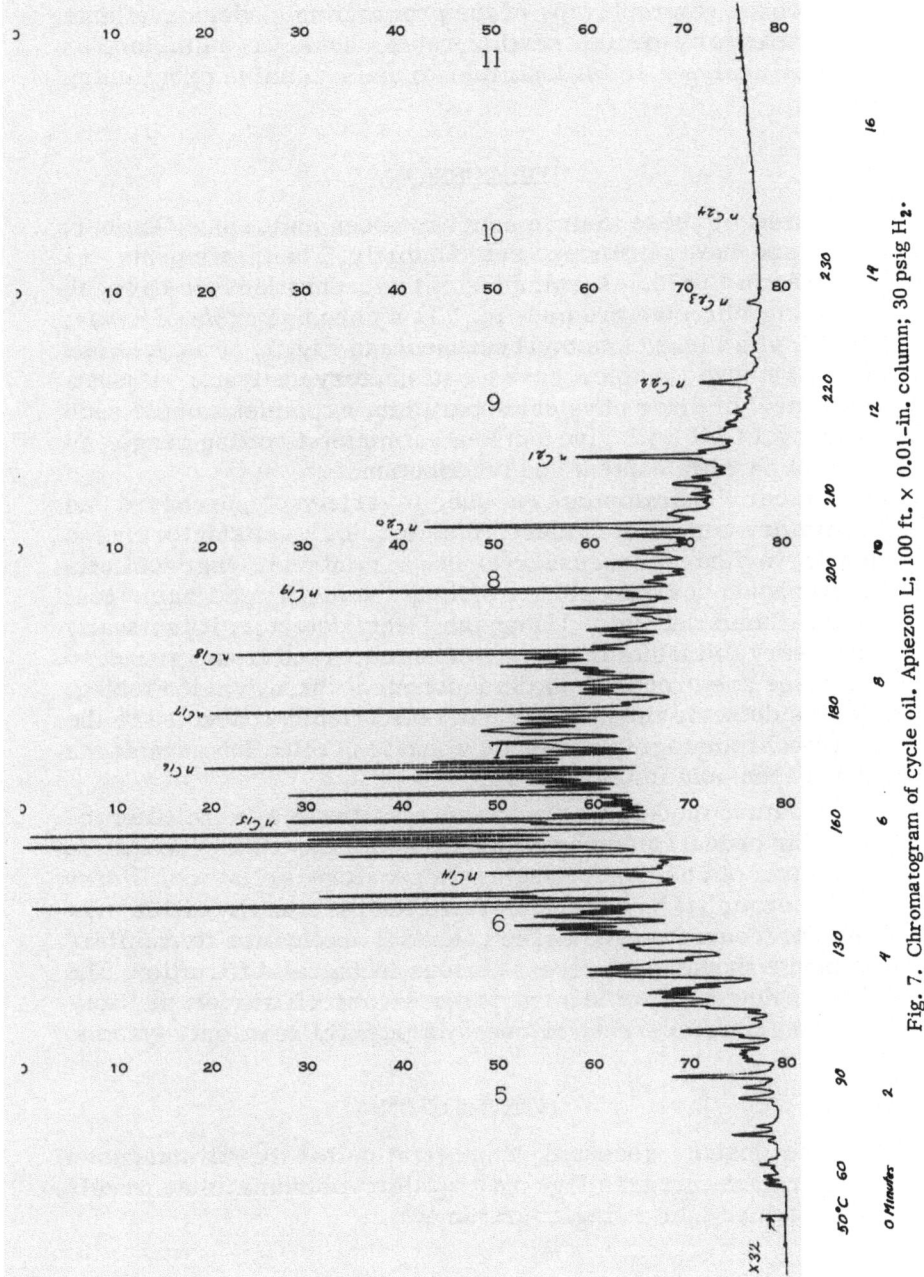

Fig. 7. Chromatogram of cycle oil, Apiezon L; 100 ft. × 0.01-in. column; 30 psig H₂.

been used satisfactorily with the temperature programer of an F & M chromatograph. Any of the programing devices available for linear or special heating rates should be suitable. For general analysis it is desirable to use a positive programing device.

## DISCUSSION

Three of these instruments have been built since October, 1961, and have performed satisfactorily. The instruments are pictured in Fig. 4. As examples of the separations we have the following chromatograms: Fig. 5 is a chromatogram of a wax; Fig. 6, of a blend of methyl benzenes; and Fig. 7, of a cycle oil. The first two samples gave a satisfactory analysis; without a great deal of prior physical separation, a complex sample such as a cycle oil will give only an estimate of boiling range and normal paraffin content and distribution.

In our laboratories we use a variety of purchased and laboratory constructed electrometers. For a satisfactory noise level, we find it necessary to use a minimum length of anti-microphonic coaxial cable to pick up the high impedance signal of the flame detector. Using the Desty detector, it is usually necessary to isolate the entire chromatograph from ground, so all gases are brought into the instrument through nylon tubing. (In this detector the positive jet is electrically connected to the entire chromatograph; gas flow and ion collection advantages are responsible for the design.)

In future models, it would seem worthwhile to include programing of dual columns to compensate for base line rise caused by elution of column coating as temperature is elevated. Emery and Koerner [4] have demonstrated the practicality of this system for conventional packed columns; application to capillary columns should not present serious technical difficulties. The second detector would also provide cancellation of the background flame current for use with integral read-out systems.

## CONCLUSION

The design presented demonstrates that the advantages of temperature programing and capillary columns may readily be combined into a single instrument.

## ACKNOWLEDGMENT

Special thanks are expressed to Mr. Lester Morse for construction of the chromatographs and for providing much of the detailed design.

### REFERENCES

1. R. Teranishi, C. C. Nimmo, and J. Corse, Anal. Chem. 32, 1384 (1960).
2. D. H. Desty, C. J. Geach, and A. Goldup, in R. P. W. Scott (ed.): Gas Chromatography, Butterworths, London (1960), p. 46.
3. R. A. Dewar, J. Chrom. 6, 312 (1961).
4. E. M. Emery and W. E. Koerner, Anal. Chem. 33, 523 (1961).

# A TECHNIQUE FOR THE IDENTIFICATION OF VOLATILE FLAVOR COMPONENTS IN FOODS

Reed Jensen* and Stuart W. Leslie

Armed Forces Food and Container Institute
Chicago, Illinois

---

Gas chromatography has been successfully employed in the isolation and preliminary identification of volatile components from foods. Pure flavor components have been prepared using a two-step distillation followed by two gas chromatographic units in series. As pure components are eluted from this apparatus, they can be collected for infrared analysis.

The problem of obtaining pure flavor components from foods in quantities suitable for infrared analysis is complicated by the minute amount of volatiles present and the desirability of keeping the starting material weight below 1000 g. This paper describes an economical, rapid, and efficient method of obtaining such samples.

The technique can be explained most clearly by pointing out that there are two separate steps which are necessary before the sample containing volatiles is ready for collection in the cold trap of the first gas chromatographic unit. The first step is a vacuum distillation of the starting material through a 40-cm Vigreux column until 50 ml of distillate have been collected (in a flask containing 50 g of salt) and cooled in liquid nitrogen. This step separates the volatiles from the solids, which are then isolated from the system. The next step is to release the volatiles from the frozen aqueous distillate and collect them in the gas-sampling trap of the first instrument. This is done by heating the salt-saturated distillate until the organic materials are evolved. Helium is passed through this solution by means of a capillary and the volatiles are carried to the cooled trapping system of the first chromatographic unit. It is convenient to use the helium from the instrument exhaust,

*Present address: Brigham Young University, Provo, Utah.

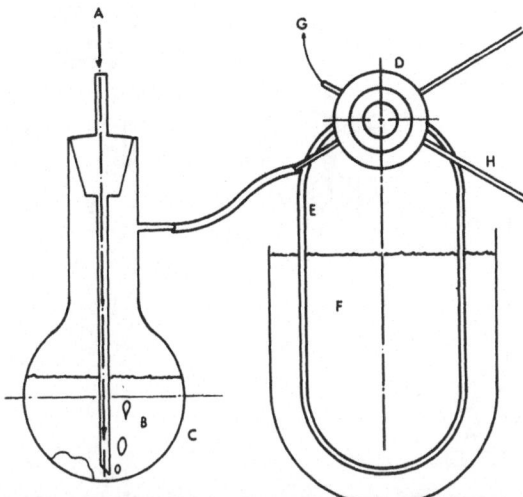

Fig. 1. Simple trapping device used in early experiments. Legend: (A) Helium inlet and capillary bubbler; (B) salt-saturated volatiles solution; (C) boiling brine solution; (D) Perkin-Elmer rotary gas valve; (E) stainless steel collection trap, $\frac{1}{8}$ -in. diameter; (F) liquid nitrogen bath; (G) helium exhaust port; (H) carrier gas to detector.

since it is controlled as to purity and flow rate. The apparatus for this is shown in Fig. 1.

The helium and the volatiles are then carried through the tube to the trap of the gas-sampling valve; the volatiles are condensed in the trap, while the helium escapes.

Doing the process in two steps has the obvious disadvantage of exposing the sample to the air and allowing part of the volatiles to escape while preparing the sample for the next distillation. This also allows oxygen to re-enter the flask, causing great difficulty when working at liquid nitrogen temperatures. It was therefore thought that an apparatus capable of combining these steps would be desirable. The apparatus shown in Fig. 2 was designed to meet these requirements. Figure 3 is an enlarged view of the intermediate collection flask and Fig. 4 is an enlarged view of the infrared gas cell designed in this laboratory.

During the first operation the stopcock is open, and for the second it is closed. It has been found necessary to apply a vacuum to the whole system at the gas-sampling valve trap in order to make the distillation proceed, since the constant-

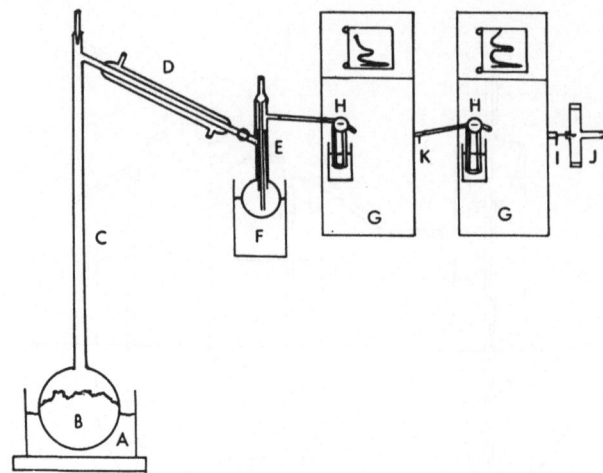

Fig. 2. Assembled apparatus for collecting volatiles. Legend:
(A) Steam bath; (B) sample; (C) Vigreux column; (D) water-cooled condenser; (E) Volatiles-collecting trap (see Fig. 3 for details); (F) liquid nitrogen cooling bath; (G) gas chromatograph; (H) gas-sampling valve and trap; (I) three-way valve with hypodermic needle; (J) sample-collecting cell for infrared analysis (see Fig. 4); (K) three-way valve.

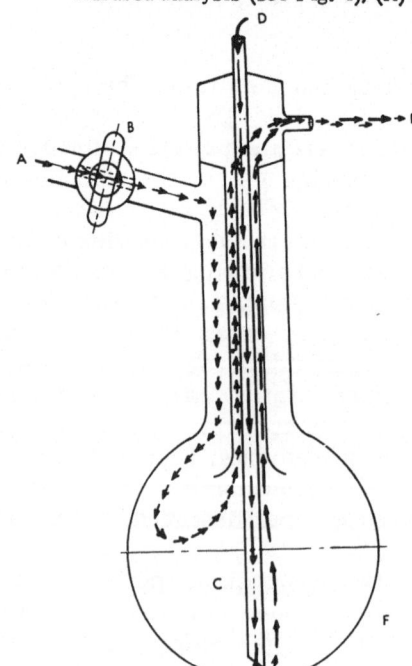

Fig. 3. Detail of collecting flask. Legend: (A) Volatiles inlet; (B) stopcock; (C) volatiles condensate; (D) helium inlet and capillary; (E) helium and volatiles outlet; (F) liquid nitrogen or boiling brine bath.

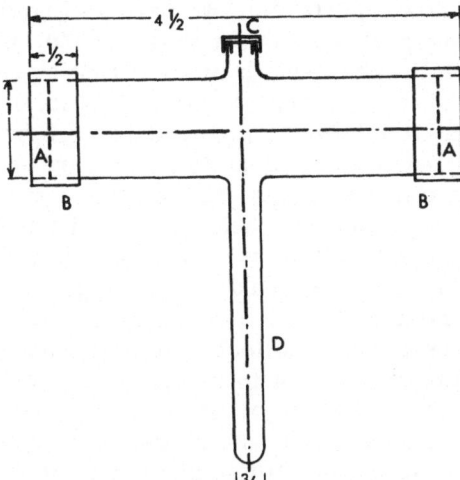

Fig. 4. Detail of infrared gas cell. Legend: (A) Sodium chloride windows ($^1/_4$ in. × 1 in. diam); (B) rubber tubing sleeve; (C) rubber serum cap; (D) cooling trap for condensable compounds.

temperature bath on the 2-liter flask is kept at 100°C. A vacuum of 650 mm is generally used. The key to balancing temperature gradients and gas flow in the whole system is the three concentric tubes in the neck of the intermediate flask as shown in Fig. 3. Note that in the first step the hot gases from the condenser must go down into the flask itself and go around the bell-shaped second tube before they can continue on to the top of the gas sampling valve. This assures excellent condensation of volatiles, including water, and keeps the gas flow to a minimum. That is, there is little gas to flow on to the vacuum pump.

After the first step is completed the stopcock is closed and the collection flask is heated in boiling brine while the trap of the gas sampling valve is in liquid $N_2$. During this operation the apparatus functions just as the simple apparatus of Fig. 1. As previously mentioned, helium from the instrument exhaust is circulated through the system to carry the volatiles to the trap to be condensed.

From here it is a simple matter to heat the trap to 200°C, turn the valve, and admit the sample to the first gas chromatograph. In order to assure that enough sample is collected to obtain a good spectrogram, the peak under study is accumulated from several identical runs in the gas-sampling valve of the second instrument by connecting the Swagelock valve from the exhaust of the first instrument to the trap of the second instrument. That is, successive runs are made, each time collecting only the peak of interest. We have determined that $1\lambda$ of liquid, or any peak equivalent to thirty (30) units (peak height in tenths of total chart width times retention time in minutes) on a 5-mv recorder with a chart speed of $\frac{1}{2}$ in./min, will provide sufficient material for an infrared spectrogram.

While the sample is being trapped at the exit port of the first instrument, the infrared gas cell is evacuated to 1 mm of pressure. The accumulated sample is then introduced into the second instrument, which has a column with different retention characteristics to ensure that the peak is really pure. The second instrument has been adjusted to have a flow rate of 25 ml/min, which permits sample collection for at least two minutes in the 50-ml infrared cell. When the pen of the second instrument starts to record the major peak a hypodermic needle, attached to the instrument exhaust, is inserted through the septum of the gas cell. When the sample has been collected the cell is placed in the beam of the Perkin-Elmer Model 21 Infrared Spectrophotometer under a heat lamp to ensure that the sample does not condense.

Spectrograms obtained in this manner have been used successfully to identify unknown flavor components in both meat and fruit products.

# SOME APPLICATIONS OF GAS CHROMATOGRAPHY TO FORENSIC CHEMISTRY*

## Daniel T. Dragel,† Ed Beck,‡
## Andrew H. Principe†

† Chicago Police Crime Detection Laboratory
‡ Armour and Co., Chicago, Illinois

## INTRODUCTION

Forensic chemistry may be defined as the application of scientific chemical techniques to problems involving legal action with the purpose of aiding the administration of justice.

In a crime laboratory, this is accomplished through the assessment of physical evidence which may be plainly visible but is usually detectable only through the application of very sensitive chemical or physical techniques. Gas chromatography, now one of the most important analytical tools, has excellent capabilities for separating individuals from complex mixtures as well as the preparation of pure compounds for subsequent identification.

## EQUIPMENT

Five types of columns are now utilized, depending on the nature of the sample to be examined. These are: (1) the Craig polyester type; (2) Apiezon M; (3) a polyglycol of 20,000 average molecular weight; (4) silica gel; and (5) a silicone rubber column. The ester column (Table I) has been used primarily for fatty methyl esters. This is a 5-ft column operated at 185°C. The Apiezon column is used with flammable materials. It is 4 ft long and is operated from 90° to 215°C, depending on the volatility of the material to be separated. Alcohols normally are separated on the $6\frac{1}{2}$-ft polyglycol column at 85°C. The 6-in. silica gel column is used for gases at room temperature. Drugs are separated on the 5-ft silicone rubber column at 215°C. Either hydrogen or helium flowing at 100 ml/min serves as the

*To be published in the Journal of Criminal Law, Criminology and Police Science.
†Chicago Police Crime Detection Laboratory.
‡Armour and Co., Chicago, Illinois.

## TABLE I
### Chromatographic Conditions

| Column Packing | Craig Polyester | Apiezon M | Glycol | Silica Gel | Silicone Rubber |
|---|---|---|---|---|---|
| Column Temp. | 185°C | 90-215 | 85 | 25 | 200-250 |
| Column Length | 5 ft | 4 ft | 6½ ft | 6 ft | 5 ft |
| Preheater Temp. | 250°C | 250 | 150 | — | 275 |
| Carrier Gas | $H_2$ or He | | | | |
| Carrier Flow Rate | 100 cc min | | | | |
| Detector | Gow-Mac Thermal Conductivity Model TE-11 | | | | |
| Recorder | Leeds & Northrup Speedomax G 1-sec 1-mv | | | | |

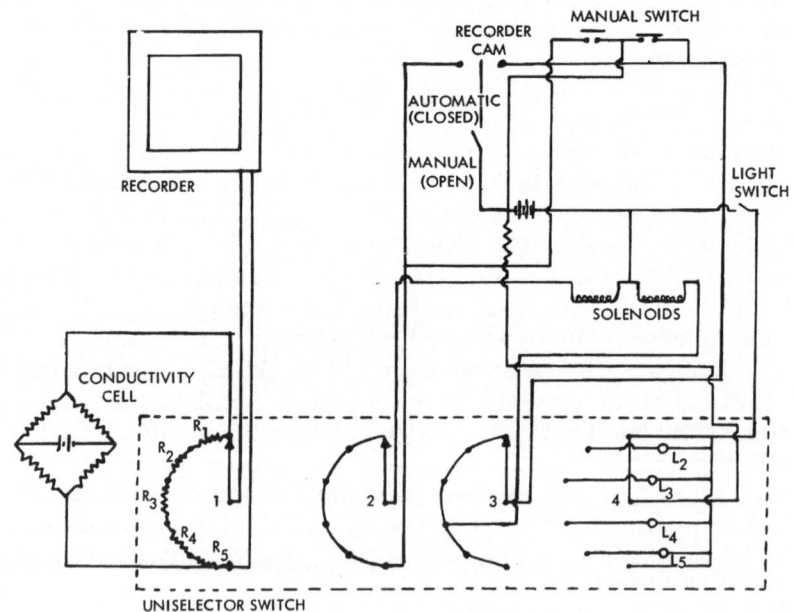

Fig. 1. Attenuator diagram.

carrier gas. The detector in this instance was Gow-Mac thermal conductivity type Model TE II and the recorder was a Leeds & Northrup 1-sec 1-mv Speedomax G.


## DISCUSSION AND RESULTS

In order to obtain positive identification of a compound or mixture, attention must be given to both major and minor peaks. In order to accomplish this, the gas chromatographic instrument must couple sensitivity with attenuation such that all peaks, whether major or minor, are shown in proper relation on the same chart. For this purpose a decimal attenuator has been constructed, the details of which are shown in Fig. 1. The heart of the attenuator is the four-gang Uniselector switch manu-factured by the General Electric Co. of England. (A similar type of switch is now available from radio supply houses.)

Gang one of the switch contains precision resistors num-bered $R_1$ through $R_5$. The values of these determine the signal suppression at any corresponding contact; in this case they are multiples of 10, thereby causing attenuation to follow a decimal pattern. The sum of values of all of the resistors should equal the input impedance of the recorder used. Gang two controls the clockwise rotation of the switch arms, which results in lesser amounts of attenuation. The most clockwise position of this gang is not wired in order to prevent the switch from moving below zero attenuation and losing its signal. Gang three controls up-scale movement, and on this gang the most counter-clockwise contact is not wired to prevent the switch from losing the signal on its high end. The contacts of gang four are wired to corresponding dial lights. These show which range or scale is functioning. The base scale, with zero attenuation, is not wired through an indicator light in order to save battery current, since most of the time the instrument operates on base range. The actual changing of ranges is accomplished by means of two solenoids in the switch—one to shift downward and the other to shift upward. These can be actuated by fiber cams on the slide-wire shaft of the recorder, or they may be operated by means of manual switches. The signal from the attenuator is recorded by a Leeds & Northrup 1-sec 1-mv Speedomax G recorder. The signal to the attenuator is furnished by a Gow-Mac Model TE II thermal conductivity cell carrying 250 ma current.

## ALCOHOL

The criminologist must characterize the chromatograms of the lower alcohols over wide concentration ranges. These extend from the 50% liquor samples to 0.1% in blood samples. In the liquor sample, automatic attenuation is especially useful since here the minor peaks can establish the brand as well as the type of beverage. The minor peaks are fermentation products other than ethanol, such as acetone, acetaldehyde, ethyl acetate, methanol, propanols, butanols, and higher alcohols.

Alcohol in the blood is a completely different problem from alcohol in the bottle. In blood, ethanol is the only peak to be determined, and since the expected concentration runs from 0.1 to 0.2%, high sensitivity must be achieved. If the blood sample is first extracted with acetone, blood solids coagulate and need not be filtered from the solution prior to injection into the chromatograph. Acetone does not interfere with the alcohol determination.

Ethanol is not the only alcohol of importance to the criminologist. Methanol also is encountered. This lightest of alcohols may be determined under the same conditions as have been outlined for ethanol and higher alcohols.

## DRUGS

The problems met in the analysis of drugs are many and varied. Some of the most frequent are: (a) Analysis of mixtures for one or more components; (b) chemical identification of individual substances; and (c) identification of small quantities.

Gas chromatography fulfills the requirements to aid the forensic chemist in all of these problems. It has the ability to separate complex mixtures, and with special techniques pure samples may be collected for positive identification. Incidentally, positive identity cannot be based on retention time alone. In order to chemically identify a substance, it must be isolated in pure form and compared with known compounds using standard physical and chemical techniques.

Parker and Kirk [1] separated and identified some 23 barbiturates, each of which produced its own individual peak. These eluted compounds may be collected by dissolving in a suitable solvent and reserving for positive chemical identification. This is accomplished employing a technique developed by Walsh and Merritt [2]. A color spot test is employed with the eluted sample at the time of collection. Spot test reagents for the various drugs are extremely sensitive, with identification limits ranging from 1 to 0.001 μg [3]. The quantity of sample dissolved from the eluted peaks may be increased if necessary by repeated injection and collection. Barbital was separated from thiopental on a 5-ft aluminum column using silicone rubber on firebrick. The sample was collected by bubbling through a test tube containing acetone and glass beads. The dissolved eluted sample was concentrated by evaporation on a steam bath and recrystallized. Evaporation should be carried out in a glass well or from a high boiling solvent to obtain well-formed crystals. At least 0.01 μg of sample must be collected for recrystallization. The recrystallized sample is examined with a polarizing microscope equipped with a Kofler Hot Stage. The properties observed on this collected sample are shown in Table II. With these chemical, physical, and optical properties we have fingerprinted this eluted peak and identified it positively as that of barbital.

## GASES

The gases most commonly encountered by the forensic chemist are carbon monoxide, carbon dioxide, light hydrocarbons, cyclopropane, and ether. Carbon monoxide and carbon dioxide are usually absorbed by blood from an environment of partial combustion. Light hydrocarbons, although much less soluble in blood, are more toxic than carbon monoxide. Cyclopropane and ether are common anesthetics and can be obtained in many clinics and hospitals. All of these gases are best detected at room temperature—the hydrocarbon types on the Apiezon column and the other gases on either molecular sieve or silica gel. Water and blood solids are irreversibly absorbed on the silica gel and molecular sieve columns, thereby making necessary frequent replacement of the column packing material. Water may be removed from samples injected into the Apiezon column using a precolumn either with calcium carbide or calcium hydride. If such a precolumn is utilized, it must be heated at least to 100°C to prevent absorption of part of the hydrocarbon sample.

## FLAMMABLES

Flammables are the number one tool of the arsonist. Fortunately, evidence of their use is not completely destroyed even

## TABLE II
### Properties of Eluted Sample

| Collected Sample | Barbital |
|---|---|
| Soluble in acetone | Soluble in acetone |
| Birefringent | Birefringent |
| Sublimes | Sublimes |
| Melting point 188°C | Melting point 188°C |
| Polymorphic | Polymorphic |
| Eutectic melting point (Salophen) 163°C | Eutectic melting point [5] (Salophen) 163°C |

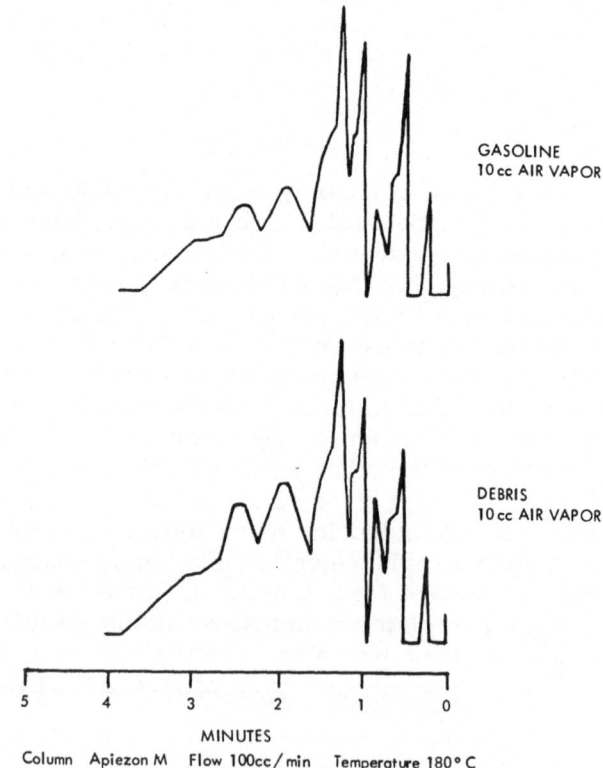

Fig. 2. Chromatograms of gasoline and of vapor sample from rags in burned debris.

for flammable material of high vapor pressure or when the environment has been at an elevated temperature for several hours. The types of materials to be expected include gasoline, paint thinner, charcoal lighter, turpentine, paint remover, fuel oil, kerosene, etc. None of these materials is a single compound but is a definite mixture, and for this reason identification may be established by preparing a chromatogram which covers not only the major ingredients but also the minor ones, including impurities. Here an automatic attenuator may be of tremendous service. Comparing entire patterns rather than single peaks eliminates much uncertainty.

The pattern shown in Fig. 2 was obtained from rags found at the scene of a fire after the fire had been extinguished. The rags were water saturated, burned, and loaded with debris from the building. Only the odor of burned building was evident to the nose. The rags were placed in a can and 10 ml of vapor was withdrawn slowly by means of a 6-in. needle inserted into the rag bulk. This vapor sample was then injected onto the Apiezon column operating at 180°C. A similar chromatogram was prepared employing a vapor sample drawn from a container of gasoline. The position and relative size of all the peaks in both chromatograms indicate that the samples used were similar. A current file of chromatograms of flammable materials can indicate to the investigator the type of material used, possible sources of the material, and in certain cases the age of the material.

TABLE III

Fatty Acid Distribution of Some Typical Fats

| Fat Source | Lauric | Myristic | Myristoleic | Palmitic | Palmitoleic | Stearic Oleic | Linoleic Linolenic |
|---|---|---|---|---|---|---|---|
| Pork | 0.4 | 0.9 | 0.1 | 31.0 | 0.1 | 66.9 | 0.5 |
| Beef | 0.3 | 3.0 | 0.2 | 29.0 | 1.4 | 65.8 | 0.3 |
| Lamb | 0.9 | 1.9 | 0.2 | 25.0 | 1.3 | 70.7 | 0 |
| Veal | 2.4 | 4.2 | 0.3 | 30.1 | 1.1 | 61.0 | 0 |
| Horse | 2.1 | 0.9 | 0 | 30.3 | 0.3 | 66.0 | 0.4 |
| N 48 | 0.9 | 1.5 | 0.8 | 19.6 | 7.3 | 58.1 | 15.5 |
| W 55 | 0.7 | 1.3 | 1.1 | 16.1 | 11.4 | 50.2 | 19.0 |
| N 32 | 1.6 | 1.3 | 1.0 | 16.3 | 9.6 | 51.2 | 19.0 |

## EXAMINATION OF FATS

Table III shows the fatty acid distribution of a few typical fat samples. The first five are from animals; the lower last three are human: a 48-year-old Negro, a 55-year-old white, and a 32-year-old Negro.

The samples were prepared for chromatography by saponifying with alcoholic KOH and refluxing for a few minutes on a steam bath with a solution of boron trifluoride in methanol according to the method of Metcalfe [4]. The fatty methyl esters so prepared were then chromatographed at 185°C on the Craig polyester column. From the fatty acid distribution, it is apparent that human fat contains more unsaturated fats than animal fat. This can be observed in the myristoleic, palmitoleic, and linoleic fractions. Also, human fat contains only about one-half as much palmitic acid as does animal fat.

Sufficient evidence is not yet available to establish definitely the relationship between different samples of human fat; however, it would appear that the lauric acid content of human fat decreases with age. It would also appear that whites have more unsaturation in their fat than Negroes.

The future of gas chromatography in forensic chemistry extends further than can be visualized at the present time. In addition to the applications mentioned in this paper, such materials as perfumes, coatings, plastics, oils, solvents, and poisons lend themselves readily to chromatographic analyses.

## REFERENCES

1. K. D. Parker and P. L. Kirk, Anal. Chem. 33, 1378 (1961).
2. T. Walsh and C. Merritt, Anal. Chem. 32, 1378 (1960).
3. C. P. Stewart and A. Stolman, Toxicology, Vol. II, Academic Press, New York (1960), p. 242.
4. L. D. Metcalfe and A. A. Schmitz, Anal. Chem. 33, 363 (1961).
5. W. C. McCrone, Fusion Methods in Chemical Microscopy, Interscience Publishers, New York (1957), p. 108.